Risk Assessment for Water Infrastructure Safety and Security

Risk Assessment for Water Infrastructure Safety and Security

Anna Doro-on

IWA Publishing

CRC Press
Taylor & Francis Group
Boca Raton London New York

CRC Press is an imprint of the
Taylor & Francis Group, an **informa** business

Co-published by IWA Publishing, Alliance House, 12 Caxton Street, London SW1H 0QS, UK
Tel. +44 (0)20 7654 5500, Fax +44 (0)20 7654 5555
publications@iwap.co.uk
www.iwapublishing.com
ISBN: 1-78040-021-7
ISBN13: 978-178040-021-1

CRC Press
Taylor & Francis Group
6000 Broken Sound Parkway NW, Suite 300
Boca Raton, FL 33487-2742

Printed in the United States of America on acid-free paper
Version Date: 20110711

International Standard Book Number: 978-1-4398-5341-2 (Hardback)

Visit the Taylor & Francis Web site at
http://www.taylorandfrancis.com

and the CRC Press Web site at
http://www.crcpress.com

Dedication

To all the heroes and victims of the terrorist attacks on September 11, 2001, including all the soldiers and civilians who risk their lives in Iraq and Afghanistan

I could not have done this book without thinking about the grief of the entire United States and remembering the most horrifying day of 9/11.

We never know what is coming,
About the things that occur,
When someone does not worry.

Nobody has the time,
To think about the crime,
That someone may do.

If each one would worry,
We would not have to look,
Death into the air.

Numerous have vanished the ones they love.
Other have lost the ones that cared about them.

Nothing would have occurred,
If everyone would just think about,
The things that are crucial.

Courageous warriors in suits of yellow and black,
Responding to a cry for help without reservation they go on to the battle.

To snatch a person from death is the goal and supreme reward.
It's what all the preparation and self sacrifice is geared toward.

They love their nation and their fellowmen the same.
It's a profession that supplies little financial gain.

Their mission is simple the American dream to keep and protect,
A life to save, a home to protect, to this they will tend.

—Tears From the Hearts of the Citizens of the United States

(Excerpted from 9/11 memorial poems created by JDK, Elizabeth, myself, and other citizens of the United States)

Contents

List of Figures

List of Tables

Preface

This book is derived from my doctoral dissertation and research papers, as a result of which I received the Young Scientist Award for 2007, given by the American Academy of Sciences and Environmental Science and Technology. The work was done under the guidance of my former mentor, Dr. Chia Shun Shih, who taught me the significance of *prospect theory* in risk analysis, which he utilized throughout his career in both the public sector and academia. This book represents a perfection of research development that has been rigorously enhanced over the past several years. The intent of presenting the materials contained within this book is to remodel and improve public perception of terrorism risk, particularly that on water infrastructure security, and to achieve incremental risk acceptability.

The primary objective of this book is to educate the reader on how to use quantitative risk assessment that analyzes the terror threats against U.S. water infrastructure and precisely defined pathways leading from the initial policy decision to the final consequences. Comparison of these risk probabilities with limits based upon the revealed preference concept and cumulative prospect theory provide the decision maker with the chance to review alternatives for acceptability. The probabilistic risk assessment methodology based on cumulative prospect theory is detailed in Chapters 9 and 11.

Since the September 11, 2001 (9/11) terrorist attacks, national concern has focused on the effectiveness of preventive measures for protecting critical infrastructure. After several years of witnessing different legislations and the implementation of several governmental programs related to infrastructure security, I still see that most of the water supply systems currently remain unprotected, including dams, reservoirs, and aqueducts. Even though there are grave threats from principal terrorist leaders against the U.S. water supply system following 9/11, it seems that the public does not feel the magnitude of the risks posed by potential attacks, particularly those against the sole source of life, the drinking water supply. Most terrorism scenarios presented by homeland security experts are focused on cyberspace and treatment facilities and not on original water sources. Meanwhile, this book graphically uncovers potential terrorism activity scenarios and bold plans of terrorist leaders that are truly considered unthinkable events by the general public, as presented in detail in Chapters 4 and 11. These presentations will not only make the homeland security decision makers and the public realize the urgent need for water infrastructure protection but also shed light on strategic improvements to water infrastructure security, to make it difficult for the terrorists to launch their attacks against the United States and to minimize the impact of the attacks that may take place.

Steve Recca of the Naval Postgraduate School's (NPS) Center for Homeland Defense and Security inspired me to prepare Chapter 11 to close this book. Chapter 11 presents a special topic on intelligence analysis for water infrastructure terrorism prevention with illustrative practical examples, terrorism warnings, risk estimation models, and risk acceptability analysis. Improvements to U.S. Intelligence agencies could make a substantial difference on keeping Americans safe and secure.

Acknowledgments

As is the case with all worthwhile and important efforts, this book was made possible only with the contribution of many talented individuals. I extend my thanks to my editor, Mark Listewnik, senior editor of the Taylor & Francis Group for his thoughtful advice, guidance, and efforts in making this book a reality. I extend my thanks also to Richard O'Hanley, publisher; to Kathryn Younce, project coordinator; George Kenney, account sales representative, Robert Sims, production project editor, Dennis Troutman, production project manager, and to all the editorial staff at Taylor & Francis for their work. I acknowledge my most sincere gratitude to my mentor Dr. Chia Shun Shih, former Associate Administrator of Research, Technology, and Analysis at the U.S. Department of Transportation, vice president of PTS International, and retired professor of the University of Texas at San Antonio, for teaching me the prospect theory in risk acceptability analysis and the application of engineering analysis for homeland security against terrorism; for working with me over the past several years, work that provided important background research for three of the chapters in this book; and for his advice on expanding my talents in different fields. Also, I would like to express my sincere gratitude to Dr. Jerome Keating, professor of the University of Texas at San Antonio, who suggested some of the vulnerability analysis methods that I use in this book. I am very grateful to Dr. Weldon Hammond, director of the Center for Water Research, University of Texas at San Antonio, for his support on my research related to groundwater security. I would like to thank Dr. Dorothy Flannagan, dean of the Graduate School; Dr. Manuel Diaz; Dr. Alberto Arroyo; Dr. Sazzad Bin-Shafiq; and all the faculty members and staff of the Department of Civil and Environmental Engineering, University of Texas at San Antonio, for their kindness and support during my stay at the university.

I would like to offer my gratitude to Steve Recca, NPS Center for Homeland Defense and Security, for his thoughtful advice, sustained efforts to keep the United States secure, and his encouragement and endeavor, which inspired me to prepare Chapter 11 to close this book. I am very grateful to Chuck Larson, of the New England Water Works Association, for providing the Automated Security Survey and Evaluation Tool (ASSET). I would like to acknowledge the Department of

Homeland Security and the Environmental Protection Agency for their efforts in providing guidelines on improving water infrastructure protection and for their contribution to my work.

I would like to acknowledge Alex Palmer and Debbie Palmer of RAMCO Companies (Alex Palmer is president of the company) for their continuous support and for the privilege of developing my skills in design and implementation of different environmental remediation technologies that were relevant to the preparation of this book; also to Mitch Kruger, computer-aided design specialist, for his help on some of my drawings at the last minute and for effectively realizing my imagination on paper; and to all the staff of RAMCO Companies for their support. I extend my thanks to Robert C. de Guzman, chairman of the Board, and the committee members of the Hispanic Engineers Business Corporation, Los Angeles, California, for their support. Also, I would like to thank Mei-Miao Kuo for her assistance, and Dr. Barry J. Hibbs, professor in Hydrogeology, for his input on groundwater resources and for his continued help and support. Further, I would like to thank Priscilla Hibbs; Debra Wong, of the Alamo Area Council of Governments; Farrah Penafiel; Destin Tianero; Edgar Lao; Loretta and Alex Shum; Kathy Pursell and Dr. Chris Pursell; and April McPherson of the Department of Homeland Security for their help and encouragement and for keeping me strong.

Finally, I express my most sincere thanks to my younger brothers, Romwel and Romcel, for their support; my mom, Marcela, for training me on artistic planning; and my dad, Romeo Doro-on, for his rigid teachings during my youth, which influenced me to prepare this book creatively. I thank my parents for the multiple talents I have and for their unconditional love and support to their daughter.

About the Author

Anna M. Doro-on is senior consultant at RAMCO Companies, Inc., an engineering and infrastructures services company in the United States. Her primary responsibilities include development and application of innovative environmental remediation technologies; civil, environmental, and water resources engineering; provision of technical oversight on projects; risk assessment and management for critical infrastructure with focus on terrorism, weapons of mass destruction, public health protection, potential reduction of contamination to water resources and the environment, natural disaster, and homeland security; technology development; and project management.

Dr. Doro-on has a PhD in environmental science and engineering from the University of Texas at San Antonio (2009), master of science (MS) in civil engineering from the University of Texas at San Antonio (2003), and bachelor of science (BS) in civil engineering specialized in structural and construction engineering at the Ateneo de Davao University, Philippines (1999). In addition to her consulting experience, Dr. Doro-on has taught courses at the University of Texas at San Antonio, Civil and Environmental Engineering Department. Her primary strength lies in developing the most cost-effective situational and site-specific solutions utilizing cutting-edge techniques and strategies. Her papers have received recognition from different professional organizations. She is a member of and is involved in the American Society of Civil Engineers–Environmental and Water Resources Institute, Association of Researchers for Construction Management Europe, Decision Making in Urban and Civil Engineering, American Academy of Science, the Society of Hispanic Professional Engineers, the Society on Underground Freight by Transportation and Other Tunnel Systems, Homeland Security and Defense Education Consortium Association, and INFORMS Computing Society.

Chapter 1

Introduction

After the September 11, 2001, series of coordinated terrorist attacks against the United States, the utmost national mission is ensuring the effectiveness of safeguard measures protecting critical national infrastructure. Security of water reserves is a matter of the highest priority for governing agencies, environmental stakeholders, and the general populace worldwide. Consequently, Executive Order (EO) 13010 designated water infrastructure as one of the eight critical national infrastructures. After the terrorist attacks, the USA PATRIOT Act of 2001 (P.L. 107–56), Homeland Security Act of 2002 (P.L. 107–296, Section 2.4), and Homeland Security Presidential Directive 7 (HSPD-7) designated water infrastructure as one of 18 separate infrastructures vital to the security of the United States. Destruction of groundwater resources and the urban water supply system—through using deadly chemical threats that are difficult to remove, blasting water supply treatment facilities, blasting reservoirs/dams, and exploding petrochemical facilities adjacent to water resources—would likely create mass casualties, cause catastrophic health effects, create chaos in regional or national security, cause irreversible damage to the water system, disrupt the downstream industry infrastructure, and cause economic destruction comparable to that from the use of a *weapon of mass destruction* (WMD). Based on review of the literature, there is inadequate protection against acts of terrorism on water infrastructure and scarce technology for ensuring safety and security. Such protective measures are urgently needed so that homeland security professionals, managers, engineers, scientists, and experts can incorporate risk assessment in policy making that provides tools for water infrastructure protection, while providing a flexible vehicle for incorporating public input.

1.1 Objective

The objective of this book is to develop a risk assessment methodology based on *cumulative prospect theory* for the analysis of threats of terrorism against water infrastructure. Model results are compared with the results of other risk and vulnerability assessment processes, formulations, and models recommended by renowned authors, private industry consultants, the U.S. Department of Homeland Security (DHS), and other government agencies for water infrastructure protection and the respective public perception of risk.

1.2 Scope

The scope of this book includes the following: development of an integrated approach of risk assessment based upon cumulative prospect theory; review of legal and regulatory requirements related to the protection policy of groundwater and water supply systems for urban areas against terrorism; application of the developed integrated model to the risk assessment of aquifers of karstic limestone producing the sole water source of large urban regions; and application of the integrated model to the risk assessment of surface water, dams, wells, wastewater treatment facilities, reservoirs, and aqueducts of large urban regions as an illustrative example for the approach.

Specifically, the following will be presented:

- Evaluation of terrorism hazards on water supply systems that affect human health and the environment
- Development of *risk estimation model* based on the event tree analysis
- Terrorism activity scenario development
- Development of *fault tree analysis* on risk estimation for potential terrorism activities, potential contamination, and vandalism of water supply systems
- Development of an integrated approach for the *risk acceptability analysis* embedded with cumulative prospect theory for acts of terrorism against the water infrastructure of urban areas
- Evaluation of hazards of prescription drugs (pharmaceuticals) and personal care products in the water supply system in addition to chemical threats against human health and the environment
- Evaluation of hazards of petrochemical facility explosions and blasting of wastewater treatment plants near water resources
- Review, evaluation, and application of standard qualitative/quantitative processes, operational formulations, and models recommended by other renowned authors, private industry consultants, DHS, and other governmental and state agencies

- Development and introduction of preventive measures, emergency preparedness, emergency response, and recovery plans
- Development of *strategic intelligence analysis* integrated with cumulative prospect theory for water infrastructure terrorism prevention

1.3 Purpose

Terrorist attacks against water infrastructure through contamination using chemicals that are difficult to treat, blasting of water supply treatment facilities, explosion of petrochemical facilities, and destruction of reservoirs/dams could impact the public in the following ways:

- Cause mass casualties and catastrophic health effects (e.g., physical and mental illness, disease, or death)
- Create chaos in regional or national security
- Cause damage to public morale and confidence
- Contaminate the water supply system and cause long-term damage to safe drinking water supplies
- Disrupt the downstream industry and commercial infrastructure that depend on safe water supplies
- Create irreversible damage to groundwater resources and water supply systems
- Cause regional or national economic and financial chaos from the loss of groundwater resources and the water supply system
- Cause damage to the environment and natural resources
- Create a need to remediate and replace portions of the water system to make it safe, which could in turn create water shortages or outages
- Result in significant costs for remediation or replacement of the water supply system, which weaken the U.S. economy

Because any of these impacts could have serious consequences, the United States should be concerned about terrorist attacks using chemical threats and other potential ways of contamination of the urban water supply system.

Accordingly, this study is crucially needed in recognizing and identifying prospective events of terrorism against water infrastructure. The uncovering of these events may lead to strategic improvements in U.S. water infrastructure security, make it more difficult for the attacks to succeed, and lessen the impact of attacks that may occur. Safeguards employed include change in policy, incorporation of intrusion detection technology, increased surveillance, and improved intelligence. In addition to strategic security enhancements, tactical security improvements to water reserves can be rapidly implemented to neutralize potential attacks.

The HSPD-7 designated the U.S. Environmental Protection Agency (EPA) as the federal lead for water infrastructure protection. Both the DHS and the EPA must highly prioritize the protection of groundwater resources and the urban water supply system because they are vulnerable to contamination from deadly agents. The destruction of groundwater resources and the urban water supply system by terrorists can cause catastrophic effects comparable to those resulting from the use of a WMD. Risk assessment can shed light on specific strategy and regulatory improvements.

Chapter 2

Acts of Terrorism and the Biological, Chemical, and Radiological Weapons Used against Water Infrastructure

2.1 Introduction

This chapter introduces the terrorist acts and the biological, radiological, and chemical weapons used for the terrorist attacks, and the new emerging threats to the drinking water supply—prescription drugs. The risks to water reserves and public health are presented herein to provide awareness not only to governmental agencies but also mainly to the general public. Reduction approaches for the contaminants are also presented in this chapter, which can be used as guidelines to water supply system recovery in the event of contamination.

The U.S. Department of Defense (DOD) has defined terrorism as the calculated use of the threat of unlawful violence to inculcate fear intended to intimidate governments or societies in the pursuit of goals that are generally political, religious, or ideological. Thus, violence, fear, and intimidation are the three key elements that produce terror in its victims. Determination of terrorist's strategies and identification of their goals in previous and or future attacks will provide guidance to strategic security enhancements for the water infrastructure in the United States.

2.2 Characterization of Terrorism

As water supply systems are essential to human life and the environment, disruption for any period could cause panic and disorder in society. Chemical threats such as arsenic and cyanide (CN) are not sufficiently understood as potential weapons for terrorism against water infrastructure. In fact, these chemicals are potent because they are not likely to be easily reduced by chlorine oxidation in municipal water-treatment facilities (Figure 4.11). Chlorination of these chemicals can produce hazardous compounds such as arsenic trichloride ($AsCl_3$) and cyanogen (CN_2). However, one of the best available techniques (BAT) identified by the U.S. Environmental Protection Agency (EPA) to treat CN is to use chlorination. Terrorists may achieve their goals of catastrophic damage against United States, gain mass media attention, and create chaos comparable or worse than 9/11 and the *suicide bombings* in Afghanistan and Iraq that generated thousands of American combat and civilian casualties, by successfully contaminating the homeland drinking water supply.

2.2.1 High-Profile Terrorism against the United States

Significant transformations in the international environment have been accompanied by technological changes that may have serious consequences for future terrorist operations against the United States. In order to maximize media attention, fear, and public anxiety, terrorists have increasingly focused their efforts on more destructive and high-profile attacks. For example, on September 11, 2001 (9/11), terrorists hijacked four U.S. commercial airliners that took off from various locations in the United States in a coordinated suicide attack. In separate attacks, two of the airplanes were crashed into the twin towers of the World Trade Center in New York City, which caught fire and eventually collapsed. Then, a third airplane crashed into the Pentagon in Washington, D.C., causing extensive damage. Casualty estimates from New York put the possible death toll at nearly 3,000, while as many as 184 people may have been lost at the Pentagon crash site. This trend toward high-profile, high-impact attacks comes at a time when interest is growing among domestic and international extremists in weapons of mass destruction (WMD).

2.2.2 Existing Regulations against Terrorism in the United States

The existing regulatory requirements for terrorism prevention and security in the United States are provided and examined in the following sections.

2.2.2.1 Critical Infrastructure Information Act

The Critical Infrastructure Information Act (CIIA), codified in United States Code §§131–134, as subtitle B of Title II of the Homeland Security Act (Public Law

107-296, 116 Stat. 2135, Sections 211–215), legalizes the disclosure of information submitted to the U.S. Department of Homeland Security (DHS) about vulnerabilities and threats to critical infrastructure. The creation of a new DHS established the safeguard of U.S. critical infrastructure such as the following: food and water systems, agriculture, health systems and emergency services, information and telecommunications, banking and finance, energy, transportation, the chemical and defense industries, postal and shipping entities, and national monuments and icons.

2.2.2.2 Freedom of Information Act

The Freedom of Information Act (FOIA) contains provisions to ensure citizen access to government information. The FOIA applies only to federal agencies and does not create a right of access to records held by Congress, the courts, or by state or local government agencies (DOJ 2010).

2.2.2.3 Chemical Facility Antiterrorism Standards: Interim Final Rule

DHS has issued an Interim Final Rule (IFR) (DHS-2006-0073, RIN 1601-AA41, 6 CFR Part 27) pursuant to Section 550, which provided the department with authority to promulgate Interim Final Regulations for the security of certain chemical facilities (DHS 2007a). The rule establishes risk-based performance standards for the security of our nation's chemical facilities (DHS 2010a). The standard requires any facility that manufactures, uses, stores, or distributes certain chemicals and their respective screening threshold quantities (STQ) to submit a chemical security assessment tool top-screen within 60 calendar days of coming into possession of the listed chemical at or above 1% by weight. Appendix A of the Chemical Facility Antiterrorism Standards (DHS 2009c) regulation lists the DHS chemicals of interest and their corresponding STQ. There are some industries and facilities that are exempt from CFATS, which include the following: (1) facilities regulated under the Maritime Transportation Safety Act of 2002; (2) public water systems, as defined in the Safe Drinking Water Act; (3) water-treatment facilities as defined in the federal Water Pollution Control Act; (4) facilities owned or operated by the DOD or the Department of Energy (DOE); and (5) facilities subject to regulation by the Nuclear Regulatory Commission (NRC). Essentially, DHS created high standards for propane, chlorine, and ammonium nitrate because of their hazardous and explosive characteristics and potential use in a terrorist attack. Acetone and urea, however, have been removed from the list, even though they can be used as precursors to explosives.

2.2.2.4 USA PATRIOT Act

The acronym PATRIOT stands for Providing Appropriate Tools Required to Intercept and Obstruct Terrorism Act of 2001 (Public Law 107-56). There are

10 USA PATRIOT Act provisions, which are all directed toward enhancing security against terrorism. The provisions include but are not limited to: (1) expanding federal agencies' powers in intercepting, sharing, and using private telecommunications, especially electronic communications, along with a focus on criminal investigations by updating the rules that govern computer crime investigations; (2) protecting the U.S. borders; (3) capturing and prosecuting of terrorists; (4) international money laundering abatement and financial antiterrorism; (5) aiding the families of Public Safety Officers who were injured or killed in terrorist attacks; (6) making grants and entering into contracts with some groups to deal with terrorist organizations that cross jurisdictional boundaries; (7) strengthening the criminal laws against terrorism and mass destruction, as well as assassination or kidnapping as a terrorist activity; (8) improving U.S. intelligence; and (9) protecting critical infrastructure. However, the PATRIOT Act limited the Director and gave him no authority to direct, manage, or undertake Foreign Intelligence Surveillance Act (FISA)-based electronic surveillance or physical search operations unless they have been authorized by statute or executive order as designated in National Security Act of 1947 (50 U.S.C. §403-3(c)).

2.2.3 International Laws and Agencies against Terrorism

Addressing the terrorism phenomenon is a very challenging task. Although condemnation of terrorist activities by the international community has been unanimous and unequivocal, efforts to regulate this phenomenon have been corrupted by differences of approach and competing concerns. A number of key issues remain unresolved, and the solution has been further complicated by the emergence of new forms of terrorism. The challenge facing the international community is translating the statements of condemnation of terrorism into definite measures that can effectively address the very negative effects and consequences of terrorist activities. The International Civil Aviation Organization, for instance, has a brief to develop agreements and recommendations on the security of air travel, including on the threat of hijackings. The International Atomic Energy Agency (IAEA) currently monitors more than 900 facilities around the world where nuclear material is stored. The United Nations Office for Drug Control and Crime Prevention (ODCCP) terrorism prevention experts have recently provided advice to the IAEA and the Organization for the Prohibition of Chemical Weapons and to many national governments seeking to incorporate international treaties against terrorism into domestic law.

2.2.4 Water Infrastructure Terrorism Attempts and Disasters in the United States

Concern over security at water reserves increased dramatically after 9/11. According to the Federal Bureau of Investigation (FBI 2001), recent terror threats against the water supply have occurred, and it specifically advised the nation's water utilities to prepare to defend against attacks on pumping stations and pipelines that deliver water to consumers. Consequently, minimal attempts at terrorism against urban water supplies in the past have created concerns among U.S. government agencies and experts.

Most literature related to water infrastructure terrorism presents different common perceptions of physical attacks, cyber terrorism, or bioterrorism (e.g., bombing of water pipelines, terrorist attacks on electronic systems controlling water operations, and injection of biological threats into hydrants and water pipes). These events are not likely considered as high-profile terrorism by principal terrorists and would not meet their high standard to inflict disaster against the United States because they do not create mega-media attention and do not cause catastrophic events and/or mass casualties comparable with the terrorist attacks on 9/11. Chapters 4 and 11 of this book presents different terrorism scenarios that the terrorist leaders might be currently planning to launch in the United States.

Biological threats against water infrastructure can be treated with tertiary treatment. Likewise, cyber terrorism of water infrastructure can cause temporary public panic but should be resolved by high-tech experts within a reasonable time. However, biological and cyber terrorism of water infrastructure should not be neglected because of its vulnerability from vandalism, criminals, amateur terrorists, or disgruntled individuals. The federal, state, and local governing agencies should acknowledge a wide range of terrorism activity scenarios and disseminate financial support equally for protection and security, and should not focus only on warfare like cyber terrorism and biological threats. Table 2.1 shows some of the major infrastructure disasters and vandalism.

Table 2.1 Major Water Infrastructure Vandalism, Terrorism, and Disasters

Date	*Description*
1889	According to Federal Emergency Management Agency (FEMA), after a night of heavy rains, the South Fork Dam had failed, sending tons of water crashing down the narrow valley. Boiling with huge chunks of debris, the wall of flood water grew at times to 60 ft high, tearing downhill at 40 miles/h and leveling everything in its path (FEMA 2010).
1911	Austin Dam, a dam in the Freeman Run Valley, Potter County, Pennsylvania, failed and destroyed the Bayless Pulp & Paper Mill and the town of Austin, and resulted in the deaths of 78 people.
1928	St. Francis Dam failed catastrophically on March 12, 1928, unleashing 12.5 billion gallons of water 140 ft high, rushing at 18 miles/h. It followed the Santa Clara River, destroying more than 1,200 homes and 10 bridges, and killing more than 500 people, on its way to the Pacific Ocean (AEG 1978).

(Continued)

Table 2.1 Major Water Infrastructure Vandalism, Terrorism, and Disasters (*Continued*)

Date	*Description*
1970	A tailings dam owned by the Buffalo Mining Company in Buffalo Creek, West Virginia failed. In a matter of minutes, 125 people were killed, 1100 people were injured, and more than 3,000 were left homeless (FEMA 2010).
1976	Teton Dam, a 123 m high dam on the Teton River in Idaho, failed, causing $1 billion in damage and leaving 11 dead. More than 4,000 homes and more than 4,000 farm buildings were destroyed as a result of the Teton Dam failure (FEMA 2010).
1977	Kelly Barnes Dam in Georgia failed, killing 39 people, most of them college students (FEMA 2010).
1985	(Failed Attempt) Law enforcement authorities discovered that a small survivalist group in the Ozark Mountains of Arkansas known as The Covenant, the Sword, and the Arm of the Lord (CSA) had acquired a drum containing 30 gallons of potassium cyanide, with the apparent intent to poison water supplies in New York, Chicago, and Washington DC.
2003	According to United Press International (UPI), Al-Qaida threatens the U.S. water supply. A U.S. intelligence official who would not comment on al-Ablaj's credibility played down the threat to U.S. water supplies in a brief interview with UPI. "It is very difficult to covertly poison a reservoir," the official said. "It would take many truckloads of poison" which would make it difficult to do secretly. (Maxnews 2003)
2003	Four incendiary devices were found in the pumping station of a Michigan water-bottling plant. The Earth Liberation Front (ELF) claimed responsibility, accusing Ice Mountain Water Company of "stealing" water for profit (Al-Rodhan 2007).
2005	The levee and flood wall failures caused flooding in 85% of New Orleans and 100% of St. Bernard. Millions of gallons of water spilled into vast areas of New Orleans, flooding thousands of homes and businesses with 10 ft or more of water.
2010	Explosion of the Deepwater Horizon, in which 11 people died, led to an oil spill that devastated a vast area of the U.S. marine environment and had a serious impact on the local fishing industry.

2.3 Chemical Terrorism Acts

Water infrastructure can easily be destroyed by terrorists due to lack of protection and deficiency in technology at original water sources (e.g., recharge system areas), during treatment, in pipelines that distribute water to points of use, or in storage systems. The U.S. antiterrorism agencies like DHS or the Central Intelligence Agency (CIA) provide reference guidelines and manuals for chemical, biological, and radiological terror threats. These reference manuals are intended to supply information on evaluating and taking action against possible chemical, radiological, or biological terrorism. In addition, this chapter presents, identifies, and characterizes such chemicals used for terrorism based on manuals and protocols provided by the Interagency Intelligence Committee on Terrorism (IICT), Environmental Protection Agency, and other governmental agencies will be used as reference on tactical security improvements and developing ways of neutralizing potential attacks on water infrastructure.

2.3.1 Characterization of Chemical Threats

The specification and characterization of chemical threats are provided herein.

2.3.1.1 Chemical Threats

Chemical threats are characterized by the rapid onset of medical symptoms and easily observed signatures such as colored residue, dead foliage, pungent odor, and dead insect and animal life, according to information from the CIA. Some of the common chemical terms are enumerated in Table 2.2.

Table 2.2 Chemical Terms

Chemical Terms	*Description*
Acetylcholinesterase	An enzyme that hydrolyzes the neurotransmitter acetylcholine. The action of this enzyme is inhibited by nerve agents.
Aerosol	Fine liquid or solid particles suspended in a gas (e.g., fog or smoke).
Atropine	A compound used as an antidote for nerve agents.
Casualty (toxic) agents	Produce incapacitation, serious injury, or death. They can be used to incapacitate or kill victims. These agents are choking, blister, nerve, and blood agents.

(*Continued*)

Table 2.2 Chemical Terms (*Continued*)

Chemical Terms	*Description*
Choking agents	Substances that cause physical injury to the lungs. Exposure is through inhalation. In extreme cases, membranes swell and lungs become filled with liquid. Death results from lack of oxygen.
Blister agents	Substances that cause blistering of the skin. Exposure is through liquid or vapor contact with any exposed tissue (eyes, skin, and lungs).
Nerve agents	Substances that interfere with the central nervous system. Exposure is primarily through contact with the liquid (skin and eyes) and secondarily through inhalation of the vapor. Three distinct symptoms associated with nerve agents are pinpoint pupils, an extreme headache, and severe tightness in the chest.
Blood agents	Substances that injure a person by interfering with cell respiration.
Chemical agent	A chemical substance that is intended for use in military operations to kill, seriously injure, or incapacitate people through its physiological effects. The agent may appear as a vapor, aerosol, or liquid; it can be either a casualty/toxic agent or an incapacitating agent.
Cutaneous	Pertaining to the skin.
Decontamination	The process of making any person, object, or area safe by absorbing, destroying, neutralizing, making harmless, or removing the hazardous material.
G-series nerve agents	Chemical agents of moderate to high toxicity developed in the 1930s. Examples are tabun (GA), sarin (GB), and soman (GD).
Incapacitating agents	Produce temporary physiological and/or mental effects through action on the central nervous system. Effects may persist for hours or days, but victims usually do not require medical treatment. However, such treatment speeds recovery.
Vomiting agents	Produce nausea and vomiting effects, can also cause coughing, sneezing, pain in the nose and throat, nasal discharge, and tears.

Table 2.2 Chemical Terms (*Continued*)

Chemical Terms	*Description*
Tear agents	Produce irritating or disabling effects that rapidly disappear within minutes after exposure ceases.
Central nervous system depressants	Compounds that have the predominant effect of depressing or blocking the activity of the central nervous system. The primary mental effects include the disruption of the ability to think, sedation, and lack of motivation.
Central nervous system stimulants	Compounds that have the predominant effect of flooding the brain with too much information. The primary mental effect is loss of concentration, causing indecisiveness and the inability to act in a sustained, purposeful manner.
Industrial agents	Chemicals developed or manufactured for use in industrial operations or research by industry, government, or academia. These chemicals are not primarily manufactured for the specific purpose of producing human casualties or rendering equipment, facilities, or areas dangerous for use by man. HCN, cyanogen chloride, phosgene, chloropicrin, and many herbicides and pesticides are industrial chemicals that also can be chemical agents.
Liquid agent	A chemical agent that appears to be an oily film or droplets. The color ranges from clear to brownish amber.
Nonpersistent agent	An agent that on release loses its ability to cause casualties after 10–15 min. It has a high evaporation rate and is lighter than air and will rapidly disperse. It is considered to be a short-term hazard. However, in small unventilated areas, the agent will be more persistent.
Organophosphorous compound	A compound, containing the elements phosphorus and carbon, whose physiological effects include inhibition of acetylcholinesterase. Many pesticides (malathione and parathion) and virtually all nerve agents are organophosphorous compounds.
Percutaneous agent	Able to be absorbed by the body through the skin.

(*Continued*)

Table 2.2 Chemical Terms (*Continued*)

Chemical Terms	*Description*
Persistent agent	An agent that on release retains its casualty-producing effects for an extended period of time, usually anywhere from 30 min to several days. A persistent agent usually has a low-evaporation rate, and its vapor is heavier than air. Therefore, its vapor cloud tends to hug the ground. It is considered to be a long-term hazard. Although inhalation hazards are still a concern, extreme caution should be taken to avoid skin contact as well.
V-series nerve agents	Chemical agents of moderate to high toxicity developed in the 1950s. They are generally persistent. Examples are VE, VG, VM, VS, and VX.
Vapor agent	A gaseous form of a chemical agent. If heavier than air, the cloud will be close to the ground. If lighter than air, the cloud will rise and disperse more quickly.

Source: Data from U.S. Central Intelligency Agency. 1998. *Terrorist CBRN: Materials and Effects*. https://www.cia.gov/library/reports/general-reports-1/cbr_handbook/cbrbook.htm#6; U.S. Central Intelligency Agency. 2010. *Terrorist CBRN: Materials and Effects*. https://www.cia.gov/library/reports/general-reports-1/terrorist_cbrn/terrorist_CBRN.htm.

2.3.1.2 Potential Chemical Threats

Terrorists have considered a wide range of toxic chemicals for attacks. Typical plots focus on poisoning foods or spreading an agent on surfaces to poison through skin contact, but some also include broader dissemination techniques. This book focuses on CN and arsenic compounds used for WMD through water supply contamination as practical examples; the chemical threats from CIA guidelines are also listed in Sections 2.3.1.3 through 2.3.1.12. The blasting of water facilities, the explosion of major petroleum refineries near water resources as depicted in Figure 4.10, and biological and radiological threats will also be analyzed in conjunction with chemical terrorism using CN and arsenic compounds.

2.3.1.3 Cyanide

Terrorists may likely consider using a number of toxic CN compounds against the United States. For immediate result, sodium or potassium cyanides are suitable weapons for terrorist attacks on groundwater resources and urban water supply systems because they can be affordably acquired from within U.S. borders, can be purchased as CN-based pesticides, or can be stolen from chemical plants, gold mining, and electroplating sites. Thus, they can be disseminated as contact poisons when mixed with chemicals that enhance skin penetration. Exposure to CN may produce nausea, vomiting, palpitations, confusion, hyperventilation, anxiety, and vertigo that may progress to agitation, stupor, coma, and death. At high doses, it causes immediate collapse.

2.3.1.4 Mustard Agents

According to the CIA (2010), a mustard agent is a blister agent that poses a contact and vapor hazard. Its color ranges from clear to dark brown depending on purity, and it has characteristic garlic-like odor. Mustard agents are not commercially available, but synthesis does not require significant expertise. In fact, the principal terrorist enemies of United States are typically being taught how to synthesize this agent. They can generate catastrophe by spraying within a crowded area. It causes damage to the lungs, and death by suffocation in severe cases due to water accumulation in the lungs. Medical treatments for exposure are very limited.

2.3.1.5 Nerve Agents

Nerve agents such as sarin, tabun, and VX disrupt a victim's nervous system and cause convulsions that can lead to death. Currently, these agents—sarin, tabun, and VX—are not commercially available and are less likely to be used against water supplies, but there are commercially available chemicals with similar properties.

2.3.1.6 Toxic Industrial Agents

There is a wide range of toxic industrial chemicals that are not as toxic as CN, mustard, or nerve agents that can be used in much larger quantities to compensate for their lower toxicity. Moreover, the effects of industrial agents such as chlorine, organophosphate pesticides, and phosgene are similar to those of mustard agenta. According to the CIA, while organophosphate pesticides are much less toxic, their effects and medical treatments are the same as for military-grade nerve agents.

2.3.1.7 Arsenic

Historically, arsenic has been used as a poison in wars, agriculture, and for household use. There are arsenic-based insecticides, such as those that control fire ants, which can be used for water poisoning. Arsenic is a group I or class-A human carcinogen on the lists of the EPA and International Agency for Research on Cancer (IARC). Also, it cannot be easily destroyed. It simply changes its form and moves around in the environment (ODHS 2002). Major uses of arsenic in the United States have been rodent poisons, insecticides, biocides, and weed killer containing arsenic in both organic and inorganic forms. In pure form, arsenic is a tasteless, odorless white powder or clear crystals. Ingestion of 2 g or more may be lethal in a very short time. Arsenic disrupts *adenosine-5'-triphosphate* (ATP) production through several mechanisms. ATP is a multifunctional nucleotide, which plays an important role in cell biology as a coenzyme that is the molecular unit of currency of intracellular energy transfer (Knowles 1980). Knowles pointed out that, at the level of the citric acid cycle, arsenic inhibits *pyruvate dehydrogenase*, and by competing with phosphate, it uncouples oxidative phosphorylation, thus inhibiting energy-linked reduction of *nicotinamide adenine dinucleotide* (NAD+), mitochondrial respiration, and ATP synthesis. These metabolic interferences lead to death from multisystem organ failure, probably from necrotic cell death (Klaassen and Watkins 2003).

2.3.1.8 Pesticides and Herbicides

Pesticides and herbicides are chemical compounds used for preventing or controlling any insects, rodents, fungi, parasites, and unwanted species of plants or animals from causing harm during or otherwise interfering with the production, processing, storage, transport, or marketing of food, agricultural commodities, wood and wood products or animal feeds, or substances that may be administered to animals for the control of insects, arachnids, or other pests. Many pesticides can be generally grouped as organochlorines, organophosphates, and carbamates. Organochlorine hydrocarbons (e.g. dichlorodiphenyltrichloroethane) could be separated into dichlorodiphenylethanes, cyclodiene compounds, and other related compounds. They operate by disrupting the sodium/potassium balance of the nerve fiber, forcing the nerve to continuously transmit. Their toxicities vary greatly, but they have been phased out because of their persistence and potential to bioaccumulate.

Organophosphate and carbamates have largely replaced organochlorines at present. Organophosphates, which are similar to nerve agents, have been replaced by less toxic carbamate. Thiocarbamate and dithiocarbamates are subclasses of carbamates. Carbamate compounds tend to be soluble in water and weakly adsorbed by soil. Thus, they easily migrate to groundwater through runoff and they have been known to be a major concern for water supply contamination. There are other groups of pesticides (including arsenic and cyanide based pesticides) that can potentially contaminate groundwater supply. The migration of pesticides through soil and water depends on their chemical properties.

2.3.1.9 Gasoline Additive: Methyl Tert-Butyl Ether

Methyl tert-butyl ether (MTBE) is a volatile, flammable, and colorless liquid with a minty odor, is immiscible with water, and is used for a gasoline additive as an oxygenate. Regulatory requirements for MTBE have became more stringent in some states in recent years and its use has declined in the United States. Studies with rats and mice suggest that drinking MTBE may cause gastrointestinal irritation, liver and kidney damage, and nervous system effects (U.S. Agency for Toxic Substances and Disease Registry [ATSDR] 2010a). The EPA concluded that available data are not adequate to quantify the health risks of MTBE at low exposure levels in drinking water, but that the data support the conclusion that MTBE is a potential human carcinogen at high doses (U.S. Environmental Protection Agency [USEPA] 2010).

2.3.1.10 Gasoline Additive: Ethanol

Ethanol is manufactured by fermenting and distilling starch crops, such as corn. The use of ethanol can reduce our dependence on foreign oil and reduce greenhouse gas emissions. Earlier, ethanol was originally reported not to pollute groundwater. Recently, the Government Accountability Office (GAO) reported that it could contaminate groundwater and surface water. There are no precise evaluations of health effects from ethanol exposures. An extensive series of toxicity and exposure assessment studies is currently in progress as part of the EPA 211(b) of the Clean Air Act testing program. Ethanol blends minimize carbon monoxide emissions, making it beneficial in parts of the United States that exceed EPA air quality standards, particularly in winter months.

2.3.1.11 Inorganic Contaminants

Some inorganic chemicals are man-made and some occur naturally. Table 2.3 introduces the health hazards of inorganic chemicals.

Table 2.3 Inorganic Contaminants' Health Effects

Contaminant	*Health Effects*
Antimony	Antimony has been shown to decrease longevity, and alter blood levels of cholesterol and glucose in laboratory animals such as rats exposed to high levels during their lifetimes.
Asbestos	Studies have shown that asbestos has produced lung tumors in laboratory animals. The available information on the risk of developing gastrointestinal tract cancer associated with the ingestion of asbestos from drinking water is limited. Ingestion of intermediate-range chrysotile asbestos fibers greater than 10 μm in length is associated with causing benign tumors in male rats. Chrysotile was the predominant type of asbestos detected in a national survey of the water supplies of 77 communities in North America.
Barium	Barium may damage the heart and cardiovascular system, and is associated with high blood pressure in laboratory animals such as rats exposed to high levels during their lifetimes.

(*Continued*)

Table 2.3 Inorganic Contaminants' Health Effects (*Continued*)

Contaminant	*Health Effects*
Beryllium	Beryllium compounds have been associated with damage to the bones and lungs and induction of cancer in laboratory animals such as rats and mice when the animals are exposed at high levels over their lifetimes. Chemicals that cause cancer in laboratory animals also may increase the risk of cancer in humans who are exposed during long periods of time.
Cadmium	Cadmium has been shown to damage the kidneys in animals such as rats and mice when the animals are exposed at high levels over their lifetimes. Some industrial workers who were exposed to relatively large amount of this chemical during their working careers also suffered damage to the kidneys.
Chromium	Chromium has been shown to damage the kidneys, nervous system, and the circulatory system of laboratory animals such as rats and mice when the animals are exposed at high levels. Some humans who were exposed to high levels of this chemical suffered liver and kidney damage, dermatitis, and respiratory problems.
Cyanide	See Section 2.4.2.1
Fluoride	Exposure to drinking water levels above 4.0 mg/L for many years may result in some cases of crippling skeletal fluorosis, which is a serious bone disorder.
Mercury	Mercury has been shown to damage the kidneys of laboratory animals, such as rats, when the animals are exposed at high levels during their lifetimes.
Nickel	Nickel has been shown to damage the heart and liver in laboratory animals when the animals are exposed to high levels over their lifetimes.
Nitrate	Excessive levels of nitrate in drinking water have caused serious illness and sometimes death in infants less than 6 months of age. The serious illness in infants is caused because nitrate is converted to nitrite in the body. Nitrite interferes with the oxygen carrying capacity of the child's blood. This is an acute disease in that symptoms can develop rapidly in infants. In most cases, health deteriorates over a period of days. Symptoms include shortness of breath and blueness of the skin. Clearly, expert medical advice should be sought immediately if these symptoms occur.

Table 2.3 Inorganic Contaminants' Health Effects (*Continued*)

Contaminant	*Health Effects*
Nitrite	Although excessive levels of nitrite in drinking water have not been observed, other sources of nitrite have caused serious illness and sometimes death in infants less than 6 months of age. The serious illness in infants is caused because nitrite interferes with the oxygen-carrying capacity of the child's blood. This is an acute disease in that symptoms can develop rapidly. However, in most cases, health deteriorates over a period of days. Symptoms include shortness of breath and blueness of the skin. Clearly, expert medical advice should be sought immediately if these symptoms occur.
Selenium	In humans, exposure to high levels of selenium over a long period of time has resulted in a number of adverse health effects, including a loss of feeling and control in the arms and legs.
Thallium	This chemical has been shown to damage the kidneys, liver, brain, and intestines of laboratory animals when the animals are exposed at high levels during their lifetimes.

Source: Data from U.S. Agency for Toxic Substances and Disease Registry (ATSDR), http://www.atsdr.cdc.gov/toxfaqs/tf.asp?id=227&tid=41, 2010; California Environmental Protection Agency-Office of Environmental Health Hazard Assessment (Cal. EPA-OEHHA), http://oehha.ca.gov/water/phg/, 2010; Standard Material Safety and Data Sheet. 2010. Source of the chemicals or chemical compounds, MSDS, http://www.msdsonline.com, 2010; Oxford University, http://msds.chem.ox.ac.uk/HY/, 2010; U.S. Environmental Protection Agency, http://www.epa.gov/safewater/pdfs/factsheets/ioc/tech/cyanide.pdf, 2010; U.S. Environmental Protection Agency, http://water.epa.gov/drink/contaminants/#Inorganic, 2010.

2.3.1.12 Organic Contaminants

Organic compounds are chemicals constructed of molecules that possess carbon-based atoms. Many organic liquid compounds are characterized as immiscible or have a very low solubility in water. Nonaqueous phase liquids (NAPLs) are hydrocarbons that exist as a separate, immiscible phase when in contact with water. Differences in the physical and chemical properties of water and NAPL result in the formation of a physical interface between the liquids, which prevents the two fluids from mixing with each other. NAPLs are typically classified as either light nonaqueous phase liquids (LNAPLs; e.g., petroleum products) that have densities less than that of water, or dense nonaqueous phase liquids (DNAPLs; e.g., chlorinated solvents) that have densities greater than that of water. Refined petroleum products are generally complex mixtures of a variety of organic compounds with minor fractions of organic and inorganic additives that fall into a number of chemical classes. Chlorinated solvents are generally released to the environment in a more or less pure form as opposed to a complex mixture. Table 2.4 presents a list of organic contaminants in the environment.

Table 2.4 Organic Contaminants

Contaminant	*Description*
Acrylamide	Acrylamide is used in wastewater treatment, papermaking, ore processing, and the manufacture of permanent press fabrics. It also occurs in many cooked starchy foods, such as potato chips, French fries, and bread that has been heated.
Alachlor	Alachlor is an herbicide and is used to control annual grasses and broadleaf weeds.
Atrazine	Atrazine is a widely used herbicide to mitigate broadleaf and grassy weeds in major crops.
Benzene	Benzene is used as an additive in gasoline and is an important solvent and precursor in the production of drugs, plastics, synthetic rubber, and dyes. It is discharged from factories and leaches from gas storage tanks and landfills.
Benzo(a)pyrene (PAHs)	PAHs are produced as byproducts of fuel burning. PAHs are also found in foods (e.g., cereal, oils, and fats).
Carbofuran	Carbofuran is one of the most toxic carbamate pesticides. It is used to control insects in a variety of field crops including rice, potatoes, corn, soybeans, and alfalfa.
Carbon tetrachloride	Discharge from chemical plants and other industrial activities.
Chlordane	Residue of banned termiticide.
Chlorobenzene	Discharge from chemical and agricultural chemical factories.
2,4-D	Runoff from herbicide used on row crops.
Dalapon	Runoff from herbicide used on rights of way.
1,2-Dibromo-3-chloropropane (DBCP)	Runoff/leaching from soil fumigant used on soybeans, cotton, pineapples, and orchards.

Table 2.4 Organic Contaminants (*Continued*)

Contaminant	*Description*
o-Dichlorobenzene	Discharge from industrial chemical factories.
p-Dichlorobenzene	Discharge from industrial chemical factories.
1,2-Dichloroethane	Discharge from industrial chemical factories.
1,1-Dichloroethylene	Discharge from industrial chemical factories.
Cis-1,2-Dichloroethylene	Discharge from industrial chemical factories.
Trans-1,2-Dichloroethylene	Discharge from industrial chemical factories.
Dichloromethane	Discharge from drug and chemical factories.
1,2-Dichloropropane	Discharge from industrial chemical factories.
Di(2-ethylhexyl) adipate	Discharge from chemical factories.
Di(2-ethylhexyl) phthalate	Discharge from rubber and chemical factories.
Dinoseb	Runoff from herbicide used on soybeans and vegetables.
Dioxin (2,3,7,8-TCDD)	Emissions from waste incineration and other combustion; discharge from chemical factories.
Diquat	Runoff from herbicide use.
Endothall	Runoff from herbicide use.
Endrin	Residue of banned insecticide.
Epichlorohydrin	Discharge from industrial chemical factories; an impurity of some water treatment chemicals.
Ethylbenzene	Discharge from petroleum refineries.
Ethylene dibromide	Discharge from petroleum refineries.
Glyphosate	Runoff from herbicide use.
Heptachlor	Residue of banned termiticide.
Heptachlor epoxide	Breakdown of heptachlor.
Hexachlorobenzene	Discharge from metal refineries and agricultural chemical factories.

(*Continued*)

Table 2.4 Organic Contaminants (*Continued*)

Contaminant	*Description*
Hexachlorocyclopentadiene	Discharge from chemical factories.
Lindane	Runoff/leaching from insecticide used on cattle, lumber, gardens.
Methoxychlor	Runoff/leaching from insecticide used on fruits, vegetables, alfalfa, livestock.
Oxamyl (Vydate)	Runoff/leaching from insecticide used on apples, potatoes, and tomatoes.
Polychlorinated biphenyls (PCBs)	Runoff from landfills; discharge of waste chemicals.
Pentachlorophenol	Discharge from wood preserving factories.
Picloram	Herbicide runoff.
Simazine	Herbicide runoff.
Styrene	Discharge from rubber and plastic factories; leaching from landfills.
Tetrachloroethylene	Discharge from factories and dry cleaners.
Toluene	Discharge from petroleum factories.
Toxaphene	Runoff/leaching from insecticide used on cotton and cattle.
2,4,5-TP (Silvex)	Residue of banned herbicide.
1,2,4-Trichlorobenzene	Discharge from textile finishing factories.
1,1,1-Trichloroethane	Discharge from metal-degreasing sites and other factories.
1,1,2-Trichloroethane	Discharge from industrial chemical factories.
TCE	Discharge from metal-degreasing sites and other factories.
Vinyl chloride	Leaching from PVC pipes; discharge from plastic factories.
Xylenes (total)	Discharge from petroleum factories; discharge from chemical factories.

Source: Data from U.S. Environmental Protection Agency, http://water.epa.gov/drink/contaminants/#Organic, 2010.

2.4 Potential Hazards of Chemical Threats

Terrorist attacks using chemical threats against groundwater can produce catastrophe. Even a relatively inept attack with limited mortality and property damage could accomplish the terrorists' goal of demoralization. Based on the U.S. DOD CBRN (2008), chemical substances that are used in terrorism are intended to kill, seriously injure, or incapacitate humans through their physiological effects. The quantity of chemicals needed for the terrorist to generate mass casualties is normally based on the LD_{50} of the chemical or chemical compound. The LD_{50} of chemical threats are presented herein.

2.4.1 *Chemicals' LD_{50}*

LD stands for *lethal dose*, and LD_{50} is the amount of a chemical(s), given all at once, which causes the death of 50% of a group of test animals. It is a standard quantification and basis of acute toxicity that is stated in milligrams (mg) of chemical or contaminant per kilogram (kg) of body weight or *water*. LC stands for *lethal concentration*, and it usually refers to the concentration of a chemical in air. Because of the changeability of dose–response effects between individual beings, the toxicity of a substance is typically expressed as the concentration or dose that is lethal to 50% of the exposed population (LC_{50} or LD_{50}). It represents the dose required to kill 50% of a population of test animals (e.g., rats, rabbits, mice). LD_{50} or LC_{50} values are standard measurements, so it is possible to verify relative toxicities among chemical substances. Hence, the lower the LD_{50} dose, the more toxic the contaminants or chemicals, which is the primary basis for terrorists to succeed in attacking a water supply.

2.4.1.1 *Cyanide's LD_{50}*

The LC_{50} for gaseous hydrogen cyanide (HCN) is 100–300 parts per million (ppm). Inhalation of CN in this range results in death within 10–60 minutes, with death coming more quickly as the concentration increases. Inhalation of 2,000 ppm of HCN causes death within 1 minute. The LD_{50} for ingestion is 50–200 mg, or 1–3 mg per kilogram (kg) of body weight, calculated as HCN. Meanwhile, for contact with unabraded skin, the LD_{50} is 100 mg as HCN per kg of body weight (ICMC 2009). The LD_{50} of sodium cyanide is 6.4 mg/kg (oral-rat).

2.4.1.2 *Arsenic's LD_{50}*

Organic forms of arsenic appear to have a lower toxicity than inorganic forms of arsenic. Research has shown that arsenites (trivalent forms) have a higher acute toxicity than arsenates (pentavalent forms) (Kingston, Hall, and Sioris 1993). The acute minimal lethal dose of arsenic in adults is estimated to be 70–200 mg or 1 mg/kg/day (Dart 2004).

2.4.1.3 Pesticides and Herbicides' LD_{50}

A mysterious secret of the chemical industry is that the *inert* ingredients, which are the carriers or bulking agents for pesticides, are usually more toxic than the *active* ingredients, yet this information is not required to be detailed or posted on the labels. The general public, and even professional applicators, usually have no idea as to these chemical contaminants. Approximately 3,700 chemicals can legally not be revealed to the public in pesticides, which comprise up to 97% of products like weed and ant killers. It is apparent that some corporations or organizations may be using this opportunity as an economical form of hazardous waste discharge. Pesticide ways of entry include the respiratory system, digestive system, and skin. The greatest hazard is by pesticide entry through the respiratory system. Absorption through the digestive tract is the second most hazardous pathway for poisoning. The skin provides an effective barrier against pesticides poisoning. However, there is substantial variation in the rate of penetration through the skin by different substances.

2.4.1.4 Gasoline Additive: MTBE's LD_{50}

MTBE exhibits low acute toxicity through the oral, dermal, and inhalation routes in mammals. According to the EPA, in rats, the average oral LD_{50} is 4,000 mg/kg. Dermal LD_{50} is more than 10,000 mg/kg and LC_{50} by inhalation is approximately 100 mg/L. MTBE is respectively regarded a skin irritant but is not showing any indication that it can irritate the lungs and the eyes, while the EPA continues to further their research. In repeated dose-toxicity studies, the principal affected organs are the liver and the kidneys, mainly at inhaled concentrations of 3,000 ppm and above or at oral doses of 250 mg/kg or higher.

MTBE is not totally standardized under the federal drinking water regulations, and some states and local agencies such as in California and New York have started to provide stringent regulations, while other states have lesser requirements for MTBE. The California Department of Health Services (CDHS) recently established a secondary maximum contaminant level (MCL) for MTBE as 0.05 mg/L (5 μg/L or 5 ppb) based on taste and odor effective January 7, 1999 (22 CCR Section 64449). An interim nonenforceable action level (AL) of 0.035 mg/L (35 μg/L or 35 ppb) in drinking water was established by CDHS in 1991 to protect against adverse health effects.

2.4.1.5 Gasoline Additive: Ethanol's LD_{50}

Ethanol is commonly used in the medical arena as a hypnotic or depressant. It depresses activity in the upper brain and is also toxic if ingested in sufficiently large quantities, but it is much less toxic than methanol or gasoline. In rats, the lethal dose of ethanol is 13.7 g/kg of body weight (Solomons 1988) or 20–40 times greater than that for methanol.

2.4.1.6 Inorganic and Organic Contaminants' LD_{50}

The inorganic and organic chemicals' LD_{50} are shown in Tables 2.5 and 2.6, and terrorist would likely use these LD_{50} data to design their attacks against U.S. water supply.

Table 2.5 Inorganic Chemicals' LD_{50}

Contaminant	*LD_{50}*
Antimony	Oral rat LD_{50}: 4,480 mg/kg (antimony acetate) Oral LD_{50}: 115 mg/kg (antimony potassium tartrate) Oral LD_{50}: 20,000 mg/kg (antimony(III)oxide)
Asbestos	Although asbestos is a known human carcinogen by the inhalation route, available epidemiological studies do not support the hypothesis that an increased cancer risk is associated with the ingestion of asbestos in drinking water.
Barium	Oral rat LD_{50}: 355 mg/kg (barium nitrate) Oral rat LD_{50}: 118 mg/kg (barium chloride) LD_{50} for rats: 630 mg/kg (barium carbonate) LD_{50} for rats: 921 mg/kg (barium acetate)
Beryllium	Typical oral mouse LD_{50}: between 0.5 and 5 mg/kg
Cadmium	Acute oral toxicity-rat (LD_{50}): 890 mg/kg
Chromium	Human 0.5–1 g, oral – lethal (potassium chromate) Rat LD_{50}: 1,800 mg/kg, oral (chromium(III)chloride) LD_{50}: 3,250 mg/kg, oral (chromium(III)nitrate)
Cyanide	See Section 2.4.1.1
Fluoride	Oral LD_{50}: 60 mg F/kg body weight to 172 mg F/kg (fluoride) Oral rat LD_{50}: 125 g/kg (sodium fluorosilicate), corresponding to 12.5 g for a 100 kg adult
Mercury	Oral rat LD_{50}: 170 mg/kg (anhydrous) 182 mg/kg (dehydrate) Oral rat LD_{50}: 1 mg/kg ((mercury II) chloride) Oral rat LD_{50}: 18 mg/kg (mercury oxide) Oral rat LD_{50}: 46 mg/kg (mercuric thiocyanate)
Nickel	Oral rat LD_{50}: >5 g/kg

(Continued)

Table 2.5 Inorganic Chemicals' LD_{50} (*Continued*)

Contaminant	*LD_{50}*
Nitrate	Oral rat LD_{50}: 200 mg/kg
Nitrite	Oral rat LD_{50}: 300 mg/kg
Selenium	Oral rat LD_{50}: Acute: 6,700 mg/kg
Thallium	Oral rat LD_{50}: 0.002 mg/kg

Source: Data from U.S. Agency for Toxic Substances and Disease Registry (ATSDR), http://www.atsdr.cdc.gov/toxfaqs/tf.asp?id=227&tid=41, 2010; California Environmental Protection Agency-Office of Environmental Health Hazard Assessment (Cal. EPA-OEHHA), http://oehha.ca.gov/water/phg/, 2010; Standard Material Safety and Data Sheet. 2010. Source of the chemicals or chemical compounds, MSDS, http://www.msdsonline.com, 2010; Oxford University, http://msds.chem.ox.ac.uk/HY/, 2010; U.S. Environmental Protection Agency, http://www.epa.gov/safewater/pdfs/factsheets/ioc/tech/cyanide.pdf, 2010; U.S. Environmental Protection Agency, http://water.epa.gov/drink/contaminants/#Inorganic, 2010.

Table 2.6 Organic Chemicals' LD_{50}

Contaminant	*LD_{50}*
Acrylamide	Oral rat LD_{50}: 124 mg/kg Skin rat LD_{50}: 400 mg/kg
Alachlor	Oral rat LD_{50}: 930 mg/kg and 1,350 mg/kg
Atrazine	Oral rat LD_{50}: 672–3,000 mg/kg Oral mouse LD_{50}: 850–1,750 mg/kg
Benzene	Oral rat LD_{50}: 930 mg/kg
Benzo(a)pyrene (PAHs)	Oral rat LD_{50}: 50 mg/kg
Carbofuran	Dermal rabbit LD_{50}: 6,783 mg/kg Oral rat LD_{50}: 7.34 mg/kg Inhalation rat LC_{50}: 0.10 mg/L/1 h
Carbon tetrachloride	Oral human LD_{LO}: 429 mg/kg Oral rat LD_{50}: 2,350 mg/kg Skin rabbit LD_{50}: >20,000 mg/kg
Chlordane	Oral rat LD_{50}: 200–700 mg/kg Oral mouse LD_{50}: 145–430 mg/kg

Table 2.6 Organic Chemicals' LD_{50} (*Continued*)

Contaminant	*LD_{50}*
Chlorobenzene	Oral rat LD_{50}: 1,110 mg/kg Oral mouse LD_{50}: 2,300 mg/kg
2,4-D	Oral rat LD_{50}: 375–666 mg/kg Oral mouse LD_{50}: 370 mg/kg
Dalapon	Oral rat LD_{50}: 9,330 mg/kg to 7,570 mg/kg
1,2-Dibromo-3-chloropropane (DBCP)	Oral rat LD_{50}: 170 mg/kg
o-Dichlorobenzene	Oral rat LD_{50}: 1,110 mg/kg
p-Dichlorobenzene	Oral rat LD_{50}: 500 mg/kg
1,2-Dichloroethane	Oral rat LD_{50}: 670 mg/kg Skin rabbit LD_{50}: 2,800 mg/kg Oral human LD_{50}: 286 mg/kg
1,1-Dichloroethylene	Oral rat LD_{50}: 200 mg/kg
Cis-1,2-Dichloroethylene	Oral rat LD_{50}: 770 mg/kg
Trans-1,2-Dichloroethylene	Oral rat LD_{50}: 1,235 mg/kg
Dichloromethane	Oral rat LD_{50}: 1,600 mg/kg Oral human LD_{50}: 357 mg/kg
1,2-Dichloropropane	Oral rat LD_{50}: 1,947 mg/kg
Di(2-ethylhexyl) adipate	Oral rat LD_{50}: 9,100 mg/kg
Di(2-ethylhexyl) phthalate	Oral rat LD_{50}: 30 gm/kg
Dinoseb	Oral rat LD_{50}: 25–46 mg/kg
Dioxin (2,3,7,8-TCDD)	Oral rat LD_{50}: 0.022–0.045 mg/kg Oral hamster LD_{50}: 1 mg/kg
Diquat	Oral rat (female) LD_{50}: 231 mg/kg
Endothall	Oral rat LD_{50}: single dose of 40–60 mg/kg Oral dog LD_{50}: 20 or 50 mg/kg-day dose died within 3–11 days
Endrin	Oral rat LD_{50}: 7–43 mg/kg Oral rabbit LD_{50}: 60 mg/kg
Epichlorohydrin	Oral rat LD_{50}: 90 mg/kg

(*Continued*)

Table 2.6 Organic Chemicals' LD_{50} (*Continued*)

Contaminant	*LD_{50}*
Ethylbenzene	Oral rat LD_{50}: 3,500 mg/kg
Ethylene dibromide	Oral rat LD_{50}: 108 mg/kg Oral mouse LD_{50}: 250 mg/kg Oral rabbit LD_{50}: 55 mg/kg
Glyphosate	Oral rat LD_{50}: 5,600 mg/kg
Heptachlor	Oral rat LD_{50}: 40–220 mg/kg Oral mouse LD_{50}: 30–68 mg/kg
Heptachlor epoxide	Oral rat LD_{50}: 15 mg/kg Oral rabbit LD_{50}: 144 mg/kg
Hexachlorobenzene	Oral rat LD_{50}: 10,000 mg/kg Oral guinea pig LD_{50}: 3,000 mg/kg
Hexachlorocyclopentadiene	Oral mouse LD_{50}: 505 mg/kg Oral rat LD_{50}: 200 mg/kg
Lindane	Oral human LD_{100}: 150 mg/kg 10–20 mg/kg (acute toxicity) Oral rat LD_{50}: 88–190 mg/kg
Methoxychlor	Oral rat LD_{50}: 5,000–6,000 mg/kg Oral mouse LD_{50}: 1,850 mg/kg
Oxamyl (Vydate)	Oral rabbit LD_{50}: 2,960 mg/kg
Polychlorinated biphenyls (PCBs)	Oral rat LD_{50}: 1,900 mg/kg
Pentachlorophenol	Oral rat LD_{50}: 27–211 mg/kg Oral mouse LD_{50}: 74–130 mg/kg Oral rabbit LD_{50}: 70–300 mg/kg
Picloram	Oral rat LD_{50}: 8,200 mg/kg Oral mouse LD_{50}: 1,061–4,000 mg/kg Oral rabbit LD_{50}: 2,000–3,500 mg/kg
Simazine	Oral mouse LD_{50}: >5,000 mg/kg Dermal rabbit LD_{50}: 3,100–10,000 mg/kg
Styrene	Oral rat LD_{50}: 2,650 mg/kg Oral mouse LD_{50}: 316 mg/kg

Table 2.6 Organic Chemicals' LD_{50} (*Continued*)

Contaminant	*LD_{50}*
Tetrachloroethylene	Oral rat LD_{50}: 2,629 mg/kg
Toluene	Oral rat LD_{50}: 636 mg/kg Dermal rabbit LD_{50}: 14,100 mg/kg Vapor mouse LC_{50}: 440 ppm for 24 h
Toxaphene	Oral hamster LD_{50}: 112–200 mg/kg Dermal LD_{50} for different species was determined as 300–1,000 mg/kg
2,4,5-TP (Silvex)	Oral mouse LD_{50}: 24 gm/kg
1,2,4-Trichlorobenzene	Oral rat LD_{50}: 756 mg/kg Dermal rat LD_{50}: 6,319 mg/kg
1,1,1-Trichloroethane	Oral rat LD_{50}: 9,600 mg/kg Oral rat LD_{50}: 6,000 mg/kg Dermal rabbit LD_{50}: 15,800 mg/kg Vapor rat LC_{50}: Acute: 18,000 ppm for 4 h
1,1,2-Trichloroethane	Oral rat LD_{50}: <2,500 mg/kg Dermal rat LD_{50}: <4,300 mg/kg
TCE	Oral rat LD_{50}: 5,650 mg/kg Oral mouse LD_{50}: 2,402 mg/kg
Vinyl chloride	Oral rat LD_{50}: 500 mg/kg
Xylene	Oral rat LD_{50}: 4,300 mg/kg Inhalation rat LD_{50}: Acute: 4,550 ppm for 4 h

Source: Data from U.S. Agency for Toxic Substances and Disease Registry (ATSDR), http://www.atsdr.cdc.gov/toxfaqs/tf.asp?id=227&tid=41, 2010; California Environmental Protection Agency-Office of Environmental Health Hazard Assessment (Cal. EPA-OEHHA), http://oehha.ca.gov/water/phg/, 2010; Standard Material Safety and Data Sheet. 2010. Source of the chemicals or chemical compounds, MSDS, http://www.msdsonline.com, 2010; Oxford University, http://msds.chem.ox.ac.uk/HY/, 2010; U.S. Environmental Protection Agency, http://www.epa.gov/safewater/pdfs/factsheets/ioc/tech/cyanide.pdf, 2010; U.S. Environmental Protection Agency, http://water.epa.gov/drink/contaminants/#Inorganic, 2010.

2.4.2 Characterization of Potential Hazards

The potential health effects of chemical threats are presented herein.

2.4.2.1 Cyanide

CN is an inhibitor of the enzyme *cytochrome c oxidase* in the fourth complex of the electron transport chain, and it is found in the membrane of the mitochondria of eukaryotic cells. CN attaches to the iron within this protein. The binding of CN to this *cytochrome* prevents transport of electrons from *cytochrome c oxidase* to oxygen. As a result, the electron transport chain is disrupted, meaning that the cell can no longer aerobically produce *adenosine triphosphate* for energy. Tissues that mainly depend on aerobic respiration, such as the central nervous system and the heart, are particularly affected. A fatal dose for a human can be as low as 1.5 mg/kg body weight (USEPA 1987).

According to the International Program on Chemical Safety (IPCS), sodium cyanide is a highly toxic chemical compound and a deadly human poison by ingestion, and the probable oral lethal dose in humans is less than 5 mg/kg. A taste (less than seven drops) is super toxic for a 70 kg (150 lbs) person.

2.4.2.2 Arsenic

Arsenic compounds are irritants, systemic toxins, and carcinogens in humans. The trivalent arsenic compounds are the ones most toxic to humans. Initial responses to acute poisoning include burning of the lips, constriction of the throat, and dysphagia (Hathaway et al. 1991). This is followed by excruciating pain in the abdominal region, severe nausea, vomiting, and diarrhea. Toxic effects on the liver, blood-forming organs, both central and peripheral nervous systems, and the cardiovascular system may also occur. Convulsions, coma, and death may follow within 24 hours of severe poisonings (Hathaway et al. 1991). Acute inhalation exposures to arsenic compounds may result in damage to the mucous membranes of the respiratory system (Parmeggiani 1983). Severe irritations of the nasal mucosae, larynx, and bronchi have been observed following exposures to arsenic compounds. In addition, exposed skin may become irritated; cases of dermatitis have been reported following dermal contact with arsenic compounds (Parmeggiani 1983). Conjunctivitis, visual disturbances, hyperpigmentation of the skin, and perforation of the nasal septum have been described in the literature (Hathaway et al. 1991). Chronic exposure causes damage to the nervous system, cardiovascular system, and liver (Parmeggiani 1983). Anemia and leukocytopenia have been reported to occur following chronic exposures to arsenic compounds (Parmeggiani 1983). Cancers of the skin, lungs, larynx, lymphoid system, and viscera have been identified as potential responses to arsenic poisoning (Hathaway et al. 1991). IARC has reviewed the available data and considers arsenic to be a Group 1 carcinogen with sufficient evidence of carcinogenicity in humans (IARC 1987).

2.4.2.3 Pesticides

Pesticides are labeled with signal words that indicate their level of toxicity.

Studies show that only 5% of pesticides reach target weeds. The rest runs off into water or dissipates in the air. More serious effects appear to be produced by direct inhalation of pesticide sprays than by absorption or ingestion of toxins. Many of the safety tests used to test these pesticide products are insufficient; they test for short-term effects on healthy adult animals. They test one chemical at a time, when generally people are exposed to various types of chemical compounds at once.

2.4.2.4 Gasoline Additive: MTBE

In controlled clinical tests in which healthy individuals were exposed to 5 mg MTBE/m^3 for 1 hour, no symptoms of adverse effects were observed (USEPA 1993). Recently, the EPA stated that the concentration of 5 mg/m^3 for 1 hour is roughly equivalent to a dose of 0.09 mg/kg. Complaints of headaches, eye irritation, nose and throat irritation, cough, nausea, and dizziness were recorded in two cities in Alaska following the introduction of MTBE-blended gasoline during the fall of 1992 (USEPA 1993). Information on the potential carcinogenicity of MTBE in humans is lacking and the results of animal studies are still under review within the EPA. However, the current unfinished assessment supports a hazard classification of possible human carcinogen based on the limited animal evidence (USEPA 1993).

2.4.2.5 Gasoline Additive: Ethanol

High concentrations of ethanol may create faintness, drowsiness, decreased awareness and responsiveness, euphoria, abdominal discomfort, nausea, vomiting, staggering gait, lack of coordination, and coma. Ingestion of gasoline is of low-to-moderate toxicity and is more toxic to children. Long-term, repeated oral exposure to ethanol may result in liver injury with fibrosis. Long-term exposure to methanol has been associated with headaches, giddiness, conjunctivitis, insomnia, and impaired vision. Meanwhile, gasoline is verified as an animal carcinogen by the International Study for Research in Cancer.

2.4.2.6 Inorganic and Organic Contaminants

The potential health hazards of inorganic and organic chemicals are presented in Tables 2.3, 2.7, and 2.8.

Table 2.7 Organic Contaminants' Health Effects

Contaminant	*Health Effects*
2,3,7,8-TCDD (Dioxin)	Dioxin has been shown to cause cancer in laboratory animals such as rats and mice when the animals are exposed at high levels during their lifetimes. Chemicals that cause cancer in laboratory animals also may increase the risk of cancer in humans who are exposed during long periods of time.
2,4-D	2,4-D has been shown to damage the liver and kidneys of laboratory animals such as rats exposed at high levels during their lifetimes. Some humans who were exposed to relatively large amounts of this chemical also suffered damage to the nervous system.
2,4,5-TP	2,4,5-TP has been shown to damage the liver and kidneys of laboratory animals such as rats and dogs exposed to high levels during their lifetimes. Some industrial workers, who were exposed to relatively large amount of this chemical during their working careers also suffered damage to the nervous system.
Alachlor	Alachlor has been shown to cause cancer in laboratory animals such as rats and mice when the animals are exposed at high levels during their lifetimes. Chemicals that cause cancer in laboratory animals also may increase the risk of cancer in humans who are exposed during long periods of time.
Atrazine	Atrazine has been shown to affect offspring of rats and the heart of dogs.
Benzo(a)pyrene	Benzo(a)pyrene has been shown to cause cancer in animals such as rats and mice when the animals are exposed at high levels.
Carbofuran	Carbofuran has been shown to damage the nervous and reproductive systems of laboratory animals such as rats and mice exposed at high levels during their lifetimes. Some humans, who were exposed to relatively large amount of this chemical during their working careers also suffered damage to the nervous system. Effects on the nervous system are generally rapidly reversible.

Table 2.7 Organic Contaminants' Health Effects (*Continued*)

Contaminant	*Health Effects*
Chlordane	Chlordane has been shown to cause cancer in laboratory animals such as rats and mice when the animals are exposed at high levels during their lifetimes. Chemicals that cause cancer in laboratory animals also may increase the risk of cancer in humans who are exposed during long periods of time.
Dalapon	Dalapon has been shown to cause damage to the kidneys and liver in laboratory animals when the animals are exposed to high levels during their lifetimes.
Di(2-ethylhexyl) adipate	Di(2-ethylhexyl) adipate has been shown to damage the liver and testes in laboratory animals such as rats and mice exposed to high levels. EPA has set the drinking water standard for di(2-ethylhexyl) adipate at 0.4 part per million (ppm) to protect against the risk of adverse health effects.
Di(2-ethylhexyl) phthalate	Di(2-ethylhexyl) phthalate has been shown to cause cancer in laboratory animals such as rats and mice exposed to high levels during their lifetimes.
Dibromochloropropane (DBCP)	Dibromochloropropane has been shown to cause cancer in laboratory animals such as rats and mice when the animals are exposed at high levels during their lifetimes. Chemicals that cause cancer in laboratory animals also may increase the risk of cancer in humans who are exposed during long periods of time.
Dinoseb	Dibromochloropropane has been shown to damage the thyroid and reproductive organs in laboratory animals such as rats exposed to high levels.
Diquat	Diquat has been shown to damage the liver, kidneys, and gastrointestinal tract and causes cataract formation in laboratory animals such as dogs and rats exposed at high levels during their lifetimes.

(*Continued*)

Table 2.7 Organic Contaminants' Health Effects (*Continued*)

Contaminant	*Health Effects*
Endothall	Endothall has been shown to damage the liver, kidneys, gastrointestinal tract, and reproductive system of laboratory animals such as rats and mice exposed at high levels during their lifetimes.
Endrin	Endrin has been shown to cause damage to the liver, kidneys, and heart in laboratory animals such as rats and mice when the animals are exposed at high levels during their lifetimes.
Ethylene dibromide (EDB)	Ethylene dibromide has been shown to cause cancer in laboratory animals such as rats and mice when the animals are exposed at high levels during their lifetimes. Chemicals that cause cancer in laboratory animals also may increase the risk of cancer in humans who are exposed during long periods of time.
Glyphosate	Glyphosate has been shown to cause damage to the liver and kidneys in laboratory animals such as rats and mice when the animals are exposed at high levels during their lifetimes.
Heptachlor	Heptachlor has been shown to cause cancer in laboratory animals such as rats and mice when the animals are exposed at high levels during their lifetimes. Chemicals that cause cancer in laboratory animals also may increase the risk of cancer in humans who are exposed during long periods of time.
Heptachlor epoxide	Heptachlor epoxide has been shown to cause cancer in laboratory animals such as rats and mice when the animals are exposed at high levels during their lifetimes. Chemicals that cause cancer in laboratory animals also may increase the risk of cancer in humans who are exposed during long periods of time.

Table 2.7 Organic Contaminants' Health Effects (*Continued*)

Contaminant	*Health Effects*
Hexachlorobenzene	Hexachlorobenzene has been shown to cause cancer in laboratory animals such as rats and mice when the animals are exposed to high levels during their lifetimes. Chemicals that cause cancer in laboratory animals also may increase the risk of cancer in humans who are exposed during long periods of time.
Hexachlorocyclopentadiene	This organic chemical is used as an intermediate in the manufacture of pesticides and flame retardants. It may get into water by discharge from production facilities. This chemical has been shown to damage the kidneys and the stomach of laboratory animals when exposed at high levels during their lifetimes.
Lindane	This organic chemical is used as a pesticide. When soil and climatic conditions are favorable, lindane may get into drinking water by runoff into surface water or by leaching into groundwater. This chemical has been shown to damage the liver, kidneys, nervous system, and immune system of laboratory animals such as rats, mice, and dogs exposed at high levels during their lifetimes. Some humans who were exposed to relatively large amounts of this chemical also suffered damage to the nervous system and circulatory system.
Methoxychlor	Methoxychlor has been shown to damage the liver, kidneys, nervous system, and reproductive system of laboratory animals such as rats exposed at high levels during their lifetimes. It has also been shown to produce growth retardation in rats.
Oxamyl	Oxamyl is used as a pesticide for the control of insects and other pests. It may get into drinking water by runoff into surface water or leaching into groundwater. This chemical has been shown to damage the kidneys of laboratory animals such as rats when exposed at high levels during their lifetimes.

(*Continued*)

Table 2.7 Organic Contaminants' Health Effects (*Continued*)

Contaminant	*Health Effects*
Pentachlorophenol	Pentachlorophenol has been shown to produce adverse reproductive effects and to damage the liver and kidneys of laboratory animals such as rats exposed to high levels during their lifetimes. Some humans who were exposed to relatively large amounts of this chemical also suffered damage to the liver and kidneys. This chemical has been shown to cause cancer in laboratory animals such as rats and mice when the animals are exposed to high levels during their lifetimes. Chemicals that cause cancer in laboratory animals also may increase the risk of cancer in humans who are exposed during long periods of time.
Picloram	Picloram has been shown to cause damage to the kidneys and liver in laboratory animals, such as rats, when the animals are exposed at high levels during their lifetimes.
Polychlorinated biphenyls (PCBs)	Polychlorinated biphenyls have been shown to cause cancer in laboratory animals such as rats and mice when the animals are exposed at high levels during their lifetimes. Chemicals that cause cancer in laboratory animals also may increase the risk of cancer in humans who are exposed during long periods of time.
Simazine	Simazine may cause cancer in laboratory animals such as rats and mice exposed at high levels during their lifetimes. Chemicals that cause cancer in laboratory animals also may increase the risk of cancer in humans who are exposed during long periods of time.

Table 2.7 Organic Contaminants' Health Effects (*Continued*)

Contaminant	*Health Effects*
Toxaphene	Toxaphene has been shown to cause cancer in laboratory animals such as rats and mice when the animals are exposed at high levels during their lifetimes. Chemicals that cause cancer in laboratory animals also may increase the risk of cancer in humans who are exposed during long periods of time.

Source: Data from U.S. Agency for Toxic Substances and Disease Registry (ATSDR), http://www.atsdr.cdc.gov/toxfaqs/tf.asp?id=227&tid=41, 2010; California Environmental Protection Agency-Office of Environmental Health Hazard Assessment (Cal. EPA-OEHHA), http://oehha.ca.gov/water/phg/, 2010; Standard Material Safety and Data Sheet. 2010. Source of the chemicals or chemical compounds, MSDS, http://www.msdsonline.com, 2010; Oxford University, http://msds.chem.ox.ac.uk/HY/, 2010; U.S. Environmental Protection Agency, http://www.epa.gov/safewater/pdfs/factsheets/ioc/tech/cyanide.pdf, 2010; U.S. Environmental Protection Agency, http://water.epa.gov/drink/contaminants/#Inorganic, 2010.

Table 2.8 Volatile Organic Contaminants' Health Effects

Volatile Organic Chemicals	*Potential Health Effects*
Benzene	Cancer
Carbon tetrachloride	Liver effects, cancer
Chlorobenzene	Liver, kidney, nervous system effects
o-Dichlorobenzene	Liver, kidney, blood cell effects
p-Dichlorobenzene	Kidney effects, possible carcinogen
1,2-Dichloroethane	Cancer
1,1-Dichloroethylene	Liver, kidney effects, possible carcinogen
Cis-1,2-Dichloroethylene	Liver, kidney, nervous system, circulatory system effects
Trans-1,2-Dichloroethylene	Liver, kidney, nervous system, circulatory system effects
1,2-Dichloropropane	Cancer
Ethylbenzene	Liver, kidney, nervous system effects
Methylene chloride	Cancer

(*Continued*)

Table 2.8 Volatile Organic Contaminants' Health Effects (*Continued*)

Volatile Organic Chemicals	*Potential Health Effects*
Styrene	Liver, nervous systems effects, possible carcinogen
Tetrachloroethylene (PCE)	Cancer
Toluene	Liver, kidney, nervous system, circulatory system effects
Total THMs Chloroform Bromoform Bromodichloromethane Chlorodibromomethane	Cancer
1,2,4-Trichlorobenzene	Liver, kidney effects
1,1,1-Trichloroethane	Liver, nervous system effects
1,1,2-Trichloroethane	Kidney, liver effects, possible carcinogen
TCE	Cancer
Vinyl chloride	Nervous system, liver effects, cancer
Xylenes (total)	Liver, kidney, nervous system effects
Bromate	Cancer
Chlorate	Anemia, nervous system effects
Haloacetic Acids (HAA5)	Cancer
Total TTHMs	Cancer
Acrylamide	Cancer, nervous system effects
Alachlor	Cancer
Aldicarb	Nervous system effects
Aldicarb sulfoxide	Nervous system effects
Aldicarb sulfone	Nervous system effects
Atrazine	Liver, kidney, lung, cardiovascular effects; possible carcinogen

Table 2.8 Volatile Organic Contaminants' Health Effects (*Continued*)

Volatile Organic Chemicals	*Potential Health Effects*
Benzo(a)pyrene (PAHs)	Liver, kidney effects, possible carcinogen
Carbofuran	Nervous system, reproductive system effects
Chlordane	Cancer
2,4-D	Liver, kidney effects
Di(2-ethylhexyl) adipate	Reproductive effects
Di(2-ethylhexyl) phthalate	Cancer
Dibromochloro-propane (DBCP)	Cancer
Dinoseb	Thyroid, reproductive effects
Diquat	Ocular, liver, kidney effects
Endothall	Liver, kidney, gastrointestinal effects
Endrin	Liver, kidney effects
Epichlorohydrin	Cancer
Ethylene dibromide (EDB)	Cancer
Glyphosate	Liver, kidney effects
Heptachlor	Cancer
Heptachlor epoxide	Cancer
Hexachlorobenzene	Cancer
Hexachlorocyclopentadiene (HEX)	Kidney, stomach effects
Lindane	Liver, kidney, nervous system, immune system, circulatory system effects
Methoxychlor	Developmental, liver, kidney, nervous system effects
Oxamyl (Vydate)	Kidney effects
Pentachlorophenol	Cancer
Picloram	Kidney, liver effects

(*Continued*)

Table 2.8 Volatile Organic Contaminants' Health Effects (*Continued*)

Volatile Organic Chemicals	*Potential Health Effects*
Polychlorinated biphenyls (PCBs)	Cancer
Simazine	Body weight and blood effects, possible carcinogen
2,3,7,8-TCDD (Dioxin)	Cancer
Toxaphene	Cancer
2,4,5-TP (Silvex)	Liver, kidney effects

Source: Data from U.S. Agency for Toxic Substances and Disease Registry (ATSDR), http://www.atsdr.cdc.gov/toxfaqs/tf.asp?id=227&tid=41, 2010; California Environmental Protection Agency-Office of Environmental Health Hazard Assessment (Cal. EPA-OEHHA), http://oehha.ca.gov/water/phg/, 2010; Standard Material Safety and Data Sheet. 2010. Source of the chemicals or chemical compounds, MSDS, http://www.msdsonline.com, 2010; Oxford University, http://msds.chem.ox.ac.uk/HY/, 2010; U.S. Environmental Protection Agency, http://www.epa.gov/safewater/pdfs/factsheets/ioc/tech/cyanide.pdf, 2010; U.S. Environmental Protection Agency, http://water.epa.gov/drink/contaminants/#Inorganic, 2010.

2.4.3 *Chlorine Oxidation of Chemical Threats in the Water Supply Treatment System*

2.4.3.1 *Cyanide*

CN can be oxidized by chlorine (Cl_2) as detailed in Section 2.5.2.3, or by other special treatment methods as presented in Section 2.5.2, depending on the pH level and concentration of CN. At a higher temperature and a water pH less than 10, toxic gases will evolve with cyanogen chloride. Also, if the water is acidic or slightly alkaline, it will lead to the formation of cyanogen chloride gas. These are extremely toxic gases and very dangerous to humans and other living beings. Cyanogen chloride is produced by the oxidation of sodium cyanide with chlorine (Section 2.5.2.3).

2.4.3.2 *Arsenic*

Chlorination is not effective to treat arsenic in the water supply system, and an advanced and costly treatment system (Sections 2.5.1.1 through 2.5.1.7) is needed to reduce or remove arsenic. Meanwhile, the reaction of chlorine with organic compounds present in the water may produce trihalomethanes (THMs), which are known carcinogens. Arsenic (As) is a naturally occurring contaminant found in many areas of groundwater. It generally occurs in two forms (valences or oxidation states): pentavalent arsenic (also known as As(V), As(+5), or arsenate) and trivalent arsenic (also known as As(III), As(+3), or arsenite) (USEPA 1999). Trivalent arsenic can be converted to pentavalent arsenic in the presence of an effective oxidant such as free chlorine. The arsenic in water containing detectable free chlorine or that has been treated with another effective oxidant will be in the pentavalent arsenic form (USEPA 1999).

2.4.3.3 Pesticides

Human exposure to pesticide residues in drinking water is a potential health problem. Although many rural homes and communities consume raw groundwater, the majority of the U.S. population drinks treated water, which usually includes chlorination. While controlling many pathogenic bacteria, chlorination of natural organic matter can produce undesirable compounds such as THMs and haloacetonitriles in drinking water. However, chlorination can degrade undesirable compounds such as pesticides, which are susceptible to degradation in chlorinated water. The effect of chlorination on pesticides was also evaluated at full-scale treatment plants in Ohio (Miltner, Fronk, and Speth 1987). The typical percent removal rates for some of the common pesticides initially present at parts per billion (ppb) levels (in microgram per liter) are summarized in Table 2.9. For atrazine, cyanazine, simazine, alachlor, metolachlor, and linuron, the removal efficiencies were either zero or extremely low. Slight removal was observed for carbofuran. Up to 98% removal was reported for metribuzin. However, according to several investigators, this high removal efficiency may be partly attributed to sample preparation in which no reducing agent was added to stabilize the samples. Thus, it was possible that chlorination could have continued for days prior to analysis of the samples collected.

Table 2.9 Removal of Pesticides Associated with Chlorination at Full-Scale Treatment Plants

Typical Pesticide	*Initial Concentration (μg/L)*	*Estimated % Removal*
Atrazine	15.0	0
Cyanazine	5.0	0
Metribuzin	5.0	98
Simazine	5.0	7
Alachlor	8.0	0–9
Metolachlor	14.0	3
Linuron	0.50	4
Carbofuran	0.10	24

2.4.4.4 Gasoline Additive: MTBE

In reviewing the literature, no studies have been published that reflect the effect of chlorination on water containing MTBE. Ultraviolet (UV) light when applied at the 254 nm wavelength has been found to remove 10%–30% of MTBE from water in a batch reactor during a 2-hour period (Wagler and Malley 1994).

2.5 Potential Reduction Approach for Chemical Threats

2.5.1 Arsenic Remediation

The existing arsenic removal technologies as shown in Table 2.10 are mostly at the experimental stage, and some of them have not been demonstrated at full scale.

Table 2.10 Arsenic Treatment Technology Maximum Removal Percentages

Treatment Technology	*Maximum Percent Removal*
Coagulation/filtration	95
Enhanced coagulation/filtration[a]	95
Coagulation-assisted microfiltration	90
Lime softening (pH > 10.5)	90
Enhanced lime softening[a] (pH > 10.5)	90
Ion exchange (sulfate < 50 mg/L)	95
Activated alumina	95
Reverse osmosis	>95
Greensand filtration (20:1 iron:arsenic)	80
POU-activated alumina	90
POU ion exchange	90

Source: Data from U.S. Environmental Protection Agency, Report No. EPA-815-R-00-028. Washington, DC: U.S. Environmental Protection Agency, 2000.

[a] Enhanced processes assume the existing plant can achieve 50% removal without modification. Process enhancements result in the balance to achieve the maximum removal. For example, an existing coagulation filtration facility can achieve 50% removal. Process enhancements result in an additional 45% removal for a total removal of 95%.

Although some of these processes may be technically feasible, they are expensive. Reverse osmosis (RO) is highly preferable technology to remove arsenic; however, it is expensive.

2.5.1.1 Coagulation/Filtration

Coagulation technology can successfully achieve pentavalent arsenate (As(V)) removals greater than 90% but experiences difficulties in removal of trivalent arsenite (As(III)). It can reduce arsenic (As(V)) levels to less than 5.0 g/L. Arsenic in the pentavalent arsenate form is more readily removed than trivalent arsenite form at pH 7.6 or lower iron and aluminum coagulants are of equal effectiveness. Moreover, recent studies have shown that arsenic removal is independent of initial concentration.

2.5.1.2 Iron/Manganese Oxidation

Iron/manganese (Fe/Mn) oxidation is normally used by facilities treating groundwater. The oxidation process used to remove iron and manganese leads to the formation of hydroxides that remove soluble arsenic by precipitation or adsorption reactions. Arsenic removal during iron precipitation is fairly efficient (Edwards 1994). Removal of 2 mg/L of iron achieved a 92.5% removal of As(V) from a 10 μg/L As(V) initial concentration by adsorption alone. Even removal of 1 mg/L of iron resulted in the removal of 83% of influent arsenic As(V) from a source with 22 μg/L As(V). Indeed, field studies of iron-removal plants have indicated that this treatment can feasibly reach 3 mg/L. However, the arsenic removal efficiencies achieved by iron removal are not as consistent as activated alumina or ion exchange. Thus, arsenic removal during manganese precipitation is relatively ineffective when compared with iron even when removal by both adsorption and coprecipitation are considered (e.g., precipitation of 3 mg/L manganese removed only 69% of As(V) of a 12.5 μg/L As(V) influent concentration).

2.5.1.3 Enhanced Coagulation

The enhanced process involves modifications to the existing coagulation process such as increasing the coagulant dosage, reducing the pH, or both. Cheng et al. (1994) conducted bench, pilot, and demonstration scale studies to examine As(V) removals during enhanced coagulation. Approximately 90% As(V) removal can be achieved under enhanced coagulation conditions. As(V) removals of more than 90% were easily attained under all conditions when ferric chloride was used. With an influent arsenic concentration of 5 μg/L, ferric chloride achieved 96% As(V) removal with a dosage of 10 mg/L and no acid addition. When alum was used, 90% As(V) removal could not be achieved without reducing the pH.

2.5.1.4 Lime Softening

Lime softening has been widely used in the United States for reducing hardness in large water treatment systems. Lime softening, excess lime treatment, split lime treatment, and lime-soda softening are all common in municipal water systems (MWS). As(III) or As(V) removal by lime softening is pH dependent. Oxidation of As(III) to As(V) before lime softening treatment will increase removal efficiencies if As(III) is the predominant form, and a pH of 10.2 or higher could feasibly reduce arsenic concentrations to less than 3 g/L in the treated water.

2.5.1.5 Activated Alumina

Activated alumina is a physical/chemical process by which ions in the feed water are sorbed to the oxidized activated alumina surface, although the chemical reactions involved are actually an exchange of ions (AWWA 1990). Field studies involving activated alumina indicate that this technology can feasibly achieve arsenic removal to 3 g/L (Stewart and Kessler 1991, Wang, Sorg, and Chen 2000). The process does not remove As(III).

2.5.1.6 Ion Exchange

Ion exchange can effectively remove arsenic. However, sulfate, total dissolved solids (TDS), selenium, fluoride, and nitrate compete with arsenic and can affect run length. Passage through a series of columns could improve removal and decrease regeneration frequency. Systems containing high levels of these constituents may require pretreatment. According to the EPA *Design Manual of Arsenic Removal Drinking Water by Ion Exchange* (EPA/600/R-03/080), dated June 2003, trivalent arsenite As(III) must be oxidized to As(V). It does not directly remove As(III). Excess oxidizing chemical might degrade the resin; therefore, its removal may be required before contact with the resin. There is a potential for discharge of higher arsenic concentrations in the treated water. For water supplies also containing nitrate, there is potential for discharging high concentrations of both nitrate and arsenic.

2.5.1.7 Membrane Processes

Membrane processes are often classified by pore size into four categories: microfiltration (MF), ultrafiltration (UF), nanofiltration (NF), and RO. The following processes can reduce or break down arsenic but not limited to filtration, electric repulsion, and adsorption of arsenic compounds up to 95%.

2.5.2 Cyanide Remediation

The destruction processes of CN are used to sever the carbon–nitrogen triple bond, thereby destroying the CN and producing nontoxic or less-toxic species. Because the carbon and nitrogen atoms in the CN molecule undergo changes in the oxidation state, destruction processes are commonly referred to as oxidation methods. The EPA has identified chlorination, ion exchange, and RO as the BATs for removing CN from drinking water (USEPA 2010; USDI-BOR 2001). Ozonation is another option that is effective but has not been approved by the EPA for removing CN (Faust and Aly 1998). Other options that may be effective include iron coagulation, hydrolysis, aeration, and boiling. No references were found to justify the use of chloramines (combined chlorine), chlorine dioxide, potassium permanganate ($KMnO_4$), UV radiation, powdered activated carbon (PAC), granular activated carbon (GAC), or conventional (multimedia) or ultrafiltration. The BATs identified by the EPA for CN removal are as follows.

2.5.2.1 Ion Exchange

In the CN reduction, the operation begins with a fully recharged anion resin bed, having enough negatively charged ions to carry out the anion exchange. Usually, a polymer resin bed is composed of millions of medium sand grain size, spherical beads. As water passes through the resin bed, the negatively charged ions are released into the water, being substituted or replaced with the CN anions in the water (ion exchange). When the resin becomes exhausted of negatively charged ions, the bed must be regenerated by passing a strong, usually NaCl (or KCl), solution over the resin bed, displacing the CN ions with chlorine (Cl) ions. Many different types of anion resins can be used to reduce dissolved CN concentrations. The use of anion IX to reduce concentrations of CN will be dependent on the specific chemical characteristics of the raw water.

2.5.2.2 Reverse Osmosis

Reverse osmosis (RO) is the method of removing contaminants through the application of pressure on the supply water to deliver it through a semipermeable membrane. The rejection by a RO membrane of a particular contaminant is based on size and electrical charge. RO delivers one of the highest quality waters. However, it is expensive to install and requires frequent monitoring of the rejection percentage for CN removal.

2.5.2.3 Hypochlorite

Hypochlorite (OCl^-), commonly known as free chlorination, conducted at pH > 10.0 is a treatment that can be utilized to remove CN; the resulting byproducts are bicarbonate ions and nitrogen gas. Hence, unlike chlorination used for the inactivation of pathogens where hypochlorous acid (HOCl) is the appropriate chemical, OCl^- is the appropriate chlorine type when destroying CN ions. During the CN reduction, hypochlorite reacts with the CN ion to form cyanate (CNO^-).

$$CN^- + OCl^- \rightarrow CNO^- + Cl^- \quad (2.1a)$$

The cyanate ion is then reduced to nitrogen, carbon dioxide, and water molecules (Equation 2.1b) and this reaction is irreversible (APHA, AWWA, WEF 1998).

$$2NaCNO + 3C_{12} + 4NaOH \rightarrow N_2 + 2CO_2 + 6NaCl + 2H_2O \quad (2.1b)$$

If CN reduction is not implemented at a pH greater than 10.0, the treatment effect could be less than acceptable. Equations 2.1c and 2.1d illustrate the formation and destruction of cyanogen chloride. According to Whelton et al. (2003), the more desirable reaction byproduct is the cyanate ion because cyanate can be sequentially reduced to harmless molecules. Several studies have shown that the breakdown of CNCl is both pH- and time-dependent.

$$NaCN + Cl^- \rightarrow CNCl + NaCl \quad (2.1c)$$

$$CNCl + 2NaOH \rightarrow NaCNO + H_2O + NaCl \quad (2.1d)$$

Moreover, hydrogen CN gas can be formed at a pH of less than 9; most CN is treated at pH values greater than 10. However, if the CN concentration is low, the pH restriction is less critical.

2.5.2.4 Pesticides Remediation

The conventional water treatment at most MWS, specifically coagulation-flocculation, sedimentation, and conventional filtration does not remove and transform pesticides in treated drinking water. The disinfection and water softening, which also routinely occur at MWS can, however, lead to pesticide transformation and, in some cases, pesticide degradation. This finding is important because disinfection and conventional coagulation/filtration are commonly used treatment processes at MWS in the United States.

PAC filtration, granular activated carbon (GAC) filtration, and RO have been demonstrated and verified to be highly effective processes at removing organic chemicals, including certain pesticides, but specific data on removal of most pesticides are not available. Moreover, air stripping can be utilized for volatile pesticides with a high Henry's Law Constant $>1 \times 10^{-3}$ atm m^3/mole, but this procedure is used at very few MWS. However, Speth and Miltner (1998) reported that, in general, compounds with Freundlich coefficients on activated carbon greater than 200 ug/g $(L/ug)^{1/n}$ would be amenable to removal by carbon sorption.

2.5.2.5 Powdered Activated Carbon Filtration and Granular Activated Carbon Filtration

GAC under the SDWA is the best available technology for removing synthetic organic chemicals (SOC). Other recommended BATs are aeration technologies for reduction of dibromochloropropane and chlorination or ozonation for removal of glyphosate. GAC and PAC are common sorbents. Activated carbon is composed of expanded layers of graphite, which leads to an extremely high surface area to mass ratio for sorption (USEPA 1989). The main difference between GAC and PAC is the particle size; PAC has smaller particles when compared with GAC. Other less common sorbents are activated aluminum, silica gel, synthetic aluminosilicates, polymeric resins, and carbonized resins. GAC is used as a filter adsorber and post-filter adsorbers are designed for synthetic organic removal.

According to the EPA, the adsorption capacity of activated carbon to remove pesticides is affected by concentration, temperature, pH, competition from other contaminants or natural organic matter, organic preloading, contact time, mode of treatment, and physical/chemical properties of the contaminant. GAC column effectiveness is also a function of the water loading rate and empty bed time, whereas PAC effectiveness is also a function of the carbon dosage. Generally, activated carbon has an attraction for contaminants that are hydrophobic (low solubility), although other parameters such as their density and molecular weight can be important.

Isotherm constants have been reported to be valuable for predicting whether activated carbon adsorbs a particular pesticide (Speth and Miltner 1990; Speth and Adams 1993). They reported that, in general, compounds with a Freundlich coefficients on activated carbon greater than 200 ug/g $(L/ug)^{1/n}$ would be amendable to removal by carbon sorption.

The performance of GAC in removing pesticides from raw water has been demonstrated by the studies of Miltner, Fronk, and Speth (1987).

2.5.2.6 Reverse Osmosis

Membranes are used in water treatment for desalinization, specific ion removal, and removal of color, organics, nutrients, and suspended solids. Membranes are used in RO, electrodialysis (ED), UF, MF, and NF (USEPA 1989). Ultrafiltration is considered a filtering technique because it is designed to exclude compounds with molecular weights greater than 500 g/mole. In contrast, RO and ED are designed to use a semipermeable membrane as a diffusion barrier for dissolved constituents in the water. ED is controlled by electrostatic attraction of ionic compounds to anionic and cationic electrodes across a semipermeable membrane. RO is controlled by hydrostatic pressure (300–1000 psi) to drive feed water through a semipermeable membrane. Membranes are typically composed of cellulose acetate (CA), polyamide membranes, and thin film composites. Membrane configurations for RO are spiral wound and hollow fine fiber membrane. The effectiveness of RO is dependent on membrane composition, physicochemical properties of raw water, pressure, and membrane treatment conditions.

The use of semipermeable membranes during RO treatment has been demonstrated to remove organic pollutants and pesticides from contaminated water. The membranes normally used in the past were either CA or polyamide. Later, a new type of membrane called thin film composites was introduced. These membranes could be produced from a variety of polymeric materials that were formed in situ or coated onto the surface of an extremely thin polysulfone support. Examples are NS-100 (cross-linked polyethylenimine membrane), FT-30 (cross-linked polyamide that contains a carboxylate group), and DSI (modified polyalkene on a polysulfone base with nonwoven polyester backing). Membranes operated with a lower pressure can also be used in water-treatment plants. Fronk and Baker (1990) conducted an evaluation of removing certain pesticides from groundwater using thin film composite membranes. Excellent removal (~100%) of organochlorine pesticides (chlordane, heptachlor, and methoxychlor) and an acetanilide compound (alachlor) was obtained. The removal of dibromochloropropane was not high and ethylene dibromide was not removed at all.

Another membrane process is NF. The membrane used is somewhat *more loose*, and the process is operated with lower effective pressure and without significant changes in water salinity (USEPA 2001). Several studies have shown that NF demonstrates a removal efficiency of up to 95% of pesticides. Moreover, MF with porosity nominally >0.1 μm and UF with porosity of 0.01 μm are sometimes combined with adsorbents such as PAC to form an integrated system that can be effective in removing pesticides. Clark, Fronk, and Lykins (1998) found that a UF/PAC (10 mg/L PAC) system was capable of removing cyanazine by 70% and atrazine by 61%. With higher PAC levels, better results can be obtained. The removal of atrazine was increased from 57% at 5 mg/L to 89% at 20 mg/L PAC (Claire et al. 1997). It would be expected that the integrated membrane/adsorbent system would lead to greater adsorption with an increase in the adsorbent time. Furthermore, pH, temperature, competitive adsorption from other contaminants, the type of PAC, and dose can affect the extent of adsorption.

2.5.2.7 Methyl Tert-Butyl Ether Remediation

It is difficult to treat water for MTBE. The main treatment methods that have proven to be effective in treating hydrocarbon organics such as MTBE from drinking water are presented in Sections 2.5.2.8 to 2.5.2.10.

If the concentration of the contaminants is high, two treatment systems are often installed or built. The first process is used to remove the *heavy* contaminant load, whereas the second provides a *polishing* step to assure full removal of the contaminant(s) and to address *breakthrough* in accordance with the regulatory requirements enforced by the EPA and other state or local agencies. Air stripping is often the first method used while activated carbon is often used as the polishing step.

2.5.2.8 Air Stripping Treatment: Advantages and Disadvantages

The air stripping process includes the delivery of large amounts of air through the contaminated water. The efficiency of this treatment technology is increased by breaking up the bulk of the water into multiple droplets. It allows the contaminants to volatilize into the air. When air stripping is used, one of the operational problems that could possibly happen is increased levels of iron or manganese in the water that can cause corrosion to the equipment, in which case additional treatment technology for iron or manganese removal will be required. Then, the entire treatment process will become more costly. One of the advantages of air stripping is that there is no regeneration of the treatment media needed.

2.5.2.9 Activated Carbon Treatment: Advantages and Disadvantages

Activated carbon has an enormous surface area within each carbon particle that attracts all types of organic contaminants, except vinyl chloride. There has been no report showing that activated carbon will effectively treat vinyl chloride. Once the removal capacity of the carbon is utilized, it may be returned to the manufacturer for rejuvenation. If activated carbon is considered for treatment, the radon and mineral radioactivity concentrations of the water should be determined because it potentially creates a low-level radionuclide waste and increased radiation within the home. Meanwhile, the bacteria will usually grow on the surface area of the carbon, where the organic contaminants can be used as a food source on its surface which alleviate the treatment process. Activated carbon should be monitored to avoid the possible release of contaminants after they have been initially adsorbed. The advantage of activated carbon treatment in pressure tanks compared with other methods is that the water does not need to be repressurized and is less likely to become contaminated by airborne contaminants. The disadvantage is that it has a low capacity of attracting MTBE compared with other organic compounds and must be replaced more frequently.

2.5.2.10 Other Possible Treatments for MBTE

2.5.2.10.1 Oxidation Treatment: Advantages and Disadvantages

Some organic compounds will react with oxygen and oxygen-like compounds. Once the oxidation treatment has been employed, the resultant compounds may be fully reduced or oxidized. Additional treatment technology may still be necessary, however. Oxidizing chemicals could include $KMnO_4$, hydrogen peroxide (H_2O_2), and ozone (O_3).

2.5.2.10.2 Ultraviolet/Ozone Destruction

The use of UV radiation in conjunction with ozone to break down MTBE is one of the emerging treatment processes. The ozone is injected into the water, and the mixture is passed through UV light. The UV light stimulates the ozone, generating oxidizing compounds for the reduction of MTBE.

2.5.3 Groundwater and Surface Water Remediation

Groundwater and surface water remediation are types of environmental cleanups that focus on addressing pollution of groundwater and surface water supplies. The main objective is to turn contaminated water into clean water that will not potentially create hazards to public health and the environment. Potable water is a limited resource, and cleanup of groundwater can free up supplies for irrigation or drinking, reducing strain on water supplies. Water that has been polluted with pharmaceuticals and endocrine disrupting chemicals, for example, could cause developmental abnormalities in fish, which could lead to a decline in fish populations. Paying for remediation can get very costly, because water is notoriously difficult to treat. Many remediation methods are used and the most common are as follows.

2.5.3.1 Pump-and-Treat

Pump-and-treat is a method of removing contaminated groundwater from strategically placed groundwater wells, treating the extracted water after it is on the surface to remove the contaminants of concern using mechanical, chemical, or biological methods, and discharging the treated water back to the ground, surface, or municipal sewer system in accordance with the regulatory requirements enforced by the EPA and state or local agencies. The method has several limitations as follows: (1) the effectiveness depends on the geology of the groundwater and the type or characteristics of the contaminant; (2) the treatment process is very slow; (3) it is very costly; and (4) it doesn't always work. Some contaminants stick to soil and rock (they are adsorbed) and they cannot easily be removed (desorbed). The NAPLs cannot be removed.

2.5.3.2 Hydraulic Containment

Hydraulic containment is a remediation technique of pumping water from groundwater wells can be performed in a manner that it changes the flowpath of water through an aquifer in ways to keep contaminants away from wells used for cities or farms. The treatment process works if the flow through the aquifer where the plume of contaminated groundwater does not divide into multiple paths. It is normally used in conjunction with pump-and-treat, and it has the same limitations.

2.5.3.3 Air Sparging with Soil Vapor Extraction

Air sparging is the process of injecting air directly into groundwater through wells. It remediates groundwater by volatilizing contaminants and enhancing biodegradation by creating bubbles. As the bubbles rise from the subsurface, the contaminants are removed from the groundwater by physical contact with the air (i.e., stripping) and are carried up into the unsaturated zone (i.e., soil). As the contaminants move into the soil, a soil vapor extraction (SVE) technology is usually utilized to remove notorious vapors. The application of oxygen to contaminated groundwater and soils also enhances biodegradation of contaminants, as it acts as a nutrient for bacteria.

Air sparging is, sometimes, referred to as in situ air stripping. When used in combination with SVE, air bubbles carry vapor phase contaminants to an SVE system, which removes them. One of the best SVE technologies that can be used to effectively remove the contaminated vapors in conjunction with air sparging is the cryogenic treatment process as presented in Section 2.5.3.9. It is very sustainable, efficient, and less costly when it is constructed and designed properly. The SVE system limits vapor plume migration by creating a negative pressure in the unsaturated zone through extraction wells. Using air sparging with an SVE system increases contaminant movement and enhances oxygenation in the subsurface, which intensifies the rate of contaminant extraction.

2.5.3.4 In situ Oxidation

The in situ oxidation method injects an oxidant such as H_2O_2 into the contaminated aquifer. The contaminant is oxidized, producing carbon dioxide and water. A permeable treatment zone process is designed and constructed by injecting a reducing reagent and buffers such as sodium dithionite and potassium carbonate to reduce the ferric iron in the aquifer sediments to ferrous iron.

2.5.3.5 Permeable Reactive Barriers

The permeable reactive barriers method is installed by constructing a trench backfilled with reactive material (e.g., iron filings, activated carbon, or peat), which absorb and break down the contaminants as water flows through the barrier. This method is suitable for relatively shallow aquifers.

2.5.3.6 Phytoremediation

Plants and trees remove organic contaminants through direct uptake of contaminants and subsequent accumulation of nonphytotoxic metabolites into the plant tissue and release of exudates and enzymes that stimulate microbial activity and as a result enhance microbial transformation in the rhizosphere (Schnoor et al. 1995). Some plants uptake notorious contaminants such as arsenic, lead, uranium, selenium, cadmium, and other toxins. Genetically altered cottonwood trees uptake mercury from the contaminated soil in Danbury, Connecticut. Moreover, transgenic Indian mustard plants are used to soak up dangerously high selenium deposits in California. The remediation consists of growing such plants so their roots tap the groundwater. Then, the plants are harvested and disposed. The method is limited to remediation of groundwater that is shallow enough to be reached by plant roots.

2.5.3.7 Natural Attenuation

Sometimes natural processes remove contaminants with no human involvement. The treatment may involve dilution, radioactive decay, sorption, volatilization, or natural chemical reactions that stabilize, reduce, or degrade contaminants. Natural attenuation describes the process of site assessment, date reduction, and data interpretation that is focused on the quantification of the capacity of a given aquifer system to assimilate groundwater contaminants through physical, chemical, and/or biological means (Hayman and Dupont 2001).

2.5.3.8 Intrinsic and Enhanced Bioremediation

Biodegradation is the breakdown of carbon-based contaminants by microorganisms into less toxic chemical compounds. The microorganisms basically transform the contaminants through metabolic or enzymatic processes. The process is employed by injecting organisms with special nutrients into the groundwater for remediation. Natural bioremediation is most effective in aquifers where bacteria are propagating easily, and where contaminant levels are very low.

2.5.3.9 Vapor Condensation-Cryogenic Technology

Vapor condensation-cryogenic technology, also known as the *cryogenic process*, is an off-gas treatment system in SVE that can be done in conjunction with air sparging (air stripping or bioslurping). It is a combination of cryogenic cooling using any of the cryogens (e.g., liquid helium, carbon dioxide, Freon, and liquid nitrogen), a technique that efficiently and sustainably recovers volatile organic compounds (VOCs) from the off-gas vapor stream of SVE or dual phase extraction systems. The cryogenic process is very cost effective, sustainable, and does not usually emit hazardous contaminants in the air when it is designed and planned appropriately. Designing the temperature levels of the system based on the freezing and boiling points of the contaminants of concern (COC), site characteristics, and chemical compatibility between the equipment's materials and the COC are the main key elements to achieve a very successful and safe operation. When using liquid nitrogen, it is very crucial to maintain certain limits of temperatures to prevent the condensation of the oxygen in the system so that potential explosions can be mitigated.

2.6 Biological Threats

Biological agents are organisms or toxins used to harm, kill, or incapacitate people. The three basic groups of biological agents that would likely be used as weapons by terrorists, criminals, or disgruntled individuals are bacteria, viruses, and toxins. Most biological agents are difficult to grow and maintain. They mostly break down quickly when exposed to sunlight and other environmental factors. The biological agents can be dispersed by spraying them into the atmosphere, by infecting livestock, and by contaminating food and water. Some of the pathogenic organisms may persist in water supplies. Most microbes and toxins can be killed and deactivated by cooking or boiling water. Additionally, most microbes are killed by boiling water for 1 minute, but some require more time. The biological threats in different categories are provided in Tables 2.11 through 2.13.

Table 2.11 Biological Threats Category A

Category A Means the High-Priority Agents, Which Include Organisms That Pose a Risk to National Security	
Biological Agent	*Characteristics, Hazard, and Reduction Approach*
Anthrax (*Bacillus anthracis*)	Humans can contract spores from inhalation of aerosolized anthrax released during a biological weapons attack. Conditions ideal for propagation of anthrax include soil of pH > 6.0 rich in organic matter. LD_{50} for inhalational anthrax in humans from weapons-grade anthrax is 2500–55,000 spores. According to the American Society of Microbiology, anthrax spores may survive in water with a concentration of 1 mg of chlorine per liter (typical tap water has a concentration of 1–2 mg/L). After 60 min in the water, there was no significant decrease in the number of viable spores.
Botulism (*Clostridium botulinum* toxin)	Botulism is mainly a foodborne intoxication, but it can also be transmitted through wound infections or intestinal infection in infants. An attack involving contamination of public drinking water is unlikely as botulinum toxins are inactivated by chlorinated water (most public drinking water is treated with chlorine to remove bacteria) (MDCPH 2004).

(*Continued*)

Table 2.11 Biological Threats Category A (*Continued*)

Category A Means the High-Priority Agents, Which Include Organisms That Pose a Risk to National Security	
Biological Agent	*Characteristics, Hazard, and Reduction Approach*
Plague (*Yersinia pestis*)	*Yersinia Pestis* is a serious issue because it contributes to waterborne and foodborne diseases that each year affects an estimated 76 million people in the United States. It can survive in water for 16 days and in moist soil for >60 days. It is inactivated by 1% sodium hypochlorite, but no reference to its tolerance to hypochlorite under the usual conditions of drinking water disinfection was recovered. *Yersinia* is 100% inactivated by 0.25 mg/L chlorine dioxide (Imangulov 1988).
Smallpox (*variola major*)	After an incubation period of approximately 12 days, signs and symptoms include chills, fever, prostration, headache, backache, and vomiting, as well as pustule formation, with a case fatality rate among the unvaccinated of 25% or more (Eitzen et al. 1998). It is inactivated by 1% sodium hypochlorite, but no reference to its tolerance to hypochlorite under the usual conditions of drinking water disinfection was recovered.
Tularemia (*Francisella tularensis*)	Tularemia is an epizootic disease of animals (especially rabbits and rodents), transmissible to humans, caused by the bacillus *Francisella tularensis* (formerly *Pasteurella tularensis*). *F. tularensis* has been weaponized in the aerosol form. *P. tularensis* (*F. tularensis*) is 99.6%–100% inactivated by 0.5–1.0 mg/L FAC at 10°C and pH 7 in approximately 5 min (Zilinskas 1997). However, other studies show that chlorine (0.5–2.0 mg/L) is ineffective against tularemia (Jensen et al. 1996).
Viral hemorrhagic fevers	Viral hemorrhagic fevers may have been weaponized for aerosol application, but no reference was recovered suggesting a potable water threat. Lassa fever virus is rapidly inactivated at 56°C; the other viruses require 30-min exposure at that temperature. All are inactivated by UV light (Parker et al. 1996). All of the listed VHF viruses are inactivated by 1%–2% sodium hypochlorite and/or 1% iodine (Parker et al. 1996), but no reference to their chlorine tolerance under the usual conditions of drinking water disinfection was recovered.

Table 2.12 Biological Threats Category B

Category B Means the Second Highest-Priority Agents, Which Include Organisms That Are Moderately Easy to Disseminate	
Biological Agent	*Characteristics, Hazard, and Reduction Approach*
Brucellosis (*Brucella* species)	Brucellosis may survive in soil for 7–69 days and in water for 20–72 days; it is inactivated by direct sunlight (Parker et al. 1996). It is inactivated by 1% sodium hypochlorite, but no reference was recovered to its tolerance to hypochlorite under the usual conditions of drinking water disinfection.
Epsilon toxin of *Clostridium perfringens*	*Clostridium perfringens* is presumed to be indefinitely stable in sewage. It is a spore formerly used as an indicator organism and is relatively insensitive to inactivation by chlorine. It is reduced to <1 $\log_{10}$ under treatment with a chlorine residual of 1.2 mg/L for 15 min at 20°C and a pH of approximately 7 (Tyrrell, Rippey, and Watkins 1995).
Food safety threats (e.g., *Salmonella* species, *Escherichia coli, Shigella*)	*Salmonella* survival in environmental media is 29–58 days in soil, 9 days in seawater, 8 days in fresh water, and up to 5 months in ice (White 1992). *Salmonella survival* is about the same. Because of the introduction of chlorine treatment of municipal water, waterborne typhoid has virtually disappeared in the United States. It requires a UV radiation dose of 15.2 $mW \cdot s/cm^2$ at 253.7 nm to achieve >99.9% inactivation (Science Applications International Corp. 1996).
Glanders (*Burkholderia mallei*)	Glanders may have been weaponized in aerosol form; a single reference suggesting its potential as an agent of drinking water contamination was recovered (Imangulov 1988). It survives in water at room temperature for up to 30 days in soil and for more than 27 days in water, but it is apparently not naturally found in soil or water. It is inactivated by 1% sodium hypochlorite water (Parker et al. 1996), but no reference to its tolerance to hypochlorite under the usual conditions of drinking water disinfection was recovered.
Melioidosis (*Burkholderia pseudomallei*)	The most serious form of melioidosis in humans, an acute septicemic condition with diarrhea, has a high case-fatality rate if untreated. Parker et al. state that it survives for years in soil and water. Melioisis is inactivated by 1% sodium hypochlorite (Parker et al. 1996), but no reference to its tolerance to hypochlorite under the usual conditions of drinking water disinfection was recovered.

(*Continued*)

Table 2.12 Biological Threats Category B (*Continued*)

Category B Means the Second Highest-Priority Agents, Which Include Organisms That Are Moderately Easy to Disseminate	
Biological Agent	*Characteristics, Hazard, and Reduction Approach*
Psittacosis (*Chlamydia psittaci*)	Signs and symptoms of psittacosis include chills and fever, headache, sore throat, nausea, and vomiting; the case-fatality rate is ≤10% (31). It is considered susceptible to heat, similar to *Rickettsia prowazekii*. It is inactivated by 1% sodium hypochlorite (Parker et al. 1996), but no reference to its tolerance to hypochlorite under the usual conditions of drinking water disinfection was recovered.
Q fever (*Coxiella burnetii*)	Fever, cough, and pleuritic chest pain may occur as early as 10 days after exposure from Q fever. It survives in tap water for 160 days at 20–22°C and resists heat, drying, osmotic shock, and UV radiation. Hence, Q fever was reduced to undetectable levels in water treated with the ERDLator, a now-discontinued item of army field equipment that combined ferric chloride and limestone coagulation with 0.8 mg/L residual chlorine disinfection, 20-min contact time, and diatomite filtration. Under the same conditions, but with a chlorine residual of 0.5 mg/L, inactivation of Q fever was incomplete (Lindsten and Schmitt 1975).
Ricin toxin from *Ricinus communis* (castor beans)	The oral LD_{50} for mice is 20 mg/kg (Franz 1997). A conservative NOAEL would be 2 μg/L for water consumptions of 15 L/day. It is detoxified in 10 min at 80°C (26) and in ~1 h at 50°C (pH 7.8); it is stable under ambient conditions (Warner 1990). Hence, ricin is >99.4% inactivated after 20-min treatment with FAC at 100 mg/L, but it is essentially unchanged at 10 mg/L (Wannemacher et al. 1993). Iodine has no measurable effect at 16 mg/L. RO can efficiently remove ricin up to 99.8% from product water, but coagulation/flocculation was not effective. Using charcoal treatment system may effectively remove ricin. Further, some individual or point-of-use water purifiers may provide protection.
Staphylococcal Enterotoxin B (SEB)	SEB is an incapacitating toxin, causing severe gastrointestinal pain, projectile vomiting, and diarrhea if ingested, and fever, chills, headache, muscle aches, shortness of breath, and nonproductive cough if inhaled. The disinfection efficacy of SEB is unknown. Water treatment systems using charcoal should remove SEB (McGeorge 1989); thus, some individual and point-of-use water purifiers may provide protection.

Table 2.12 Biological Threats Category B (*Continued*)

Category B Means the Second Highest-Priority Agents, Which Include Organisms That Are Moderately Easy to Disseminate	
Biological Agent	*Characteristics, Hazard, and Reduction Approach*
Typhus fever (*Rickettsia prowazekii*)	Signs and symptoms of typhus fever include high fever, chills, intense headache, back and muscle pains, and skin eruptions (Freeman et al. 1979). An infective dose of fewer than 10 organisms has been estimated, corresponding to a drinking water concentration of <1 organism per liter for consumption of either 15 L/day or 5 L/day for 7 days, if in fact epidemic typhus is transmissible through water (Parker et al. 1996). Typhus fever is inactivated by 1% sodium hypochlorite (Parker et al. 1996), but no reference to its tolerance to hypochlorite under the usual conditions of drinking water disinfection was recovered.
Viral encephalitis	Encephalomyelitis is usually arthropodborne, diseases of animals to which humans may be susceptible. No reference suggesting potential as an agent of drinking water contamination was recovered. It is inactivated by 1% sodium hypochlorite (Parker et al. 1996), but no reference to its tolerance to hypochlorite under the usual conditions of drinking water disinfection was recovered.
Water Safety Threat (*Cryptosporidium Parvum*)	The signs and symptoms of cryptosporidiosis are profuse watery diarrhea, nausea, and stomach cramps. However, it has been suggested as a potential agent for sabotaging potable water supplies by reason of its infectivity and ready availability (Burrow 1999). The RO with water purification unit will remove 100% of *Cr. parvum* oocysts, which are 3–7 μm in size. Removal of oocysts by direct filtration will approach 3 $\log_{10}$ in well-operated municipal systems (Clancy et al. 1998) and may exceed 5 $\log_{10}$ for slow sand filtration (SAIC 1996), but chlorination of the product water provides no protection if filtration performance degrades.

(*Continued*)

Table 2.12 Biological Threats Category B (*Continued*)

Category B Means the Second Highest-Priority Agents, Which Include Organisms That Are Moderately Easy to Disseminate	
Biological Agent	*Characteristics, Hazard, and Reduction Approach*
Brucellosis (*Brucella* species)	The causative agents of brucellosis are *Brucella melitensis* and *Brucella suis*; the latter has been weaponized for aerosol application. Because brucellosis is contracted through consumption of contaminated milk, it is prudent to consider water as a potential route of infection (Imangulov 1988). According to Parker, *Br. melitensis* may survive in soil for 7–69 days and in water for 20–72 days; it is inactivated by direct sunlight. *Br. melitensis* is inactivated by 1% sodium hypochlorite (Parker et al. 1996), but we did not find any reference to its tolerance to hypochlorite under the usual conditions of drinking water disinfection.
Clostridium perfringens	*Clostridium perfringens* is a common organism in secondary sewage effluent. The spores may have potential for weaponization in aerosol form. No reference suggesting potential as an agent of drinking water contamination was recovered. *C. perfringens*, a spore formerly used as an indicator organism, is relatively insensitive to inactivation by chlorine. It is reduced by <1 $\log_{10}$ under treatment with a chlorine residual of 1.2 mg/L for 15 min at 20°C and a pH of approximately 7 (Tyrrell, Rippey, and Watkins 1995).

Table 2.13 Biological Threats Category C

Category C Means the Third Highest-Priority Agents, Which Include Emerging Pathogens That Could Be Engineered for Mass Distribution	
Biological Agent	*Characteristics, Hazard, and Reduction Approach*
Nipavirus	According to Sawatsky et al. (2008), the henipaviruses are naturally harbored by Pteropid fruit bats (flying foxes) and are characterized by a large genome, a wide host range, and their recent emergence as zoonotic pathogens capable of causing illness and death in domestic animals and humans. It can be inactivated by chlorine.
Hantavirus	Symptoms begin 1–6 weeks after inhaling the virus and typically start with 3–5 days of flu-like illness including fever, sore muscles, headaches, nausea, vomiting, and fatigue. As the disease gets worse, it causes shortness of breath because of fluid-filled lungs. It can be inactivated by chlorine.

2.7 Radiological Threats

There are three kinds of potential radiological threats. A terrorist or extremist group may actually steal a nuclear weapon, they may steal radioactive materials from chemical and ammunition plants, or they may attack a nuclear plant. Security experts have tried to analyze various scenarios such as the sabotage of vulnerable areas where radiological materials are stored or used. These scenarios have led to new approaches to tightening up security and improving intrusion prevention technologies. Table 2.14 shows the radiological terms.

Table 2.14 Radiological Terms

Terms	*Description*
Acute radiation syndrome	Consists of three levels of effects: hernatopoletic (blood cells, most sensitive); gastrointestinal (GI cells, very sensitive); and central nervous system (brain/muscle cells, insensitive). The initial signs and symptoms are nausea, vomiting, fatigue, and loss of appetite. Below about 200 rems, these symptoms may be the only indication of radiation exposure.
Alpha particle (α)	The alpha particle has a very short range in air, and a very low ability to penetrate other materials, but it has a strong ability to ionize materials. Alpha particles are unable to penetrate even the thin layer of dead cells of human skin and, consequently, are not an external radiation hazard. Alpha-emitting nuclides inside the body as a result of inhalation or ingestion are a considerable internal radiation hazard.
Beta particles (β)	High-energy electrons emitted from the nucleus of an atom during radioactive decay. They normally can be stopped by the skin or a very thin sheet of metal.
Cesium-137 (Cs-137)	A strong gamma ray source that can contaminate property, entailing extensive clean-up. It is commonly used in industrial measurement gauges and for irradiation of material. Half-life is 30.2 years.
Cobalt-60 (Co-60)	A strong gamma ray source that is extensively used as a radiotherapeutic for treating cancer, food and material irradiation, gamma radiography, and industrial measurement gauges. Half-life is 5.27 years.
Curie (Ci)	A unit of radioactive decay rate defined as 3.7×10^{10} disintegrations per second.

(Continued)

Table 2.14 Radiological Terms (*Continued*)

Terms	*Description*
Decay	The process by which an unstable element is changed to another isotope or another element by the spontaneous emission of radiation from its nucleus. This process can be measured by using radiation detectors such as Geiger counters.
Decontamination	The process of making people, objects, or areas safe by absorbing, destroying, neutralizing, making harmless, or removing the hazardous material.
Dose	A general term for the amount of radiation absorbed during a period of time.
Dosimeter	A portable instrument for measuring and registering the total accumulated dose of ionizing radiation.
Gamma rays (γ)	High-energy photons emitted from the nucleus of atoms; similar to X-rays. They can penetrate deeply into body tissue and many materials. Cobalt-60 and Cesium-137 are both strong gamma-emitters. Shielding against gamma radiation requires thick layers of dense materials, such as lead. Gamma rays are potentially lethal to humans.
Half-life	The amount of time needed for half of the atoms of a radioactive material to decay.
Highly enriched uranium (HEU)	Uranium that is enriched to above 20% Uranium-235 (U-235). Weapons-grade HEU is enriched to above 90% U-235.
Ionize	To split off one or more electrons from an atom, thus leaving it with a positive electric charge. The electrons usually attach to one of the atoms or molecules, giving them a negative charge.
Iridium-192	A gamma-ray emitting radioisotope used for gamma-radiography. The half-life is 73–83 days.
Isotope	A specific element always has the same number of protons in the nucleus. That same element may, however, appear in forms that have different numbers of neutrons in the nucleus. These different forms are referred to as "isotopes" of the element. For example, deuterium (2H) and tritium (3H) are isotopes of ordinary hydrogen (H).

Table 2.14 Radiological Terms (*Continued*)

Terms	*Description*
Lethal dose (50/30)	The dose of radiation expected to cause death within 30 days to 50% of those exposed without medical treatment. The generally accepted range is from 400 to 500 REM received over a short period of time.
Nuclear reactor	A device in which a controlled, self-sustaining nuclear chain reaction can be maintained with the use of cooling to remove generated heat.
Plutonium-239 (Pu-239)	A metallic element used for nuclear weapons. The half-life is 24–110 years.
Rad	A unit of absorbed dose of radiation defined as deposition of 100 ergs of energy per gram of tissue. It amounts, approximately, to one ionization per cubic micrometer.
Radiation	High energy alpha or beta particles or gamma rays that are emitted by an atom as the substance undergoes radioactive decay.
Radiation sickness	Symptoms resulting from excessive exposure to radiation of the body.
Radioactive waste	Disposable, radioactive materials resulting from nuclear operations. Wastes are generally classified into two categories: high-level and low-level waste.
Radiological dispersal device (RDD)	A device (weapon or equipment), other than a nuclear explosive device, designed to disseminate radioactive material in order to cause destruction, damage, or injury by means of the radiation produced by the decay of such material.
Radioluminescence	The luminescence produced by particles emitted during radioactive decay.
REM	A Roentgen Equivalent Man is a unit of absorbed dose that takes into account the relative effectiveness of radiation that harms human health.
Shielding	Materials (lead, concrete, and so on) used to block or attenuate radiation for protection of equipment, materials, or people.

(*Continued*)

Table 2.14 Radiological Terms (*Continued*)

Terms	*Description*
Special nuclear material (SNM)	Plutonium and uranium enriched in the isotope Uranium-233 or Uranium-235.
Uranium-235 (U-235)	Naturally occurring uranium U-235 is found at 0.72% enrichment. U-235 is used as a reactor fuel or for weapons; however, weapons typically use U-235 enriched to 90%. The half-life is 7.04×10^8 years.
X-ray	An invisible, highly penetrating electromagnetic radiation of much shorter wavelength (higher frequency) than visible light. Very similar to gamma-rays.

Source: Data from U.S. Central Intelligence Agency. 1998. Terrorist CBRN: Materials and Effects. https://www.cia.gov/library/reports/general-reports-1/cbr_handbook/brbook.htm#6 (accessed August 20, 2009 and October 18, 2010).

2.8 Prescription Drugs (Pharmaceuticals), Personal Care Products, and Endocrine Disrupting Compounds in the Water System

Prescription drugs or pharmaceuticals, personal care products (PCPs), and endocrine disrupting chemicals or compounds (EDCs) are the emerging contaminants that have been detected in surface water, groundwater, estuarine water, and drinking water. These contaminants enter the water system through treated and untreated wastewater, and urban and agricultural runoff. The EPA has been intently working with federal, state, and local agencies to understand the implications of these emerging contaminants, particularly the prescription drugs. The EPA continues to assess and determine their way of exposure, levels of exposure, and potential effects on public health and the environment.

2.8.1 Prescription Drugs

The most prominent prescription drug in the water environment is ethynylestradiol (EE2), which is widely used as an oral contraceptive. In 1999, the first report documenting EE2 occurrence in U.S. surface water was published (Snyder et al. 1999). More importantly, the occurrence of both synthetic estrogen EE2 and the endogenous estrogen 17β-estradiol (E2) in U.S. wastewater effluents were subsequently identified as putative contaminants linked to reproductive ailments in fish

(Snyder et al. 2001). In addition, quantities of antibiotics, steroids, antidepressants, and hormones are also present in the water supply. The long-term effects are currently not clearly defined, as the EPA continues to work on the important issue of prescription drugs in water.

2.8.2 Personal Care Products

PCPs whether they be hair dye, skin care products, shampoo, conditioner, or Rogaine; perfume; toothpaste or mouthwash; antibacterial soap or hand lotion; almost all of it goes down the drain when we do laundry, wash the dishes, wash our hands, brush our teeth, bathe, or do any of the other myriad things that incidentally use household water. Recent studies have shown that dish detergents contain high concentrations of hazardous chemicals such as benzene and naphthalene. Unfortunately, most wastewater treatment facilities are not equipped to filter out PCPs, household products, and pharmaceuticals, and a large portion of the chemicals passes right into the local lakes and rivers that accept the treatment plant's supposedly clean effluent. Study of the effects of these chemicals getting into the water is just beginning, but examples of problems are now arising regularly, for example scientists are finding fragrance molecules inside fish tissue. The EPA is determined to undertake a scientific approach in evaluating the risks associated with contaminants associated with PCPs.

2.8.3 Endocrine Disrupting Compounds

The human endocrine system is a composite network of glands and hormones that regulates many of the body's functions including growth, development, and maturation, as well as the way various organs operate. An endocrine disruptor, also known as endocrine disrupting compounds (e.g., *diethylhexylphthalate*, *diethylstilbesterol*, or synthetic estrogen *ethinylestradiol* in birth control pills), is a synthetic chemical that when absorbed into the body either mimics or blocks hormones and disrupts the body's normal functions. EDCs can also be found in many PCPs like cosmetics and deodorants. Meanwhile, Research and Development's National Risk Management Research Laboratory (NRMRL) conducts research on the efficacy of existing risk-management techniques to minimize exposure to suspected EDCs and develops new risk management tools (USEPA 2010). According to the EPA, the most commonly reported EDCs in studies on the impact of wastewater treatment are reproductive steroid hormones (especially estrogens) and the estrogenic biodegradation products of alkylphenol ethoxylate surfactants. For example, a recent publication by the United States Geological Survey (USGS) showed that reproductive hormones and estrogenic alkylphenols were present in 40% and 70%, respectively, of the surveyed U.S. surface waters.

2.8.4 Potential Reduction of Prescription Drugs (Pharmaceuticals), Personal Care Products, and Endocrine Disrupting Compounds

Most prescription drugs are highly water soluble (Daughton and Ternes 1999). For prescription drugs that are synthetic organic compounds (SOCs), GAC, PAC, RO, and NF are likely to be effective. The following advanced treatment systems are suitable for treating prescription drugs, PCPs, and EDCs:

2.8.4.1 Granular Activated Carbon

The contaminants accumulate within the pores, and the greatest efficiency is attained when the pore size is only slightly larger than the material being adsorbed. The performance of GAC for specific contaminants is determined in the laboratory by trial runs and is performed one chemical at a time. Both powdered activated carbon (PAC) and GAC have been demonstrated to be effective at removing pharmaceuticals from water, with greater than 50% removal for most compounds. The Freundlich equation can be used to indicate the efficiency of GAC/PAC treatment. The Freundlich equation is expressed as: $Q_e = K \times C_e^n$, where Q_e is the equilibrium capacity of the carbon for the target compound (μg/g), C_e is the equilibrium liquid phase concentration of the target compound (μg/L), and K and $1/n$ are the Freundlich coefficients in (μg/g)(L/μg)$^{1/n}$ and dimensionless units, respectively. n is a parameter associated to both the relative magnitude and diversity of energies associated with a specific sorption method. The K values that are determined for each chemical are a means of expressing the capacity of a particular GAC to remove a chemical.

2.8.4.2 Membranes

Membrane systems can be an effective technology for reducing the concentration of a diverse set of pharmaceutical compounds during both drinking water and wastewater treatment. NF and RO were the most effective membranes. RO membranes removed more than 80% of all target compounds.

2.8.4.3 Iron–Tetra Amidomacrocyclic Ligand

The iron plus *tetra-amido macrocyclic ligand* (Fe-TAML) activators, developed by Carnegie Mellon University scientists in Pittsburgh, Pennsylvania, are made of an iron atom at the center, surrounded by four nitrogen atoms, which in turn are corralled by a ring of carbon atoms. According to Carnegie Mellon scientists, if H_2O_2 is present, it can displace a water ligand and create a catalyst that triggers oxidation reactions with other compounds in the solution. These catalysts can work with H_2O_2 to rapidly break down 17β-estradiol and 17β-ethinylestradiol within 5 min, whereas 17β-estradiol has a natural half-life of about a week and 17β-ethinylestradiol takes about twice that time to naturally degrade. Thus, Fe-TAML catalyst requires more rigorous experimental studies, while not all stakeholders totally believe such treatment is necessary. Many wastewater treatment plants rely on biological treatment systems, but this is not always effective for all new emerging contaminants and other chemicals of concern. Membrane filtration, including RO, is an effective way to remove the majority of chemicals, but it is expensive to build, maintain, and operate.

2.8.4.4 Chlorine Oxidation

Half of the target chemical compounds were highly reactive and more than 80% were removed, while the remaining compounds were removed at less than a 20% rate. The more reactive compounds generally have aromatic rings with hydroxyl, amine, or methoxy groups. Only certain pharmaceutical compounds will be removed with high efficiency. Also, free chlorine is more effective than chloramine. However, chlorine can react with natural organic and inorganic matter in the water to form carcinogens, which cause adverse reproductive or developmental effects in laboratory animals.

2.8.4.5 Ozonation

Ozonation was highly effective at removing COC and is among the most effective strategies. Oxidative treatment success has been reported for clofibric acid, ibuprofen, and diclofenac, using O_3, H_2O_2/UV, or O_3/H_2O_2. Ozonation has been reported to be effective in the breaking down of diclofenac with complete conversion of the chlorine into chloride ions. Experts have shown that oxidative treatment with both H_2O_2/UV and O_3 is effective for reducing carbamazepine, clofibric acid, diclofenac, sulfamethoxazole, ofloxacin, and propranolol. Ozone treatment of treated water by biological process from wastewater treatment plants is reported to reduce the concentration of several types of pharmaceuticals (measured by the parent compound) below detection limits.

2.9 Illustrative Example for Quantifying the Chemical Threats to Yield Mass Casualties and Acute Injuries

Terrorists can design their mission to attack U.S. drinking water supplies effortlessly through quantification of chemicals needed to yield mass casualties and acute injuries based on the LD_{50} of each chemical or chemical compound. The terrorist leaders have confidently threatened to contaminate the drinking water supply of the United States in recent years. However, many people are still not fully convinced of the potential of these threats, particularly attacking the reservoirs, aqueducts, or an aquifer. The DHS and EPA has focused on protecting water and wastewater treatment plants including hydrants, cyberspace for water facilities, and water tanks but not the original water sources such as the aquifer recharge zone and aqueducts. In fact, once the raw water is totally contaminated with chemicals that are difficult to remove by the traditional water treatment systems, the *denial to water service* can immediately take place. Chapter 4 presents terrorism activity scenarios, Chapters 8 through 9 present the risk estimation model on water supply contamination, and Chapter 11 illustrates the bold planning of terrorist leaders. These presentations are intended to make the general public imagine the possible horrific events that could happen if the terrorists successfully attack the water infrastructure. Revealing these events may shed light on strategic improvements to water infrastructure security to make it difficult for the attacks to succeed and reduce the impact of the attacks that may occur. This section presents the amount of chemicals needed to destroy the water supplies. Accordingly, this is not intended to promote harm against human health and the environment but to drive the general public in acknowledging and realizing the probability or feasibility of water source contamination.

2.9.1 Example of Water Contamination

In the San Antonio region, the Edward Aquifer average daily pumpage is 136.5 million gallons per day (MGD) or 418 acre-ft (515,595,415.52 Liters, L). (1) From 1934 through 1994, the average recharge to the Edwards aquifer was 676,600 acre-feet (834,573,823,300 L); (2) decades later in 1992, following a year of record rainfall, recharge to the aquifer was the highest ever recorded at 2,486,000 acre-ft (3,066,435,892,300 L).

Assume Edwards Aquifer maintains a volume of 676, 600 acre-feet or 834, 573,823,300 L in a year. CN-based pesticide (or sodium cyanide) with an LD_{50} of 6.4 mg/kg (6.4mg/L) will be used as the chemical threat (CT):

Option 1: CN-based pesticide

$$CT_{Cyanide} = (6.4\,\text{mg/L}) \times 834{,}573{,}823{,}300\ \text{L}$$

$$CT_{Cyanide} = 5.3 \times 10^{12}\ \text{mg}$$

$$CT_{Cyanide} = 5.3 \times 10^{12}\ \text{mg} \approx 5{,}300{,}000\ \text{kg} \approx 11{,}684{,}500\ \text{lbs}$$

Option 2: It is very typical that the commercially available pesticide has an approximate LD_{50} of 10 mg/L (Note: See Section 2.4.1.3, according to several resources a mysterious secret of the chemical industry is that inert ingredients, which are the bulking agents for pesticides are often more toxic, however, this information is not required on the labels.)

$$CT_{\text{Pesticide}} = (10\,\text{mg/L}) \times 834{,}573{,}823{,}300\ \text{L}$$

$$CT_{\text{Pesticide}} = 8.3 \times 10^{12}\ \text{mg}$$

$$CT_{\text{Pesticide}} = 8.3 \times 10^{12}\ \text{mg} \approx 8{,}300{,}000\ \text{kg} \approx 18{,}298{,}368\ \text{lbs}$$

The terrorist group requires 5,300,000 kg or 11,684,500 lbs of sodium cyanide to destroy Edwards Aquifer. The terrorist group also has an option to use the traditional pesticides available in the market, if it is difficult to acquire sodium cyanide or CN-based pesticide in some areas. They require 18,298,368 lbs of pesticide to generate mass casualties. Based on the illustration in Figure 4.2 in Chapter 4, terrorists can set up various undetectable stations to implement their plan. Aside from foreign support, they can use credit cards and loans within the United States to financially support the attack.

Meanwhile, one trailer dolly or pick-up truck has the capacity to carry 600 lbs of chemical threats. A terrorist can acquire 3,000 lbs per day (600 lbs × 5 trips from local department stores—open 24 hours) or 21,000 lbs per week (3,000 lbs × 7 days) using a trailer dolly or pick-up truck. In another scenario, a terrorist can use a larger truck with a capacity to haul 6,000 lbs of pesticide or 24,000 lbs per day (6,000 lbs × 4 trips from local department stores—open 24 hours) to acquire a minimum of 168,000 lbs per week (24,000 lbs × 7 days). Further, the terrorist group needs a minimum of 24 stations [(18,298,368 ÷ (21,000 + 168,000) = 24] or 24 *points of target*. Two terrorists can be involved in one station between San Antonio and Austin, Texas to entirely contaminate the Edwards Aquifer within a single month. If there are four or more additional terrorists assigned for this mission, it could be less than a month before terrorists can achieve their ultimate goal without exhausting themselves or being caught by authorities or intelligence agents. Therefore, aquifer destruction can possibly be executed anytime by terrorists based on their extreme hatred and aim to inflict catastrophe against United States. A similar process to the one presented above in option 2 can be utilized to contaminate aqueducts, water tanks, and reservoirs by quantifying the amount of chemical threats using the chemical's LD_{50}.

This chapter sufficiently identified and presented the chemical threats against water infrastructure including some of the effective treatment technologies, which can be used to treat the water supply in the event of contamination. Fundamentally, identifying the potential threats and their hazards is very crucial in optimizing procedures for the protection of the homeland water infrastructure. The following chapter introduces the most accessible explosives and blasting agents used for terrorist attacks. Thus, blasting of water supply facilities, explosion of dams and reservoirs, and the creation of leaks or explosions to petrochemical refineries or chemical plants near water resources could create further damage. The explosive materials or blasting agents potentially used in terrorism are detailed in Chapter 3.

References

Al-Rodhan, N. 2007. *Policy Briefs on the Transnational Aspects of Security and Stability*. New Brunswick, NJ: Transaction Publishers.

America Public Health Association. 1998. American Water Works Association (AWWA), Water Environment Federation (WEF). *Standard Methods for Water and Wastewater Examination*. 20th ed. Washington, DC: America Public Health Association.

American Water Works Association (AWWA). 1990. *Water Quality and Treatment: A Handbook of Community Water Systems*. New York: McGraw-Hill Publishing Company.

Association Engineering Geologists. 1978. Special Publication Commemorating the 50th Anniversary of Failure of the St. Francis Dam, San Francisquito Canyon, near Saugus, California. California: Association Engineering Geologists Southern California Section.

Burrows, W. D. and Renner S.E.1999. Biological warfare agents as threats to potable water. *Environ Health Perspective*, 107(12):975–984.

California Environmental Protection Agency. 2000. *Public Health Goal for Chemicals in Drinking Water, Prepared by Pesticides and Environmental Toxicology Section of Environmental Health Hazard Assessment California Environmental Agency*. California: California Environmental Protection Agency.

California Environmental Protection Agency-Office of Environmental Health Hazard Assessment (Cal. EPA-OEHHA). 2010. All Public Health Goals for Contaminants in Drinking Water. http://oehha.ca.gov/water/phg/ (accessed between April 1, 2010 and August 30, 2010).

Cheng, R. C., S. Liang, H.-C. Wang, and J. Beuhler. 1994. Enhanced coagulation for arsenic removal. *J Am Water Works Assoc* 9:79–90.

Clark, R. M., C. A. Fronk, and B. W. Lykins Jr., 1988. Removing organic contaminants. *Environ Sci Technol* 22(10):1126–30.

Claire, D., S. Randtke, P. Adams, and S. Shreve. 1997. Microfiltration of a high-turbidity surface water with post-treatment by nanofiltration and reverse osmosis. In Proceedings, AWWA Membrane Technology Conference, New Orleans, LA, February 23–26.

Clancy, J. L., and C. Fricker. 1998. Control of cryptosporidium: How effective is drinking water treatment? *Water Qual Int* (July/August):37–41.

Corso, P. S., M. H. Kramer, K. A. Blair, D. G. Addiss, J. P. Davis, and A. C. Haddix. 2003. Cost of illness in the 1993 waterborne Cryptosporidium outbreak, Milwaukee, Wisconsin. *Emerging Infectious Disease Journal*, v9(4):426–31.

Dart, R. C. 2004. *Medical Toxicology*, 1393–1401. Philadelphia: Williams & Wilkins.

Daughton, C. G., and T. A. Ternes. 1999. Pharmaceuticals and Personal Care Products in the Environment: Agents of Subtle Change? *Environ Health Perspect* 107(suppl 6):907–38.

Edwards, M. A. 1994. Chemistry of Arsenic removal during coagulation and Fe-Mn oxidation. *J Am Water Works Assoc* v86(9):64–77.

Eitzen, E., J. Pavlin, T. Cieslak, G. Christopher, and R. Culpepper, 1998. *Medical Management of Biological Casualties Handbook*. 3rd ed. Fort Detrick, MD: U.S. Army Medical Research Institute of Infectious Diseases.

Faust, S. D., and O. M. Aly. 1998. *Chemistry of Water Treatment*. Chealsea, MI: Ann Arbor Press.

Federal Bureau of Investigation. 2001. *Terrorism: Are America's Water Resources and Environment at Risk*. http://www.fbi.gov/news/testimony/terrorism-are-americas-water-resources-and-environment-at-risk (accessed September 15, 2007).

Franz, D. R. 1997. Defense against toxin weapons. In *Medical Aspects of Chemical and Biological Warfare*, ed. F. R. Sidell, E. T. Takafugi, and D. R. Franz, 603–19. Washington, DC: TMM Publications.

Freeman, B. A. 1979. *Burrows Textbook of Microbiology*. 21st ed. Philadelphia, PA: WB Saunders.

Fronk, C. A., and D. Baker. 1990. Pesticide removal by membrane processes. Paper presented at the 1990 Annual Conference of AWWA, Cincinnati, OH, June 17–21.

Hathaway, G. J., N. H. Proctor, J. P. Hughes, and M. L. Fischman. 1991. *Proctor and Hughes' Chemical Hazards of the Workplace*. 3rd ed. New York: Van Nostrand Reinhold.

Hartman, M. J., L. F. Morasch, and W. D. Webber, eds. 2001. *Hanford Site Groundwater Monitoring for Fiscal Year 2000*. Prepared for U.S. Department of Energy. March 2001. Richland, WA: Pacific Northwest National Laboratory.

Hayman, M, and R. R. Dupont. 2001. *Groundwater and Soil Remediation: Process Design and Cost Estimating of Proven Technologies*. Reston, Virginia: ASCE Press.

Imangulov, R. J. 1988. *Medical Service and Protection from Weapons of Mass Destruction in the Subunits*, 1–159. ICD 91-25. Aberdeen Proving Ground, MD: U.S. Army Research Institute of Chemical Defense.

International Agency for Research on Cancer (IARC). 1987. *Overall Evaluations of Carcinogenicity: An Updating of IARC Monographs*, Vol. 1–42. IARC Monographs on the Evaluation of Carcinogenic Risks to Humans, supplement 7. Geneva: International Agency for Research on Cancer.

International Cyanide Management Code for the Gold Mining Industry (ICMC). 2009. *Environmental and Health Effects of Cyanide*. Washington DC: International Cyanide Management Code for the Gold Mining Industry.

Jensen J.G., Vanderveer D.E, and Greer W.T. 996. *Analysis for the Automatic In-line Detection of Chemical and Biological Agents in Water Systems*. Int Rpt AL/CF-TR-1996-0181. Wright-Patterson AFB, OH: Chemical Biological Defense Division, Armstrong Laboratory, 1996.

Kingston, R. L., S. Hall, and L. Sioris. 1993. Clinical observations and medical outcome in 149 cases of arsenate and killer ingestion. *J Toxicol Clin Toxicol* 31(4):581–91.

Klaassen, C., Watkins, J. 2003. *Casarett and Doull's Essentials of Toxicology*. McGraw Hill. p. 512. ISBN 978-0071389143.

Knowles, J. R. 1980. Enzyme-catalyzed phosphoryl transfer reactions. *Annu Rev Biochem* 49:877–919.

Lindsten, D. C., and R. P. Schmitt. 1975. *Decontamination of Water Containing Biological Warfare Agents*. Rpt no 2131, AD A014983. Fort Belvoir, VA: U.S. Army Mobility Equipment Research and Development Center.

Maxnews. 2003. *Al Qaeda Threatens Water Supply*. http://archive.newsmax.com/archives/articles/2003/5/29/94821.shtml (accessed January 17, 2008).

McGeorge, H. J. II. 1989. *Chemical/Biological Terrorism Threat Handbook*. Rpt DAALO3-86-D-0001. Aberdeen Proving Ground, MD: Chemical Research, Development and Engineering Center.

Miltner, R. J., C. A. Fronk, and T. F. Speth. 1987. Removal of Alachlor from Drinking Water. In *Proceedings of the Nat'l Conference on Environmental Engineering*, July 1987, Orlando, FL: ASCE.

Michigan Department of Community Health Office. 2004. Botulism: Information for the public. http://www.michigan.gov/documents/Botulism_fact_sheet_7-22-04_162864_7.pdf (accessed May 20, 2010).

Oregon Department of Human Service (ODHS). 2002. *Arsenic (As)*. Oregon: Office of Environmental Health, Environmental Toxicology Section, Oregon Department of Human Service.

Oxford University. 2010. Physical chemistry department standard material and data sheet. http://msds.chem.ox.ac.uk/HY/ (accessed August 2009 through October 2010).

Parker, A. C., J. Kirsi, W. H. Rose, and D. T. Parker. 1996. *Counter Proliferation—Biological Decontamination*. Rpt DAAD09-92-D-0004. Aberdeen Proving Ground, MD: U.S. Army Test and Evaluation Command.

Parmeggiani, L. 1983. *Encyclopedia of Occupational Health and Safety*. 3rd rev. ed. Geneva, Switzerland: International Labour Organisation.

Sawatsky, B., C. Ranaheera, H. M. Weingartl, and M. Czub. 2008. *Hendra and Nipah Virus. Animal Viruses: Molecular Biology*. Norwich, UK: Caister Academic Press.

Schnoor, J. L., L. A. Licht, S. C. McCutcheon, N. L. Wolfe, and L. H. Carreira. 1995. Phytoremediation of organic and nutrient contaminants. *Environ Sci Technol* 29(7):318A–323A.

Science Applications International Corp. 1996. *Ultraviolet Light Disinfection Technology in Drinking Water Application. An Overview*. EPA/811/R-96/002. Washington, DC: U.S. Environmental Protection Agency.

Snyder, S. A., T. L. Keith, D. A. Verbrugge, E. M. Snyder, T. S. Gross, K. Kannan and J. P. Giesy. 1999. Analytical methods for detection of selected estrogenic compounds in aqueous mixtures. *Environ Sci Technol* 33(16):2814–20.

Snyder, S. A., D. L. Villeneuve, E. M. Snyder, and J. P. Giesy. 2001. Identification and quantification of estrogen receptor agonists in wastewater effluents. *Environ Sci Technol* 35(18):3620–5.

Solomons, T. W. G. 1988. *Organic Chemistry*. 4th ed. New York: John Wiley and Sons, Inc.

Speth, T. F., and J. Q. Adams. 1993. GAC and air stripping design support for the safe drinking water act. In *Strategies and Technologies for Meeting SDWA Requirements*, ed. R. Clark and R. S. Summers, 47–89. Ann Arbor, MI: Lewis Publishers.

Speth, T. F., and R. J. Miltner. 1990. Technical note: Adsorption capacity of GAC for synthetic organics. *J Am Water Works Assoc* 82(2):72–5.

Speth, T. F., and R. J. Miltner. 1998. Technical note: Adsorption capacity of GAC for synthetic organics. *J Am Water Works Assoc* 90(4):171–4.

Standard Material Safety and Data Sheet. 2010. Source of the chemicals or chemical compounds' MSDS. http://www.msdsonline.com (accessed 2008 through 2010).

Stewart, H. T., and K. J. Kessler. 1991. Evaluation of arsenic removal by activated alumina filtration at a small community public water supply. *J New Eng Water Works Assoc* 105:179–99.

Tyrrell, S. A., S. R. Rippey, and W. D. Watkins. 1995. Inactivation of bacterial and viral indicators in secondary sewage effluents, using chlorine and ozone. *Water Res* 29(11):2483–90.

U.S. Agency for Toxic Substances and Disease Registry (ATSDR). 2010a. Toxic substances portal: methyl tert-butyl ether (MTBE). http://www.atsdr.cdc.gov/toxfaqs/tf.asp?id=227&tid=41 (accessed July 29, 2010).

U.S. Agency for Toxic Substances and Disease Registry. 2010b. Toxic substances portal. http://www.atsdr.cdc.gov/substances/index.asp (accessed July 28, 2010).

U.S. Department of the Interior-Bureau of Reclamation (USDI-BOR). 2001. Cyanide fact sheet. http://www.usbr.gov/pmts/water/publications/reportpdfs/Primer%20Files/08%20-%20Cyanide.pdf (accessed June 25, 2010).

U.S. Central Intelligency Agency. 1998. Terrorist CBRN: Materials and effects. https://www.cia.gov/library/reports/general-reports-1/cbr_handbook/cbrbook.htm#6 (accessed August 20, 2009 and October 18, 2010).

U.S. Central Intelligency Agency. 2010a. Terrorist CBRN: Materials and effects. https://www.cia.gov/library/reports/general-reports-1/terrorist_cbrn/terrorist_CBRN.htm (accessed August 20, 2009 and October 18, 2010).

U.S. Central Intelligence Agency. 2010b. CIA and war on terrorism. https://www.cia.gov/news-information/cia-the-war-on-terrorism/ (accessed October 12, 2010).

U.S. Department of Defense. 2002. *Environmental Assessment for Pulse Fast Neutron Analysis Cargo Inspection Test Facility at Ysleta Port of Entry Commercial Cargo Facility, El Paso Texas*. Dahlgren, Virginia, USA: Counterdrug Technology Development Program Office, Naval Surface Warfare Center.

U.S. Department of Homeland Security (DHS). 2002. *The Homeland Security Act 2002*. Public Law 107-296, 116 Stat. 2135, Sections 211–215. Washington DC: U.S. Department of Homeland Security.

U.S. Department of Homeland Security. 2003. *Homeland Security Presidential Directive 7: Critical Infrastructure Identification, Prioritization, and Protection*. Washington DC: U.S. Department of Homeland Security.

U.S. Department of Homeland Security. 2005. *Interim National Infrastructure Protection Plan*. Washington, DC: U.S. Department of Homeland Security.

U.S. Department of Homeland Security. 2007a. *Appendix to Chemical Facility Anti-Terrorism Standards; Final Rule*, 6 CFR Part 26, Federal Register/ Vol. 72, No. 223. Washington DC: U.S. Department of Homeland Security.

U.S. Department of Homeland Security. 2007b. *Chemical Facility Anti-terrorism Standards; Final Rule*, 6 CFR Part 27, DHS-2006-0073, Federal Register/Vol. 72, No. 67. Washington DC: U.S. Department of Homeland Security.

U.S. Department of Homeland Security. 2007c. *Homeland Security Appropriation Act 2007*, Public Law 109-295. Washington DC: U.S. Department of Homeland Security.

U.S. Department of Homeland Security. 2009a. *Interagency Security Committee Use of Physical Security Performance Measures*. Washington DC: U.S. Department of Homeland Security.

U.S. Department of Homeland Security. 2009b. *National Infrastructure Protection Plan, Water Sector*. Washington DC: U.S. Department of Homeland Security.

U.S. Department of Homeland Security. 2009c. *Identifying Facilities Covered by the Chemical Security Regulation*. http://www.dhs.gov/files/programs/gc_1181765846511.shtm (accessed May 5, 2010).

U.S. Department of Homeland Security. 2010a. *Homeland Security Presidential Direction 10: Biodefense for the 21st Century*. Washington, DC: U.S. Department of Homeland Security.

U.S. Department of Homeland Security. 2010b. *Risk for Chemical Anti-terrorism Standards*. Washington, DC: U.S. Department of Homeland Security.

U.S. Department of Justice (DOJ). 2010. Freedom of Information Act. http://www.justice.gov/oip/ (accessed June 2, 2010).

U.S. Environmental Protection Agency. 1987. *Drinking water criteria document for cyanide. Prepared by the Office of Health and Environmental Assessment, Environmental Criteria and Assessment Office, Cincinnati, OH, for the Office of Drinking Water*. Washington, DC: U.S. Environmental Protection Agency.

U.S. Environmental Protection Agency. 1989. *Technologies for Upgrading Existing or Designing New Drinking Water Treatment Facilities.* Report No. EPA/625/4-89/023. Washington, DC: U.S. Environmental Protection Agency.

U.S. Environmental Protection Agency. 1993. *Technical Information Review. Methyl tertiary Butyl Ether.* CAS No. 1634-04-4. Washington, DC: Office of Pollution Prevention and Toxics, U.S. Environmental Protection Agency.

U.S. Environmental Protection Agency. 1999. *Technologies and Costs for Removal of Arsenic from Drinking Water.* Report No. EPA 815-P-01-001. http://water.epa.gov/drink/info/arsenic/upload/2005_11_10_arsenic_techcosts.pdf (accessed April 29, 2009).

U.S. Environmental Protection Agency. 2000. *Technologies and Cost for Removal of Arsenic from Drinking Water, Targeting and Analysis Branch Standards and Risk Management Division Office of Groundwater and Drinking Water,* Report No. EPA-815-R-00-028. Washington, DC: U.S. Environmental Protection Agency.

U.S. Environmental Protection Agency. 2001. *The Incorporation of Water Treatment Effects on Pesticide Removal and Transformations in Food Quality Protection Act (FQPA).* Drinking Water Assessments. http://www.bvsde.paho.org/bvsacg/i/fulltext/pesticide/pesticide.pdf (accessed January 15, 2009).

U.S. Environmental Protection Agency. 2002a. *Factoids: Drinking Water and Ground Water Statistics for 2002.* Washington, DC: U.S. Environmental Protection Agency.

U.S. Environmental Protection Agency. 2002b. *National Primary Drinking Water Regulations: Long Term 1 Enhanced Surface Water Treatment Rule; Final Rule,* 40 CFR Parts 9, 141, and 142. Washington, DC: U.S. Environmental Protection Agency.

U.S. Environmental Protection Agency. 2002c. *The Clean Water and Drinking Water Infrastructure Gap Analysis Report,* 50. Report No. EPA 816-R-02-020. Washington, DC: U.S. Environmental Protection Agency.

U.S. Environmental Protection Agency. 2003. *Design Manual of Arsenic Removal Drinking Water by Ion Exchange.* Report No. EPA/600/R-03/080. Washington, DC: U.S. Environmental Protection Agency.

U.S. Environmental Protection Agency. 2004. *National Public Water Systems Compliance Report,* 96. Report No. EPA 305-R-04-001. Washington DC: U.S. Environmental Protection Agency.

U.S. Environmental Protection Agency. 2004b. *Water Security Legislation and Directives.* Washington DC: U.S. Environmental Protection Agency.

U.S. Environmental Protection Agency. 2010a. Technical factsheet on cyanide. http://www.epa.gov/safewater/pdfs/factsheets/ioc/tech/cyanide.pdf (accessed January 5, 2010).

U.S. Environmental Protection Agency. 2010b. Organic chemicals. http://water.epa.gov/drink/contaminants/#Organic (accessed January 2010).

U.S. Environmental Protection Agency. 2010c. Inorganic chemicals. http://water.epa.gov/drink/contaminants/#Inorganic (accessed January 2010).

U.S. Federal Emergency Management Agency. 2010. Why dams fail. http://www.fema.gov/hazard/damfailure/why.shtm (accessed October 17, 2010).

Wagler, J. L., and J. P. Malley Jr. 1994. The removal of MTBE from a model groundwater using UV/peroxide oxidation. *J New Eng Water Works Assoc* v67, 236–43.

Wang, L., T. Sorg, and A. Chen. 2000. Arsenic removal by full scale Ion exchange and activated Alumina treatment systems. Paper presented at the American Water Works Association Inorganic Contaminants Workshop, Albuquerque, NM, February 27–29.

Wannemacher, R.W. Jr., R. E. Dinterman, W. L. Thompson, M. O. Schmidt, and W. D. Burrows. 1993. *Treatment for Removal of Biotoxins from Drinking Water*. Report no. TR9120, AD A275958. Fort Detrick, MD: US Army Biomedical Research and Development Laboratory.
Warner, J. S. 1990. *Review of Reactions of Biotoxins in Water*. Rpt CBIAC Task 152. Fort Detrick, MD: U.S. Army Medical Research and Development Command.
Whelton, A. J., J. Jensen, T. Richards, and R. M. Valvidia. 2003. The Cyanic Threat to Potable Water. Paper presented at the Annual Conference and Exposition, AWWA Proceedings, Anaheim, California, June 14–17.
White, G. C. 1992. *The Handbook of Chlorination and Alternative Disinfectants*. 3rd ed. New York: Van Nostrand Reinhold.
Zilinskas, R. A. 1997. Iraq's biological weapons. The past as future? *JAMA* 278(5):418–24.

Chapter 3

Explosives Used Against Water Infrastructure

3.1 Introduction

This chapter introduces some of the explosives or *blasting agents* that can be easily created as components of improvised explosive devices (IEDs) used against water infrastructure. IEDs are responsible for numerous American combat casualties including civilians in Iraq and Afghanistan. The Department of Homeland Security (DHS) has no specific and detailed information to indicate that IEDs are currently being planned for use in the United States. This chapter provides the basics of explosives and their design, specifications, and characteristics as weapons usually used by terrorists. Warnings against terrorism and appropriate preventive measures can be determined by having an idea of the materials used in explosives preparation and their specifications. In Section 3.3.12.1, a process is suggested to calculate safe distances from a potential explosion. This method can be used to help design and install perimeters and emergency response stations for major assets.

3.2 Characterization of Explosive Materials

Explosive materials include explosives, blasting agents, and detonators. A list of explosive materials determined to be within the coverage of 18 U.S.C. Chapter 40, Importation, Manufacture, Distribution, and Storage of Explosive Materials is issued at least annually by the Director of the Bureau of Alcohol, Tobacco, Firearms, and Explosives (ATF) of the Department of the Treasury. The U.S. Department of Transportation (DOT) classifications of explosive materials used in commercial blasting operations are not identical with the statutory definitions of the Organized Crime Control Act of 1970, Title 18 U.S.C., Section 841. To achieve uniformity in transportation, the definitions of the DOT in Title 49 Transportation Code of Federal Regulations (CFR), Chapter 1, subdivides these materials into class A explosives (detonating, or otherwise maximum hazard), class S explosives (flammable hazard), class C explosives (minimum hazard), and oxidizing material (a substance that yields oxygen readily to stimulate the combustion of organic matter). Hence, there are several categories of explosive materials; however, the ones described in Sections 3.2.1 through 3.2.19 are the most inexpensive and easy to acquire for terrorist attacks against water infrastructure.

3.2.1 Acetone Peroxide

Acetone peroxide (tricycloacetone peroxide; Chemical Abstract Service [CAS] number 17088-37-8) is formed from acetone in sulfuric acid solution when acted upon by 45% hydrogen peroxide. Its properties are comparable to those of primary explosives. Acetone peroxide is not used in practice because of its tendency to undergo sublimation. It is highly susceptible to heat, friction, and shock. However, it can be a powerful weapon for destroying water infrastructure and the chemical compositions are commercially viable. The cyclic dimer ($C_6H_{12}O_4$), an open monomer, and a dimer are also formed, but under special conditions the cyclic trimer ($C_9H_{18}O_6$) is the primary product. The criteria and characteristics of acetone peroxide include but are not limited to the following: oxygen balance, –151.3%; lead block test, 250 cm^3/10 g; melting point, 91°C; impact sensitivity, 0.3 N·m; and friction sensitivity, 0.1 N. Acetone peroxide was used as the explosive in the July 2005 London bombings. There was a series of coordinated terrorist bomb blasts that hit London's public transport system during the morning rush hour. Three bombs exploded within 50 seconds of each other on the London Underground trains. It should be noted that the terrorists usually launch their attacks during rush hours. Detection method and emergency response techniques based on the timing and distance of the explosion may need to be analyzed closely.

3.2.2 Ammonium Nitrate

Ammonium nitrate (NH_4NO_3; AN; CAS number 6484-52-2) is hygroscopic and water soluble. AN fertilizer does great things for the agriculture industry. It is made of a chemical compound of 27% nitrogen and 8% calcium carbonate, is typically affordable, and does an impressive duty bolstering all kinds of harvests. Hence, it is very accessible to terrorists and criminals as a weapon of destruction, and it was used in the notorious 1995 federal building bombing in Oklahoma City. Meyer, Kohler, and Homburg (2007) indicated that the product shows a great tendency to cake and the resulting difficulties are avoided by transformation into *prills* (prills are small aggregate materials formed from a melted liquid). AN is commercially sold as dense prills and as porous prills employed for industrial explosives such as mining after milling.

Meyer, Kohler, and Homburg (2007) specified that ammonium nitrate explosives composed of AN with carbon carriers (e.g., coal or wood meal) and sensitizers such as *nitroglycol* or TNT with aluminum powder in it for producing a stronger explosion.

According to Meyer, Kohler, and Homburg (2007), AN tends to be very difficult to detonate, and another high explosive or strong industrial or military blasting cap—a #8 cap and 12 oz of *pentaerythritol tetranitrate* (PETN) or hexahydro-1,3,5-trinitro-1,3,5-triazine (RDX)—is needed to detonate it; it melts at 180°C, holds 378 cal/g of energy, and has a detonating velocity of 3,460 m/s. Table 3.1 provides the characteristics and specifications of AN.

Table 3.1 Characteristics and Specifications of Ammonium Nitrate

Color: colorless crystals
Molecular weight: 80
CAS number: 6484-52-2
Energy of formation: −4,428.0 kJ/kg
Enthalpy of formation: −4,567.0 kJ/kg
Oxygen balance: +19.99%
Nitrogen content: 34.98%
Volume of explosion gases: 980 L/kg
Heat of explosion (H_2O liq.): 479 kJ/kg, (H_2O gas): 1,441 kJ/kg
Melting point: 169.6°C = 337.3°F
Lead block test: 180 cm^3/10 g
Deflagration point: starts decomposition at the melting point of 169.6°C; completes approximately at 210°C (boiling point).
Impact sensitivity: up to 50 N·m
Friction sensitivity: 353 N pistil load no reaction
Critical diameter of steel sleeve test: 1 mm

Sources: Data from U.S. Environmental Protection Agency, http://www.epa.gov/ttn/chief/ap42/ch08/final/c08s03.pdf, 2010; Meyer, R., J. Kohler, and A. Homburg, *Explosives*, 5th ed., New York: Wiley, 2002; Akhavan, J., *The Chemistry of Explosives*, Cambridge, UK: Royal Society of Chemistry, 2004; U.S. Department of Treasury's Bureau of Alcohol, Tobaco, Firearms, and Explosives (ATF), *Definition and test procedures of Ammonium Nitrate Fertilizer*, U.S. government publications, 1984; U.S. Department of Treasury's Bureau of Alcohol, Tobaco, Firearms, and Explosives (ATF), http://searchjustice.usdoj.gov/search?q=explosives+&site=default_collection&sort=date%3AD%3AL%3Ad1&output=xml_no_dtd&ie=UTF-8&oe=UTF-8&client=atf&proxystylesheet=atf, 2010.

3.2.3 Ammonium Nitrate–Fuel Oil

ANFO is a tertiary explosive composed of AN and liquid hydrocarbons. The application technique of these mixtures has now become much easier owing to the fact that the material, which has a strong tendency to agglomerate, is commercially produced as porous prills (Meyer, Kohler, and Homburg 2007). These porous prills have the capacity to approximately absorb 6% of the oil, which is the quantity required to maintain an oxygen balance that will generate a higher explosion with igniting by a powerful primer.

3.2.4 Cyclonite (RDX)

Cyclonite $(CH_2\text{-}N\text{-}NO_2)_3$ is a white, crystalline solid used in mixture with other explosives/blasting agents and plasticizers, phlegmatizers, or desensitizers. RDX can be dissolved in acetone, ether and ethanol, but it is insoluble in water. *Cyclohexanone, nitrobenzene*, and *glycol* are solvents at elevated temperatures (Meyer, Kohler, and Homburg 2007). Its *detonation velocity* at a density of 1.76 g/ cm^3 is quantified as 8750 m/s. The chemical reaction of concentrated nitric acid with hexamine produces RDX. The decomposition temperature is 170°C and the melting point is 204°C. Based on testimony in the U.S. Court of Appeals for the Ninth Circuit, dated February 2, 2010, Ahmed Ressam, the Al-Qaeda millennium bomber, used a small quantity of RDX as one of the components in the explosives that he used to bomb Los Angeles International Airport on New Year's Eve 1999; the combined explosives could have produced a blast greater than that produced by a devastating car bomb. RDX was the main component used in the 2006 Mumbai, India, train bombings. It is also believed to be the explosive used in the 2010 Moscow Metro, Russia, bombings. Table 3.2 presents the characteristics and specifications of RDX.

Table 3.2 Characteristics and Specifications of RDX

Color: colorless crystals
Empirical formula: $C_3H_6N_6O_6$
CAS number: 121-82-4
Molecular weight: 222.1
Energy of formation: +401.8 kJ/kg
Enthalpy of formation: +301.4 kJ/kg
Optimum oxygen balance: –21.6%
Optimum nitrogen content: 37.84%

Table 3.2 Characteristics and Specifications of RDX (*Continued*)

Volume of explosion gases: 903 L/kg
Empirical formula: $C_3H_6N_6O_6$
Molecular weight: 222.1
Energy of formation: +401.8 kJ/kg
Enthalpy of formation: +301.4 kJ/kg
Optimum oxygen balance: –21.6%
Optimum nitrogen content: 37.84%
Volume of explosion gases: 903 L/kg
Heat of explosion (H_2O liq.): 5,647 kJ/kg, (H_2O gas): 5,297 kJ/kg
Heat of detonation (H_2O liq.): 6,322 kJ/kg
Specific energy: 1,375 kJ/kg
Density: 1.82 g/cm^3
Melting point: 204°C
Heat of fusion: 161 kJ/kg
Lead block test: 480 cm^3/10 g
Detonation velocity, confined: 8,750 m/s
Impact sensitivity: 7.5 N·m
Friction sensitivity: 120 N

Sources: Data from U.S. Environmental Protection Agency, http://www.epa.gov/ttn/chief/ap42/ch06/final/c06s03.pdf, 2010; U.S. Environmental Protection Agency, http://nlquery.epa.gov/epasearch/epasearch?querytext=Explosives&typeofsearch=epa&sort=term_relevancy&results_per_page=10&doctype=all&originalquerytext=Ammonium+Nitrate&areaname=&faq=no&filterclause=&sessionid=981E2B6116865D395E59B9D752558E2C&referer=http%3A%2F%2Fepa.gov%2F&prevtype=epa&result_template=epafiles_default.xsl&areasidebar=search_sidebar&areapagehead=epafiles_pagehead&areapagefoot= epafiles_pagefoot&stylesheet=http%3A%2F%2Fwww.epa.gov%2Fepafiles% 2Fs%2Fepa. css&po=3333, 2010; Meyer, R., J. Kohler, and A. Homburg, *Explosives*, 5th ed., New York: Wiley, 2002; U.S. Department of Treasury's Bureau of Alcohol, Tobacco, Firearms, and Explosives (ATF), http://searchjustice.usdoj.gov/search?q=explosives+&site=default_collection&sort=date%3AD%3AL%3Ad1&output=xml_no_dtd&ie=UTF-8&oe=UTF-8&client=atf&proxystylesheet=atf, 2010; Akhavan, J., *The Chemistry of Explosives*, Cambridge, UK: Royal Society of Chemistry, 2004.

3.2.5 Dingu and Sorguyl

Dingu and *sorguyl* were introduced by the Soiete Nationale Des Poudres Et Explosivs, Sogues, France (Meyer, Kohler, and Homburg 2002). The reaction between glyoxal and urea produces *glycolurile*. Once glycolurile undergoes the dinitration process, it produces dingu. The characteristics of dingu are presented in Table 3.3. It is easily decomposed by alkaline hydrolysis. It is stable in contact with neutral or acid water. It is insoluble in molten TNT but soluble in *dimethyl sufoxide*. Nitration with nitric acid and nitrogen pentoxide can generate sorguyl. Nitro derivatives of glycolurile have recently attracted renewed interest because sorguyl has proved to be one of the most powerful modern explosives (Boileau, Emeury, and Keren 1975). Sorguyl has high density and high detonation velocity, which can be used by terrorists for attacking dams. Dingu and sorguyl are not very common in the United States and not easily detected onsite. Sorguyl is not hygroscopic, decomposes easily by hydrolysis (decomposes when mixed with molten TNT), and is insoluble in both hydrocarbons and chlorinated hydrocarbons. The characteristics and specifications of dingu and sorguyl are presented in Tables 3.3 and 3.4.

3.2.6 Hexamethylenetetramine Dinitrate

Hexamethylenetetramine dinitrate is soluble in water but insoluble in alcohol, ether, chloroform, and acetone. It is usually made from hexamethylenetetramine and nitric acid; it is an important precursor of primary explosives. Table 3.5 provides the characteristics and specifications of hexamethylenetetramine dinitrate.

Table 3.3 Characteristics and Specifications of Dingu

Color: colorless
Empirical formula: $C_4H_4N_6O_6$
Molecular weight: 232.1
Oxygen balance: –27.6%
Nitrogen content: 36.2%
Density: 1.94 g/cm^3
Detonation velocity, confined: 7,580 m/s
Impact sensitivity: 5–6 N·m
Decomposition: begins at 266°F
Friction sensitivity: up to 300 N

Sources: Data from Agrawal, J. P., and R. D. Hodgson, *Organic Chemistry of Explosives*, West Sussex, UK: John Wiley & Sons, 2007; Meyer, R., J. Kohler, and A. Homburg, *Explosives*, 5th ed., New York: Wiley, 2002; Boileau, J., J. M. Emeury, and J. P. Keren, German Patent 2,435,651, 1975; Emeury, J. L., and H. H. Girardon, U.S. Patent 4,211, 874, 1980.

Table 3.4 Characteristics and Specifications of Sorguyl

Color: colorless
Empirical formula: $C_4H_2N_8O_{10}$
Molecular weight: 322.1
Oxygen balance: +5%
Nitrogen content: 34.79%
Density: 2.01 g/cm^3
Detonation velocity, confined: 9,150 m/s
Impact sensitivity: 1.5–2 N·m
Deflagration point at 459°F

Sources: Data from Agrawal, J. P., and R. D. Hodgson, *Organic Chemistry of Explosives*, West Sussex, UK: John Wiley & Sons, 2007; Meyer, R., J. Kohler, and A. Homburg, *Explosives*, 5th ed., New York: Wiley, 2002; Boileau, J., J. M. Emeury, and J. P. Keren, German Patent 2,435,651, 1975.

3.2.7 Hexanitroazobenzene

2,2',4,4',6,6',-Hexanitroazobenzene (CAS number 19159-68-3) is normally created from *dinitrochlorobenzene* and *hydrazine*. Also, oxidation and nitration of tetranitrohydrazobenzene can produce *hexanitroazobenzene* (Table 3.6). It is a threat to be used for destroying dams, underground sewer pipelines, and aqueducts because it is a considerably powerful underwater explosive.

3.2.8 Hexanitrodiphenylamine

Hexanitrodiphenylamine is mostly toxic and a poisonous underwater explosive with TNT and aluminum powder. It is less powerful than *hexanitroazobenzene* but can still destroy large dams when it is appropriately installed near dam abutments on unstable ground. Additionally, it is insoluble in water and most organic solvents and forms acid-sensitive salts. It is prepared by nitration of *asym-dinitrodiphenylamine* and formed by condensation of dinitrochlorobenzene with aniline (Meyer, Kohler, and Homburg 2007). As stated by Meyer, Kohler, and Homburg (2007), it is an explosive with a relatively low sensitivity to heat and has been used as a precipitant for potassium. The characteristics and specifications of hexanitrodiphenylamine are presented in Table 3.7.

3.2.9 Hexanitrohexaazaisowurtzitane

As indicated by Meyer, Kohler, and Homburg (2007), *hexanitrohexaazaisowurtzitane* or CL-20 (CAS number 135285-90-4) is obtained by condensing glyoxal with

Table 3.5 Characteristics and Specifications of Hexamethylenetetramine Dinitrate

Color: colorless crystals
Empirical formula: $C_6H_{14}N_6O_6$
CAS number: 100-97-0
Molecular weight: 266.2
Energy of formation: −1,296.6 kJ/kg
Enthalpy of formation: −1,417.7 kJ/kg
Oxygen balance: +78.3%
Nitrogen content: 31.57%
Volume of explosion gases: 1,081 L/kg
Heat of explosion (H_2O liq.): 2,642 kJ/kg, (H_2O gas): 2,434 kJ/kg
Melting point: 169.6°C = 316°F
Lead block test: 220 cm^3/10 g
Begins decomposition at melting point: completes at 316°F
Impact sensitivity: up to 50 N·m no reaction
Friction sensitivity: 240 N pistil load reaction

Sources: Data from Meyer, R., J. Kohler, and A. Homburg, *Explosives*, 5th ed., New York: Wiley, 2002; U.S. Department of Treasury's Bureau of Alcohol, Tobaco, Firearms, and Explosives (ATF), http://searchjustice.usdoj.gov/search?q=explosives+&site=default_collection&sort=date%3AD%3AL%3Ad1&output=xml_no_dtd&ie=UTF-8&oe=UTF-8&client=atf&proxystylesheet=atf, 2010.

Table 3.6 Characteristics and Specifications of Hexanitroazobenzene

Color: orange red
Empirical formula: $C_{12}H_4N_8O_{12}$
Molecular weight: 452.2
CAS number: 19159-68-3
Oxygen balance: −49.7%
Nitrogen content: 24.78%
Melting point: 430°F

Sources: Data from U.S. Naval Technical Mission to Japan, http://www.fischertropsch.org/primary_documents/gvt_reports/USNAVY/USNTMJ%20Reports/USNTMJ_toc.htm, 1945; Meyer, R., J. Kohler, and A. Homburg, *Explosives*, 5th ed., New York: Wiley, 2002; U.S. Department of Treasury's Bureau of Alcohol, Tobaco, Firearms, and Explosives (ATF), http://searchjustice.usdoj.gov/search?q=explosives+&site=default_collection&sort=date%3AD%3AL%3Ad1&output=xml_no_dtd&ie=UTF-8&oe=UTF-8&client=atf&proxystylesheet=atf, 2010.

Table 3.7 Characteristics and Specifications of Hexanitrodiphenylamine

Color: yellow crystals
Empirical formula: $C_{12}H_5N_7O_{12}$
Molecular weight: 439.2
CAS number: 131-73-7
Energy of formation: +162 kJ/kg
Density: 1.64 g/cm^3
Enthalpy of formation: +94.3 kJ/kg
Specific energy 1,098 kJ/kg
Optimum oxygen balance: +52.8%
Detonation velocity: 7,200 m/s at a density of 1.6 g/cm^3
Optimum nitrogen content: 22.33%
Volume of explosion gases: 791 L/kg
Heat of explosion (H_2O liq.): 4,075 kJ/kg, (H_2O gas): 4,004 kJ/kg
Melting point: 464°F–466°F
Lead block test: 325 cm^3/10 g
Temperature of decomposition: completes at 316°F
Impact sensitivity: 7.5 N·m pistil load no reaction
Friction sensitivity: 353 N pistil load reaction
Acetone mixture: not more than 0.1%
Insoluble in 1:3 pyridine (C_5H_5N, an important solvent and reagent)

Sources: Data from U.S. Army and Air Force (USAAF), http://www.lexpev.nl/downloads/oldchemicalweapons1994.pdf, 1994; Meyer, R., J. Kohler, and A. Homburg, *Explosives*, 5th ed., New York: Wiley, 2002; U.S. Department of Treasury's Bureau of Alcohol, Tobaco, Firearms, and Explosives (ATF), http://searchjustice.usdoj.gov/search?q=explosives+&site=default_collection&sort=date%3AD%3AL%3Ad1&output=xml_no_dtd&ie=UTF-8&oe=UTF-8&client=atf&proxystylesheet=atf, 2010; U.S. Naval Technical Mission to Japan, http://www.fischertropsch.org/primary_documents/gvt_reports/USNAVY/USNTMJ%20Reports/USNTMJ_toc.htm, 1945.

benzylamine to produce *hexabenzylhexaazaisowurtzitane*. It is one of the most energetic organic explosives due to its high density and detonation velocity that is near to or can exceed ±10,000 m/s. It can be used as one of the IED's components to destroy a portion of a large metropolitan area when planned carefully. The characteristics and specifications of CL-20 are presented in Table 3.8.

Table 3.8 Characteristics and Specifications of CL-20

Empirical formula: $C_6H_6N_{12}O_{12}$
Color: white crystalline solid
Molecular weight: 438.19
Energy of formation: +1,005.3 kJ/kg
Density: 2.04 g/cm^3
Enthalpy of formation: +920.5 kJ/kg
Specific energy: 1323 kJ/kg
Optimum oxygen balance: –10.95%
Optimum nitrogen content: 38.3%
Heat of explosion (H_2O liq.): 6,314 kJ/kg, (H_2O gas): 6,084 kJ/kg
Melting point: 195°C
Impact sensitivity: 4 N·m
Friction sensitivity: 48 N

Sources: Data from Simpson, R. L., P. A. Urtiew, D. L. Ornellas, G. L. Moody, K. J. Scribner, and D. M. Hoffman, *Propellant, Explosives, Pyrotechnics* 22(5):249–55, October 1997; Meyer, R., J. Kohler, and A. Homburg, *Explosives*, 5th ed., New York: Wiley, 2002; Lee, K. E., R. L. Hatch, M. Mezger, and S. Nicolich, US Patent 6,214,137 B1, April 10, 2001; Lee, J. S., and K. S. Jaw, *J Therm Anal Calorim* 85:463–7, 2006.

3.2.10 Lead Azide

Lead azide ($Pb(N_3)_2$; CAS number 13424-46-9) is poisonous, insoluble in water, and resistant to heat and moisture. It is prepared by reacting sodium azide and lead nitrate; production of large crystals may occur and should be mitigated for safety from potential explosion during the preparation by precipitation with dextrin or polyvinyl alcohol. Flammability can be improved by adding flammable additive, such as *lead trinitroresorcinate* (lead trinitroresorcinate is a slurry or wet mass of orange–yellow crystals, which is a weak but highly sensitive explosive). Hence, water does not reduce this explosive's impact sensitivity, therefore, it can be a component of an IED for devastation of water infrastructure. The characteristics and specifications of lead azide are presented in Table 3.9.

Table 3.9 Characteristics and Specifications of Lead Azide

Color: colorless crystals
Molecular weight: 291.3
Energy of formation: +1,663.3 kJ/kg
Density: 4.8 g/cm^3
Enthalpy of formation: +1,637.7 kJ/kg
Net content as $PbCrO_4$: not less than 91.5%
Optimum oxygen balance: +5.5%
Detonation velocity: depends on the density (e.g., 4,500 m/s at a density of 3.8 g/cm^3)
Optimum nitrogen content: 28.85%
Volume of explosion gases: 231 L/kg
Explosion heat: 1,638 kJ/kg
Melting point: 464°F–466°F
Lead block test: 325 cm^3/10 g
Temperature of decomposition: completes at 316°F
Impact sensitivity: 2.5–4 N·m (pure); 3.0–6.5 N·m (dextrinated)
Friction sensitivity: 0.1–1 N
Moisture content: 0.3%
Water solubility: 1%
Lead content: 68%
Deflagration point: 350°C
Bulk density: 1.1 g/cm^3

Sources: Data from Verneker, V. R., and A. C. Forsyth, *J Phys Chem* 72:111, 1968; Meyer, R., J. Kohler, and A. Homburg, *Explosives*, 5th ed., New York: Wiley, 2002; U.S. Department of Treasury's Bureau of Alcohol, Tobaco, Firearms, and Explosives (ATF), http://searchjustice.usdoj.gov/search?q=explosives+&site=default_collection&sort=date%3AD%3AL%3Ad1&output=xml_no_dtd&ie=UTF-8&oe=UTF-8&client=atf&proxystylesheet=atf, 2010; Mcnicol, L. J. P., U.S. Patent 3,264,150, August 2, 1966.

3.2.11 Lead Styphnate

Lead styphnate (or lead trinitroresorcinate; CAS number 15245-44-0) is sensitive to fire, can readily ignite by static discharges from the human body, is insoluble in water, and is moderately soluble in acetone and ethanol. Consequently, it is usually employed as an initiating explosive with lead azide forming the detonator charge;

this is unlikely to be utilized by terrorists in attacking large urban areas due to its very high ignition sensitivity that requires safe handling and transportation procedures, as authorities are vigilant to suspicious activities. Table 3.10 shows the characteristics and specifications of lead styphnate.

Table 3.10 Characteristics and Specifications of Lead Styphnate

Color: orange-yellow to dark brown crystals
Empirical formula: $C_6H_3N_3O_9Pb$
CAS number: 15245-44-0
Molecular weight: 468.3
Energy of formation: –1,747.2 kJ/kg
Density: 3.0 g/cm^3
Enthalpy of formation: –1,786.9 kJ/kg
Optimum oxygen balance: –18.8%
Detonation velocity: 5,200 m/s at a density of 2.9 g/cm^3
Optimum nitrogen content: 8.97%
Volume of explosion gases: 231 L/kg
Lead block test: 130 cm^3/10 g
Impact sensitivity: 2.5–5.0 N·m
Moisture content: 0.15%
Net content: 98%
Water solubility: not more than 1%
Lead content: 43.2% – 44.3%
Ca, Mg: 0.5%
Na: 0.07%
pH: 5–7
Deflagration point: 518°F
Bulk density: 1.3–1.5 g/cm^3

Sources: Data from Jiang, Z., In *Proceedings of the 24th International Symposium on Shock Waves*, Beijing, China, July 11–14, 2004, 1:984, Berlin, Heidelberg: Springer, 2005; Meyer, R., J. Kohler, and A. Homburg, *Explosives*, 5th ed., New York: Wiley, 2002; U.S. Department of Treasury's Bureau of Alcohol, Tobaco, Firearms, and Explosives (ATF), http://searchjustice.usdoj.gov/search?q=explosives+&site=default_collection&sort=date%3AD%3AL%3Ad1&output=xml_no_dtd&ie=UTF-8&oe=UTF-8&client=atf&proxystylesheet=atf, 2010; Ledgard, J. B., *Preparatory Manual of Explosives: A Comprehensive Laboratory Manual*, South Bend, IN: Paranoid Publications Group, 2002.

3.2.12 Mercury(II) Fulminate

Mercury(II) fulminate (CAS number 628-86-4) is toxic and highly sensitive to friction and shock. It is water insoluble and can be phlegmatized by the addition of fats, oils, or paraffin. Meyer, Kohler, and Homburg (2007) specified that it is prepared by dissolving mercury in nitric acid with 95% ethanol. Then, energetic gas and crystals are produced, the crystals are filtered by suction and washed until they become neutral after the reaction. The mercury(II) fulminate product is obtained as a small brown to grey pyramid-shaped crystal, the color of which is caused by the presence of colloidal mercury. It is normally stored under water and dried at 104°F shortly before use (Meyer, Kohler, and Homburg 2007). Its characteristics and specifications are presented in Table 3.11.

3.2.13 Nitrocellulose

Nitrocellulose (CAS number 9004-70-0) is prepared by the action of a nitrating mixture on high-quality cellulose prepared from wood pulp. The crude nitration product is first centrifuged to remove the bulk of the acid, after which it is stabilized

Table 3.11 Characteristics and Specifications of Mercury (II) Fulminate

Color: colorless
Empirical formula: $Hg(CNO)_2$
CAS number: 628-86-4
Molecular weight: 284.6
Energy of formation: +958 kJ/kg
Density: 4.42 g/cm^3
Enthalpy of formation: +941 kJ/kg
Optimum oxygen balance: −11.2%
Nitrogen content: 9.84%
Impact sensitivity: 1.0–2.0 N·m
Net content: not less than 98%
Water solubility: not more than 1%
Deflagration point: 330°F

Sources: Data from Beck, W., J. Evers, M. Göbel, G. Oehlinger, and T. M. Klapötke, *Zeitschrift für anorganische und allgemeine Chemie* 663(9):1417–22, 2007; Meyer, R., J. Kohler, and A. Homburg, *Explosives*, 5th ed., New York: Wiley, 2002; Perry, D. L., and S. L. Phillips, *Handbook of Inorganic Compounds*. Boca Raton, FL: CRC Press, 1995.

by preliminary and final boiling operations (Meyer, Kohler, and Homburg 2002). Also, the nitration processes are resumed while a measured amount of nitric acid and anhydrous sulfuric acid are applied to regulate the spent acid. Standard nitrocellulose types are manufactured and blended to the desired nitrogen content. Blasting soluble nitrocotton (dynamite nitrocotton; 12.3% nitrogen) is held at a high viscosity to maintain good gelatinizing properties and all nitrocelluloses are soluble in acetone. In addition, it is a highly flammable compound formed by nitration of cellulose. Most airport X-ray machines may not be able to detect nitrocellulose, although another type of technology called a *trace detection* machine can. Most underdeveloped countries may not have the technology to detect nitrocellulose; terrorists will be able to pass through their security system and can hijack the airlines heading to the United States for a series of attacks. Table 3.12 shows the characteristics and specifications of nitrocellulose.

Table 3.12 Characteristics and Specifications of Nitrocellulose

Color: white fibers
Empirical formula of the structural unit: $C_{12}H_{14}N_6O_{22}$
Nitration grade = 14.14%
CAS number: 9004-70-0
Optimum nitrogen content: 13.4% or 13.5% with anhydrous phosphoric acid
Molecular weight of the structure unit: 324.2 + % N/14.14270
Optimum oxygen balance: –28.7%
Volume of explosion gases: 871 L/kg
Heat of explosion (H_2O liq.): 4,312 kJ/kg, (H_2O gas): 3,991 kJ/kg
Density: 1.67 g/cm^3, by pressing: 1.3 g/cm^3
Lead block test: 370 cm^3/10 g
Impact sensitivity: 3 N·m
Friction sensitivity: up to 353 N
Ashes: not more than 0.4%
Insoluble in acetone: not more than 0.4%
Alkali, as $CaCO_3$: not more than 0.05%
Sulfate, as H_2SO_4: not more than 0.05%
$HgCl_2$: none

Table 3.12 Characteristics and Specifications of Nitrocellulose (*Continued*)

Nitrocellulose for gelatinous explosives must gelatinize nitroglycerine completely within 5 minutes at 60°C.
Linters (cotton fibers) as raw material
Properties $(C_6H_{10}O_5)_n$
White fibers
Molecular weight of structural unit: 162.14
Specifications ⟨-cellulose content
(insoluble in 17.5% NaOH): at least 96%
Fat; resin (soluble in CH_2C_{l2}): not more than 0.2%
Optimum moisture: not more than 7.0%
Optimum ash content: not more than 0.4%

Sources: Data from U.S. Environmental Protection Agency, http://www.epa.gov/ttn/chief/ap42/ch08/final/c08s03.pdf, 2010; Meyer, R., J. Kohler, and A. Homburg, *Explosives*, 5th ed., New York: Wiley, 2002; U.S. Department of Treasury's Bureau of Alcohol, Tobaco, Firearms, and Explosives (ATF), http://searchjustice.usdoj.gov/search?q=explosives+&site=default_collection&sort=date%3AD%3AL%3Ad1&output=xml_no_dtd&ie=UTF-8&oe=UTF-8&client=atf&proxystylesheet=atf, 2010.

3.2.14 Nitroglycerin

Nitroglycerin (CAS number 55-63-0) is an oily, colorless liquid, and a high explosive that is so unstable that the slightest jolt, impact, or friction can cause it to spontaneously detonate. Because the molecule contains oxygen, nitrogen, and carbon, when it explodes a large amount of energy is released and the rate of decomposition reaction makes it a violent explosive. It forms new molecules as depicted in the following chemical equation:

$$4C_3H_5N_3O_9(s) \rightarrow 6N_2(g) + 12CO(g) + 10H_2O(g) + 7O_2(g)$$

High explosives are decomposed instantaneously by a supersonic shock wave passing through the material. Based on the chemical equation above, 4 moles of nitroglycerin generates 35 moles of hot gases, which makes it as one of the most powerful explosives. The characteristics and specifications of nitroglycerin are provided in Table 3.13.

3.2.15 Octagen (HMX)

Octagen, or *octahydro-1,3,5,7-tetranitro-1,3,5,7-tetrazocine* (HMX), is a powerful and relatively insensitive *nitroamine* high explosive, chemically related to RDX.

Table 3.13 Characteristics and Specifications of Nitroglycerin

Color: yellow oil	
Empirical formula: $C_3H_5N_3O_9$	
Molecular weight: 227.1	
CAS number: 55-63-0	
Energy of formation: –1,539.8 kJ/kg	
Enthalpy of formation: –1,632.4 kJ/kg	
Optimum oxygen balance: +3.5%	
Optimum nitrogen content: 18.50%	
Volume of explosion gases: 716 L/kg	
Heat of explosion (H_2O liq.): 6,671 kJ/kg, (H_2O gas): 6,214 kJ/kg	
Specific energy: 1,045 kJ/kg	
Density: 1.591 g/cm^3	
Solidification point: +13.2°C (stable modification), +2.2°C (unstable modification)	
Specific heat: 1.3 kJ/kg	
Lead block test: 520 cm^3/10 g	
Detonation velocity, confined: 7,600 m/s	
Impact sensitivity: 0.2 N · m	
Friction sensitivity: up to 353 N	
Nitroglycerine as a Component of Explosives	
Nitrogen content	Not less than 18.38%
Abel test at 82.2°C	Not less than 10 minutes
Glycerol as a raw material	
Smell	Not offensive; pungent
Color	Clear, as pale as possible
Reaction to litmus	Neutral
Nitroglycerine as a Component of Explosives	
$AgNO_3$ test: traces only	Fatty acids: traces only
Ash content: maximum 0.03%	Water content: maximum 0.50%
Refractive index (nD)	20: 1.4707–1.4735

Table 3.13 Characteristics and Specifications of Nitroglycerin (*Continued*)

Acidity: not more than 0.3 mL n/10	NaOH/100 mL
Alkalinity: not more than 0.3 mL n/10	HCl/100 mL

Sources: Data from U.S. Environmental Protection Agency, http://www.epa.gov/ttn/chief/ap42/ch06/final/c06s03.pdf, 2010; U.S. Environmental Protection Agency, http://nlquery.epa.gov/epasearch/epasearch?querytext=Explosives&typeofsearch=epa&sort=term_relevancy&results_per_page=10&doctype=all&originalqueryt ext=Ammonium+Nitrate&areaname=&faq=no&filterclause=&sessionid=981E2B6116865D395E59B9D752558E2C&referer=http%3A%2F%2Fepa.gov%2F&prevtype=epa&result_template=epafiles_default.xsl&areasidebar=search_sidebar&areapagehead=epafiles_pagehead&areapagefoot=epafiles_pagefoot&stylesheet=http%3A%2F%2Fwww.epa.gov%2Fepafiles%2Fs%2Fepa.css&po=3333, 2010; Meyer, R., J. Kohler, and A. Homburg, Explosives, 5th ed., New York: Wiley, 2002; U.S. Department of Treasury's Bureau of Alcohol, Tobaco, Firearms, and Explosives (ATF), http://searchjustice.usdoj.gov/search?q=explosives+&site=default_collection&sort=date%3AD%3AL%3Ad1&output=xml_no_dtd&ie=UTF-8&oe=UTF-8&client=atf&proxystylesheet=atf, 2010.

Octagen is insoluble in water and is made by the nitration of hexamine with ammonium nitrate and nitric acid in an acetic acid. It can be used to manufacture *cyclotrimethylene-trinitramine* (RDX), another high explosive similar in structure to HMX. The characteristics and specifications of HMX are shown in Table 3.14.

3.2.16 Pentaerythritol Tetranitrate

PETN is an explosive with high brisance; it is very stable; Meyer, Kohler, and Homburg (2007) indicated that it is insoluble in water; sparingly soluble in alcohol, ether, and benzene; and soluble in acetone and methyl acetate. Pentaerythrol is mixed into concentrated nitric acid with efficient stirring and cooling to produced PETN. An approximate optimum fraction of 70% HNO_3 shall be attained to precipitate the residue of the product, then acetone is used for finishing. PETN is more complex to detonate than TNT but it has a higher level of shock and friction sensitivity; an explosion will not be produced by dropping or igniting. A deflagration to detonation transition can take place in some cases. PETN is used to avoid the need for primary explosives; the energy needed for an immediate initiation of PETN by an electric spark is approximately up to 60 mJ. It is a major ingredient of the plastic explosive *semtex*. PETN can only become a violent explosive when it is mixed with other explosives. The explosive that almost brought down Northwest Airlines Flight 253 was extremely powerful, allowing terrorists to use only small quantities to cause enormous damage. PETN crystals are hard to detect if carried in a sealed container. These PETN crystals can be used by terrorists entering foreign international airports, particularly in underdeveloped countries where they may not have sophisticated detection systems; terrorists can hijack or use their airlines to attack the United States and

Table 3.14 Characteristics and Specifications of HMX

Color: colorless crystals
Empirical formula: $C_4H_8N_8O_8$
Molecular weight: 296.2
Energy of formation: +353.6 kJ/kg
Enthalpy of formation: +60.5 kcal/kg = +253.3 kJ/kg
Optimum oxygen balance: −21.6%
Optimum nitrogen content: 37.83%
Volume of explosion gases: 902 L/kg
Heat of explosion (H_2O gas): 5,249 kJ/kg, (H_2O liq.): 5,599 kJ/kg
Specific energy: 1,367 kJ/kg
Density: α-modification: 1.87 g/cm^3 β-modification: 1.96 g/cm^3 γ-modification: 1.82 g/cm^3 δ-modification: 1.78 g/cm^3 Melting point: 275°C
Modification transition temperatures: $\alpha \rightarrow \delta$: 193°C–201°C $\beta \rightarrow \delta$: 167°C–183°C $\gamma \rightarrow \delta$: 167°C–182°C $\alpha \rightarrow \beta$: 116°C $\beta \rightarrow \gamma$: 154°C
Transition enthalpies: $\alpha \rightarrow \delta$: 25.0 kJ/kg $\beta \rightarrow \delta$: 33.1 kJ/kg $\gamma \rightarrow \delta$: 9.46 kJ/kg $\beta \rightarrow \gamma$: 23.6 kJ/kg $\alpha \rightarrow \gamma$: 15.5 kJ/kg $\alpha \rightarrow \beta$: 8.04 kJ/kg

Table 3.14 Characteristics and Specifications of HMX (*Continued*)

Specific heat, β-modification: 0.3 kcal/kg at 80°C	
Lead block test: 480 cm^3/10 g	
Detonation velocity, confined, β-modification: 9100 m/s	
Deflagration point: 287°C	
Impact sensitivity: 7.4 N·m	
Friction sensitivity: At 120 N	
CAS number: 2691-41-0	
Grade A	Not less than 93%
Grade B	Not less than 98%
Melting point	Not less than 270°C
Acetone-insoluble	Not more than 0.05%
Ashes	Not more than 0.03%
Acidity, as Ch_3COOH	Not more than 0.02%

Sources: Data from Gibbs, T. R., and Poppolato, A.; *LASL Explosive Property Data.* Berkeley: University of California Press, 1980; Meyer, R., J. Kohler, and A. Homburg, *Explosives,* 5th ed., New York: Wiley, 2002; U.S. Department of Treasury's Bureau of Alcohol, Tobaco, Firearms, and Explosives (ATF), http://searchjustice.usdoj.gov/search?q=explosives+&site=default_collection&sort=date%3AD%3AL%3Ad1&output=xml_no_dtd&ie=UTF-8&oe=UTF-8&client=atf&proxystylesheet=atf, 2010; Akhavan, J., *The Chemistry of Explosives,* Cambridge, UK: Royal Society of Chemistry, 2004.

to successfully create massive explosions near or on critical national infrastructure. The characteristics and specifications are shown in Table 3.15.

3.2.17 Picric Acid

2,4,6-trinitrophenol (TNP) is commonly known as *picric acid*; it is soluble in hot water, alcohol, ether, benzene, and acetone. Also, it has a greater magnitude of explosion than TNT. It is generated by dissolving phenol during nitration of the resulting *phenoldisulfonic acid* with nitric acid. The crude product is purified by washing in water and needs an elevated pouring temperature (Meyer, Kohler, and Homburg 2007). But, the solidification point can be reduced by using *nitronaphthalene* or *dinitrobenzene*. The characteristics and specifications of picric acid are presented in Table 3.16.

Table 3.15 Characteristics and Specifications of PETN

Color: colorless crystals
Empirical formula: $C_5H_8N_4O_{12}$
Molecular weight: 316.1
Energy of formation: –1,610.7 kJ/kg
Enthalpy of formation: –1,704.7 kJ/kg
Optimum oxygen balance: –10.1%
Optimum nitrogen content: 17.72%
Volume of explosion gases: 780 L/kg
Heat of explosion (H_2O gas): 5,850 kJ/kg, (H_2O liq.): 6,306 kJ/kg
Heat of detonation (H_2O liq.): 6,322 kJ/kg
Specific energy: 1,205 kJ/kg
Density: 1.76 g/cm^3
Melting point: 141.3°C
Heat of fusion: 152 kJ/kg
Specific heat: 1.09 kJ/kg
Lead block test: 523 cm^3/10 g
Detonation velocity, confined: 8,400 m/s = 27,600 ft/s at ⟩ = 1.7 g/cm^3
Deflagration point: 202°C = 396°F
Impact sensitivity: 3 N·m
Friction sensitivity: 60 N
CAS number: 78-11-5

Sources: Data from Lee, J. S., and K. S. Jaw, *J Therm Anal Calorimetry* 93:953–7, 2008; Meyer, R., J. Kohler, and A. Homburg, *Explosives*, 5th ed., New York: Wiley, 2002; U.S. Department of Treasury's Bureau of Alcohol, Tobaco, Firearms, and Explosives (ATF), http://searchjustice.usdoj.gov/search?q=explosives+&site=default_collection&sort=date%3AD%3AL%3Ad1&output=xml_no_dtd&ie=UTF-8&oe=UTF-8&client=atf&proxystylesheet=atf, 2010; Akhavan, J., *The Chemistry of Explosives*, Cambridge, UK: Royal Society of Chemistry, 2004.

Table 3.16 Characteristics and Specifications of Picric Acid

Color: yellow crystals
Empirical formula: $C_6H_3N_3O_7$
Molecular weight: 229.1
Energy of formation: –1,014.5 kJ/kg
Enthalpy of formation: –1,084.8 kJ/kg
Detonation velocity: 7,350 m/s at a density of 1.7 g/cm^3
Deflagration point: 570°F
Solidification point: not less than 240°F
Moisture content: not more than 0.1%
Benzene-insoluble: not more than 0.15%
Ash content: not more than 0.1%
Lead content: not more than 0.0004%
Optimum oxygen balance: –45.4%
Iron content: not more than 0.005%
Insolubility in water: not more than 0.15%
Optimum nitrogen content: 18.34%
Volume of explosion gases: 826 L/kg
Heat of explosion (H_2O gas): 3,437 kJ/kg, (H_2O liq.): 3,350 kJ/kg
Specific energy: 995 kJ/kg
Density: 1.76 g/cm^3
Heat of fusion: 76.2 kJ/kg
Impact sensitivity: 7.4 N·m
Friction sensitivity: 353 N
Specific heat: 1.065 kJ/kg
CAS number: 88-89-1

Sources: Data from U.S. Naval Technical Mission to Japan, http://www.fischertropsch.org/primary_documents/gvt_reports/USNAVY/USNTMJ%20Reports/USNTMJ_toc.htm,1945; Meyer, R., J. Kohler, and A. Homburg, *Explosives*, 5th ed., New York: Wiley, 2002; U.S. Department of Treasury's Bureau of Alcohol, Tobaco, Firearms, and Explosives (ATF), http://searchjustice.usdoj.gov/search?.q=explosives+&site=default_collection&sort=date%3AD%3AL%3Ad1&output=xml_no_dtd&ie=UTF-8&oe=UTF-8&client=atf&proxystylesheet=atf, 2010; Akhavan, J., *The Chemistry of Explosives*, Cambridge, UK: Royal Society of Chemistry, 2004.

3.2.18 Plastic Explosives

Plastic explosives (e.g., gelignite, composition 4 or C-4, and plastrite) are commonly called semtex, and they contain high-brisance crystalline explosives, such as octagen and RDX with petroleum jelly (Vaseline) or gelatinized liquid nitro compounds in poly-additive plastics (e.g., polysulfides, polybutadiene, acrylic acid). These explosives are easy to use by terrorists or disgruntled individuals.

3.2.19 2,4,6-Trinitrotoluene

2,4,6-trinitrotoluene is a yellow, odorless solid and is commonly known as TNT; it is usually used in military bombs, and grenades; for industrial uses; and in underwater blasting. The production of TNT in the United States occurs solely at military arsenals. It can be produced pure and mixed with ammonium nitrate, with aluminum powder, with RDX, and in other combinations. It is one of the most highly used explosives in the military because it is neutral and very stable. Table 3.17 defines the specifications and characteristics of TNT.

Table 3.17 Characteristics and Specifications of TNT

Color: pale yellow crystals; flakes if granulated
Empirical formula: $C_7H_5N_3O_6$
Molecular weight: 227.1
Energy of formation: –219.0 kJ/kg
Enthalpy of formation: –295.3 kJ/kg
Optimum oxygen balance: –73.9%
Optimum nitrogen content: 18.50%
Volume of explosion gases: 825 L/kg
Heat of explosion (H_2O gas): 3,646 kJ/kg, (H_2O liq.): 4,564 kJ/kg
Specific energy: 92.6 mt/kg = 908 kJ/kg
Density, crystals: 1.654 g/cm^3
Density, molten: 1.47 g/cm^3
Solidification point: 80.8°C
Heat of fusion: 96.6 kJ/kg
Specific heat at 20°C: 1.38 kJ/kg
Lead block test: 300 cm^3/10 g
Detonation velocity, confined: 6,900 m/s

Table 3.17 Characteristics and Specifications of TNT (*Continued*)

Deflagration point: 300°C
Impact sensitivity: 1.515 N·m
Friction sensitivity: up to 353 N
CAS number: 118-96-7

Sources: Data from U.S. Environmental Protection Agency, http://www.epa.gov/ttn/chief/ap42/ch06/final/c06s03.pdf, 2010; U.S. Environmental Protection Agency, http://nlquery.epa.gov/epasearch/epasearch?querytext=Explosives&typeofs earch=epa&sort=term_relevancy&results_per_page=10&doctype=all&originalquerytext=Ammonium+Nitrate&areaname=&faq=no&filterclause=&sessionid=981E2B6116865D395E59B9D752558E2C&referer=http%3A%2F%2Fepa.gov%2F&prevtype=epa&result_template=epafiles_default.xsl&areasidebar=search_sidebar&areapagehead=epafiles_pagehead&areapagefoot=epafiles_pagefoot&stylesheet=http%3A%2F%2Fwww.epa.gov%2Fepafiles%2Fs%2Fepa.css&po=3333, 2010; Meyer, R., J. Kohler, and A. Homburg, Explosives, 5th ed., New York: Wiley, 2002; U.S. Department of Treasury's Bureau of Alcohol, Tobaco, Firearms, and Explosives (ATF), http://searchjustice.usdoj.gov/search?q=explosives+&site=default_collection&sort=date%3AD%3AL%3Ad1&output=xml_no_dtd&ie=UTF-8&oe= UTF-8&client=atf&proxystylesheet=atf, 2010; Akhavan, J., The Chemistry of Explosives, Cambridge, UK: Royal Society of Chemistry, 2004; U.S. Naval Technical Mission to Japan, http://www.fischertropsch.org/primary_documents/gvt_reports/USNAVY/USNTMJ%20Reports/USNTMJ_toc.htm, 1945.

3.3 Components and Applications of Explosive Materials

In recent years, there have been almost constant terrorist attack warnings and bomb threats and anyone who could be affected by these dangers should have knowledge of the typical explosive components and devices used as weapons. Moreover, this knowledge could help prevent disaster by timely detection of these potential attacks. Terrorists aim to inflict mass casualties and cause maximum loss of life and property damage, and explosives are typically their first weapons of choice. Some of the common components and considerations to bomb making that can destroy water infrastructure are presented in Sections 3.3.1 through 3.3.17.

3.3.1 Alginates

Alginates are anionic polysaccharides that are capable of binding 200–300 times their own volume of water. They can be used as swelling agents to explosive mixtures in order to improve the resistance of such explosives to moisture.

3.3.2 Aluminum Powder

Aluminum powder is usually a crucial ingredient to explosives for producing heat explosion, and as a result a higher temperature is shared to the fumes. If the proportion of aluminum in the explosive formulation is extremely high, a gas impact effect results, since successive combining of the unreactive parts of the fumes with atmospheric oxygen may produce a delayed second explosion.

3.3.3 Base Charge

The base charge is normally the finishing component of any blasting detonator. It is composed of a secondary nitramine explosive (Ledgard 2002).

3.3.4 Blasting Caps

Blasting caps are made of cylindrical copper or aluminum capsules, which are utilized as initiators of explosive charges. PETN or another type of secondary charge is added to achieve a higher brisance. A blasting cap can be ignited by the flame of a safety fuse or ignited electrically. The normal size should only be used with a slow fuse. Currently, number 8 blasting caps are commercially available, for all practical purposes. As stated by Meyer, Kohler, and Homburg (2007), the number 8 blasting cap consists of a 300 mg primary charge and an 800 mg secondary charge, and is 0.4–5 cm in length and 0.7 cm in external diameter.

3.3.5 Blasting Galvanometer

An instrument that is used for testing electric blasting circuits, enabling the blaster to locate breaks, short circuits, or faulty connections before an attempt is made to fire the shot.

3.3.6 Blasting Machine

According to Meyer, Kohler, and Homburg (2007), two blasting machines exist: (1) one with direct energy supply and equipped with a self-induction or a permanent magnet generator, and (2) one with an indirect energy supply, in which the generated electrical energy is stored in a capacitor and, after the discharge voltage has been attained, the breakthrough pulse is sent to a blasting train. In order to ignite *bridgewire detonators*, they need to be installed and connected in parallel, the output of the machines shall be higher since more than 95% of the electrical energy is lost in the blasting circuit.

3.3.7 Blast Meters and Boosters

Blast meters are simple devices used to measure the range of pressure created by a shock wave. A booster can be a cap-sensitive cartridge or a press-molded cylinder for the initiation of non-cap-sensitive charges, for example, blasting agents or cast TNT (Meyer, Kohler, and Homburg 2007).

3.3.8 Bridgewire Detonator

Bridgewire detonators are used in the industrial detonation of explosive charges (Meyer, Kohler, and Homburg 2002). They contain a bridge made of thin resistance wire with an igniting pill built around the wire and immersed in pyrotechnical substance after drying, which will glow by using an electric pulse. The delayed-action detonators may be set for a delay of half a second (half-second detonators) or for a delay of 2–34 milliseconds (millisecond detonators). Hence, if multiple charges are to be detonated at the same time, the detonators need to be connected in series with the connecting wire. Special blasting machines must be used in parallel connection with detonators.

3.3.9 Brisance

Brisance is the destructive fragmentation effect of a charge on its designated and direct vicinity. The relevant parameters of explosives are detonation rate and loading density, gas yield, and heat of explosion. The higher the loading density of the explosive the higher speed of the reaction rate and intensity of the impact of the detonation. Moreover, an increase in density is in conjunction with an increase in the detonation rate of the explosive, whereas the shock wave pressure in the detonation front varies with the square of the detonation rate. Therefore, higher loading density is very significant.

3.3.10 Deflagration

Deflagration is a technical term describing subsonic combustion that usually propagates through the liberated heat of reaction. The burning of powder is a deflagration process.

3.3.11 Delay Time and Element

Delay time is the time or distance interval between the instant a device carrying the fuze is launched and the instant the fuze becomes armed (Meyer, Kohler, and Homburg 2007). Delay compositions are mixtures of substances that when pressed into delay tubes react without the progression of gaseous products and thus ensure minimum variation in the delay period. Examples of such mixtures are potassium permanganate with antimony, and redox reactions with fluorides and other halides. The delay element can be an explosive train component consisting of a primer, a delay column, and a relay detonator or transfer charge collected in that order in a single housing to produce a regulated time interval. Hence, the time or distance between the initiation of the fuze and the detonation can be designed.

3.3.12 Detonation

Detonation is a chemical reaction created by an explosive agent/material, which produces a shock wave. Increase in temperature and pressure gradients are created in the wave front, in order to initiate the chemical reaction instantaneously. Detonation speeds are in the approximate range of 1,500–9,000 m/s; slower reactions, which are propagated by thermal conduction and radiation, are known as deflagration.

3.3.12.1 Shock Wave

Shock waves are generated in nonexplosive form by a rapid change in pressure, allowing a movable piston in a tube to be suddenly accelerated from rest and then continue its motion at a constant rate. The air in front of the piston is compressed and warms up a little; the compression range is determined by the velocity of sound in the air. In addition, this allows the piston to accelerate again and continue its motion at the higher rate. The new compression is applied to the medium, part of which is already in motion; it is moving at a higher and quicker rate, the movement of the subject is superposed and the sonic velocity is intensified in a warmer medium. If the medium is an explosive gas mixture rather than air, an explosive reaction will be instantly initiated in front of the shock wave. Explosions normally produce a shock wave in the surrounding air. Hence, this compression shock is the standard principle of the long-distance effect of explosions. According to Meyer, Kohler, and Homburg (2002), if the propagation of the shock wave is nearly spherical, the compression ratio p_1/p_o decreases rapidly, and so does the p_o velocity of matter W; it becomes zero when the shock wave becomes an ordinary sound wave. If the explosion-generated shock wave is propagated in three-dimensional space, its effect decreases with the third power of the distance. This is the guideline adopted by German accident prevention regulations, in which the safety distance (in meters) is quantified from the term $f \cdot \sqrt[3]{M}$, where M is the maximum amount of explosives in kilograms, which are present in the building or asset at any time, whereas f is a factor that varies, according to the required degree of safety, from 1.5 to 8 (distance from the nondangerous part of an asset). This expression can be used to design the location of an emergency response station for a major asset, to immediately respond in the event of an extreme terrorist attack as illustrated in the scenarios in combat zones presented in Chapter 11. Meanwhile, Meyer, Kohler, and Homburg (2002) pointed out that the shock wave theory is easier to understand if we consider a planar shock wave, on the assumption that the tube is indestructible (such shock wave tubes are utilized as research instruments in gas dynamics and in solid-state physics; the shock sources are explosions or membranes bursting under pressure). Comparative treatment of the behavior of the gas in the tube yields the following relationships:

From the law of conservation of mass,

$$\rho_0 D = \rho_1 (D - W) \quad \text{or} \quad v_1 D = v_0 (D - W) \tag{3.1}$$

From the law of conservation of momentum,

$$p_1 - p_0 = \rho_0 DW \quad \text{or} \quad v_0(p_1 - p_0) = DW \tag{3.2}$$

From the law of conservation of energy,

$$p_1 W = \eta_0 D(e_1 - e_2 + W^2/2) \tag{3.3}$$

Rearrangements yield the so-called *Hugoniot* equation:

$$e_1 - e_0 = \frac{1}{2}(p_1 + p_0)(v_0 - v_1) \tag{3.4}$$

The following expressions are obtained for velocity D of the shock wave and for the velocity of matter W:

$$D = v_0 \sqrt{\frac{\rho_1 - \rho_0}{v_0 - v_1}} \tag{3.5}$$

and

$$W = \sqrt{(p_1 - p_0)(v_0 - v_1)} \tag{3.6}$$

These relationships are valid irrespective of the state of aggregation.

3.3.12.2 Detonation Wave Theory

An explosive chemical reaction is produced in the wave front because of the extreme temperature and pressure conditions. The development and transmission of the shock wave is sustained by the energy of the reaction. The equations presented in Section 3.3.12.1 are still valid; the meaning of the equation parameters is

p_1—Detonation pressure
ρ_1—Density of gaseous products in the front of the shock wave; this density is thus higher than the density of the explosive ρ_o
D—Detonation rate
W—Velocity of fumes

Equation 3.1 remains unchanged.

Since p_o is negligibly small as compared to the detonation pressure p_1, we can write Equation 3.2 as $p_1 = \rho_o DW$.

The pressure created by detonation in the wave front is proportional to the product of density, detonation rate, and fume velocity, given that fume velocity is the square of detonation rate. For a known explosive, detonation velocity increases with increasing density. As per the equation $p_1 = \rho_o DW$, detonation pressure increases noticeably if the initial density of the explosive substance can be raised to its optimum charge, for example, by casting or pressing, or if the density of the explosive substance is elevated (e.g., the density of TNT is 1.64 and octagen is 1.96). High density of the explosive is important if high *W* brisance is required. Meanwhile, the detonation pressure and rate may be reduced by decreasing ρ_o, that is, by the application of a more loosely textured explosive material. This is initiated and employed if the blasting has to be applied on softer rocks and if a weaker *thrust* effect is expected.

The determination of the maximum level of detonation pressure p_1, in Equation 3.7, has been studied by X-ray measurements. While the detonation velocity can be quantified by electronic recorders, there is no standard quantification for the fume velocity *W*; but it can be projected by the direction of angle of the fumes behind the wave front. The relation between *D* and *W* is $W = \frac{D}{\gamma+1}$; γ is presented as the *polytrop exponent* in the modified state equation.

$$p = C\rho^{\gamma}, \quad \text{while } C = \text{constant}$$

The value of γ is nearly 3, therefore p_1 is

$$p_1 = \eta_o D^2/4 \tag{3.7}$$

Equation 3.2 can be recomputed as

$$p_1 - p_2(v_0 - v_1)\rho_0^2 D^2 \tag{3.8}$$

Equation 3.4, utilized to the detonation development relating the chemical energy of reaction q, becomes

$$e_1 - e_o = \frac{1}{2}(p_1 + p_o)(v_o + v_1) + q \tag{3.9}$$

Equations 3.5 and 3.6 remain unmodified, but *D* is currently equivalent to the detonation rate, whereas *W* represents the fume velocity.

3.3.12.3 Selective Detonation

Selectivity in the course of a detonation process is noted when processes with very different sensitivities, and thus also with very different induction periods, participate in the intensive chemical reaction produced by the shock wave (Ahrens 1977). If the concentration or the amount of the shock wave is minimal as a result of external conditions—explosion in an unconfined space, for example—the induction periods of less-sensitive reactions may become infinite, that is, the reaction may fail to take place. Hence, this selectivity is important for ion exchanges. According to Meyer, Kohler, and Homburg (2002), if the explosive is detonated while unconfined the only reaction that will occur is that of the nitroglycerine–nitroglycol mixture, which is fast and is limited by its relative proportion and is thus firedamp safe.

3.3.12.4 Sympathetic Detonation

Sympathetic detonation signifies the beginning of an explosive charge without a priming mechanism by the detonation of another charge. The maximum distance between two cartridges in line is based on the flashover tests, by which the detonation is transmitted. The transmission method is complicated by shock waves, hot reaction products, and even the *hollow charge* effect. The detonation velocity is defined as the rate of propagation of a detonation in an explosive; if the density of the explosive is at its optimum charge and if the explosive is charged into columns that are considerably wider than the critical diameter, the detonation velocity is a characteristic of each individual explosive.

3.3.12.5 Detonation Development Distance

Detonation development distance is a term denoting the distance or space required for the full detonation rate to be attained. This distance is short for initiating explosives. The detonation development distance relating to less sensitive explosive materials is strongly influenced by the consistency, density, and cross-section of the charge.

3.3.13 Electro-Explosive Device

An electro-explosive device (EED) is a detonator or initiator initiated by an electric current. A *one-ampere/one-watt initiator* or EED is one that will not fire when 1 A of current at 1 W of power is delivered and given to a bridgewire for a designated time.

3.3.14 Oxidizer and Oxygen Balance of Explosives

All explosive materials contain and require oxygen to achieve an explosive reaction. In addition, oxygen can be introduced by *nitration*. The most critical solid-state oxidizers are nitrates, particularly ammonium nitrate and sodium nitrate for explosives. The quantity of oxygen, released as a result of total conversion of the explosive material to CO_2, H_2O, SO_2, Al_2O_3, etc., is called *positive* oxygen balance. If the amount of oxygen is insufficient, which is known as a *negative* oxygen balance, the deficient amount of oxygen needed to complete the reaction is designated with a negative sign. The most favorable composition for an explosive can be easily quantified from the oxygen values of its components. Commercial explosives must have an oxygen balance close to zero in order to minimize the production of toxic gases such as monoxide and nitrous gases.

3.3.15 Heat of Explosion

The heat of explosion can be calculated using theoretical principles and experimentally determined. The quantified value is the difference between the energies of formation of the explosive components and the energies of formation of the explosion products. Moreover, the values of heats of explosion can be quantified from the *partial* heats of explosion of the components of the propellant. The calculated values do not agree exactly with those obtained by experiment; if the explosion takes place in a bomb, the true compositions of the explosion products are different and, moreover, vary with the loading density. In accurate calculations, these factors must be taken into consideration.

3.3.16 Underwater Detonation

The destructive effects of underwater detonation change according to distance and closeness effects. The first effect is caused by the action of the pressure shock wave and the latter mainly by the thrust created by the increasing and intensifying gas bubble.

3.3.16.1 Shock Wave of Underwater Detonation

The adjacent layer of water is compressed under the effect of high pressure, which transmits that pressure onto the next level, and this transfers the pressure onto further levels or a chain reaction to different levels. The velocity of propagation intensifies with pressure, accordingly generating a steeply ascending pressure front, which reveals the characteristic of a shock wave to the pressure wave. At the beginning, the velocity of propagation surpasses that of the speed of sound, but the velocity declines with greater distance. Thus, the optimum pressure is directly proportional to the cube root of the charge weight and inversely proportional to the distance or space, as depicted in the following expression:

$$P_{max} = CL^{1/3}/e$$

where P is pressure in bar, L is loading weight in kilograms, e is the distance in meters, and c is the typical empirical factor of 500.

3.3.16.2 Gas Bubble

The underwater explosion created the gas that primarily penetrates the small cavity formerly filled by the explosive, thus creating a gas bubble under a high level of pressure, which then expands. Accordingly, the bubble expansion creates a water mass that progresses radially at high velocity away from the point of explosion, which is known as *thrust*. The optimum amount of kinetic energy distributed to the water during an explosion is called *thrust energy*. The gas bubble can be oscillated repeatedly several times and is forced upward toward the surface of water. The variation in pressure between the top and the bottom layers of the bubble causes the bottom layer to move at a higher velocity, propelling it upward into the bubble. It is likely that both surfaces will meet. Within a partial area, the water obtains an upward thrust, producing a *water hammer*. Effective and powerful underwater explosives with mixtures of aluminum powder are those that can generate a high-pressure gas bubble for the formation of thrust. Detection technologies may need to be developed that detect the chemical components of explosives that generate high-pressure gas bubbles upon entry to major water assets.

3.3.17 Quantification of the Amount of Explosives

One way to calculate the total quantity of explosives, which is recommended by Langefors and Khilstrom (1963), is

$$Q = 0.07B^2 + 0.4B^3 + 0.004B^4$$

where B = the burden in meters and Q = quantity of explosive in kilograms.

The first term is the explosive needed to produce surface blast design and to satisfy other dissipative processes. The second term is the principal term that relates the weight of explosive and the weight of rock. The third term, usually very small, provides the energy for the swelling and lifting of the mass.

3.4 Hazards of Explosives

High explosives are capable of severely mutilating the human body. Explosives tend to rip the body into different pieces like a shark with jagged uneven bits of body parts removed. According to Jared Ledgard (2002), other than the obvious effects of injuries caused by explosives upon the body, there are other effects known as the secondary effects of exposure to detonations, which include temporary loss of vision, hearing impairment, fragmentation wounds, burns, and inhalation and/or skin absorption of poisonous fumes. Fragmentation wounds cause a whole multitude of problems as they are like multiple gunshot wounds. Patients have died many hours later due to many types of complications arising from fragmentation wounds.

The chemicals contained in explosives are notorious contaminants to the water supply, and they are discharged from explosive and ammunition plants. There are several cases in the United States where these contaminants cannot be removed or be easily detected by municipal water treatment plants. The most notorious emerging water supply contaminant generated from explosive plants in the United States is perchlorate, which mainly affects the thyroid glands. A state of emergency was declared on November 20, 2010 in Barstow, California because the water supply was contaminated with perchlorate and other toxic chemicals used to make explosives and rocket fuels. It has been found that these toxic chemicals can cause various forms of cancer.

3.5 The Challenge of Improvised Explosive Devices in the United States

According to the U.S. Department of Defense (DOD) (DOD 2010), an IED is a device placed or fabricated in an improvised manner incorporating destructive, lethal, noxious, pyrotechnic, or incendiary chemicals and designed to destroy, incapacitate, harass, or distract people. IEDs were popularized by adversaries in Iraq, Afghanistan, and Pakistan. An IED were can be prepared almost everywhere "undetectably." When constructed creatively and intelligently, it can defeat even highly protected assets. The most dangerous IEDs that can be utilized are commercial airplanes, particularly from foreign countries, which could be used to attack explosive chemical plants and petroleum refineries near U.S. international airports (see Figure 4.10 and Figures 11.7a and b) not only to create massive destruction to water resources near the area but also to generate mass casualties and economic aftershocks. Since most of the airports within the United States are highly secured with sophisticated technologies, terrorists are looking at underdeveloped foreign countries' international airports as the best alternative to hijack commercial airplanes and use them as weapons of mass destruction (WMDs). These countries may not have equivalent security measures as the United States or other developed countries. Nevertheless, one of the most effective strategies to counter the threat of IEDs is to improve intelligence. Chapter 4 introduces water infrastructure and potential terrorism activity scenarios.

References

Agrawal, J. P., and R. D. Hodgson. 2007. *Organic Chemistry of Explosives*. West Sussex, UK: John Wiley & Sons.

Ahrens, H. 1977. Propellants, Explosives, Pyrotechnics. *An International-Peer-Reviewed Journal on Energetic Materials Official Journal of the International Pyrotechnics Society* 2(1–2):7–20.

Akhavan, J. 2004. *The Chemistry of Explosives*. Cambridge, UK: Royal Society of Chemistry.

Beck, W., J. Evers, M. Göbel, G. Oehlinger, and T. M. Klapötke. 2007. The Crystal and Molecular Structure of Mercury Fulminate (Knallquecksilber). *Zeitschrift für anorganische und allgemeine Chemie* 663(9):1417–22.

Boileau, J., J. M. Emeury, and J. P. Keren. 1975. German Patent 2,435,651 (February 6, 1975).

Bowden, F. P., and A. D. Yoffe. 1958. *Fast Reactions in Solids*. London: Butterworth Science.

Emeury, J. L., and H. H. Girardon. 1980. Continuous process for the manufacture of Dinitroglycolurile (Dingu). US Patent 4,211, 874.

Gibbs, T. R., and Poppolato, A. 1980. *LASL Explosive Property Data*. Berkeley: University of California Press.

Jiang, Z. 2005. Shock waves. In *Proceedings of the 24th International Symposium on Shock Waves*, Beijing, China, July 11–14, 2004, 1:984. Berlin, Heidelberg: Springer.

Langefors, U., and B. Khilstrom. 1963. *The Modern Technique of Rock Blasting*. New York: Wiley.

Ledgard, J. B. 2002. *Preparatory Manual of Explosives: A Comprehensive Laboratory Manual*. South Bend, IN: Paranoid Publications Group.

Lee, J. S., and K. S. Jaw. 2006. Thermal Decomposition Properties and Compatibility of CL-20, NTO with Silicon Rubber. *J Therm Anal Calorim* 85:463–7.

Lee, J. S., and K. S. Jaw. 2008. Thermal Decomposition Properties and Compatibility of CL-20, NTO with Silicon Rubber. *J Therm Anal Calorimetry* 93:953–7.

Lee, K. E., R. L. Hatch, M. Mezger, and S. Nicolich. 2001. High Performance explosive containing CL-20. US Patent 6,214,137 B1 (April 10, 2001).

Lewis, B., and G. Elbe. 1987. *Combustion, Flames and Explosives of Gases*. 3rd ed. Orlando, FL: Academic Press.

Series No. 22, Washington DC: American Chemical Society.

Mcnicol, L. J. P. 1966. Explosive Lead Azide Process. U.S. Patent 3,264,150. August 2, 1966.

Meyer, R., J. Kohler, and A. Homburg. 2002. *Explosives*. 5th ed. Wiley.

Meyer, R., J. Kohler, and A. Homburg. 2007. *Explosives*. 6th ed. Wiley.

Perry, D. L., and S. L. Phillips. 1995. *Handbook of Inorganic Compounds*. Boca Raton, FL: CRC Press.

Simpson, R. L., P. A. Urtiew, D. L. Ornellas, G. L. Moody, K. J. Scribner, and D. M. Hoffman. 1997. CL-20 performance exceeds that of HMX and its sensitivity is moderate. *Propellant, Explosives, Pyrotechnics* 22(5):249–55, October 1997.

U.S. Army and Air Force (USAAF). 1994. U.S. Army Chemical Material Destruction Agency Old Chemical Weapons: Munitions Specification Report. http://www.lexpev.nl/downloads/oldchemicalweapons1994.pdf (accessed October 29, 2010).

U.S. Department of Defense. 2010. Department of Defense Dictionary of Military and Associated Terms. http://ra.defense.gov/documents/rtm/jp1_02.pdf (accessed on March 10, 2011).

U.S. Department of Treasury's Bureau of Alcohol, Tobacco, Firearms, and Explosives (ATF). 2010. Multiple searches on the characteristics of explosives. http://searchjustice.usdoj.gov/search?q=explosives+&site=default_collection&sort=date%3AD%3AL%3Ad1&output=xml_no_dtd&ie=UTF-8&oe=UTF-8&client=atf&proxystylesheet=atf (accessed July 2010).

U.S. Department of Treasury's Bureau of Alcohol, Tobacco, Firearms, and Explosives (ATF). 1984. *Definition and test procedures of Ammonium Nitrate Fertilizer*. http://www.atf.gov/publications/download/hist/definition-and-test-procedures-for-ammonium-nitrate.pdf (accessed October 20, 2010).

U.S. Environmental Protection Agency. 2010a. Ammonium Nitrate. http://www.epa.gov/ttn/chief/ap42/ch08/final/c08s03.pdf (accessed October 4, 2010).

U.S. Environmental Protection Agency. 2010b. Explosives. http://www.epa.gov/ttn/chief/ap42/ch06/final/c06s03.pdf (accessed October 4,2010).

U.S. Environmental Protection Agency. 2010c. Multiple searches on the characteristics of explosives. http://nlquery.epa.gov/epasearch/epasearch?querytext=Explosives&typeofsearch=epa&sort=term_relevancy&results_per_page=10&doctype=all&originalquerytext=Ammonium+Nitrate&areaname=&faq=no&filterclause=&sessionid=981E2B6116865D395E59B9D752558E2C&referer=http%3A%2F%2Fepa.gov%2F&prevtype=epa&result_template=epafiles_default.xsl&areasidebar=search_sidebar&areapagehead=epafiles_pagehead&areapagefoot=epafiles_pagefoot&stylesheet=http%3A%2F%2Fwww.epa.gov%2Fepafiles%2Fs%2Fepa.css&po=3333 (accessed May 2010).

U.S. Naval Technical Mission to Japan. 1945. Japanese Bombs, Intelligence Targets Japan (DNI) OF 4 Sept. 1945, Fascicle 0–1, Target 0-23. http://www.fischertropsch.org/primary_documents/gvt_reports/USNAVY/USNTMJ%20Reports/USNTMJ_toc.htm (accessed October 26, 2010).

Verneker, V. R., and A. C. Forsyth. 1968. Mechanism for controlling the reactivity of lead azide. *J Phys Chem* 72:111.

Chapter 4

Water Infrastructure

4.1 Introduction

This chapter introduces the homeland water infrastructure, which includes groundwater, surface water, water tanks, municipal wastewater treatment plants, municipal water treatment plants, reservoirs, dams, and aqueducts. Vandalism and terrorism activity scenarios against water infrastructure will also be presented herein.

4.2 Acts of Terrorism against Water Infrastructure

The September 11, 2001 (9/11) terrorist attacks displayed a brutal execution against the United States. The terrorists and their leaders think in terms of a long time frame for achieving their goals. They focus on generating mass casualties, economic aftershocks, and fear and creating massive media attention that can exceed the level of the 9/11 attacks. Water infrastructure destruction through contamination or blasting of urban water supply systems (e.g., aqueducts, reservoirs, water tanks, and water treatment plants) and the explosion of petroleum refineries/pipelines near water resources can potentially create irreversible damage to water resources, generate mass casualties, disrupt the downstream industry, and injure the environment comparable or worse than the 9/11 attacks and the Deepwater Horizon explosion accident that occurred on April 20, 2010. Recently, the United States has been facing enormous economic challenges. Any terror attacks that strike at one of the largest metropolitan areas of the United States during this tough time could be devastating, for instance, the explosion of major petroleum refineries or creating leaks to petroleum pipelines. This not only contaminates water resources and injures the environment but also breeds absolute catastrophe. In considering this terrorist activity scenario, no nuclear bomb is necessary.

Consequently, the Homeland Security Presidential Directive-7 (HSPD-7) designated the United States Environmental Protection Agency (EPA) as the federal lead for water infrastructure protection. Both the Department of Homeland Security (DHS) and the EPA must prioritize the protection of groundwater resources and the urban water supply system because they can be defenseless from vandalism (e.g., temporary denial of service attack) and/or terrorism. Accordingly, if any of the weapons presented in Chapters 2 and 3 are used in the attack, the result could be catastrophic.

4.3 Groundwater Resources

Groundwater is located beneath the surface in soil pore spaces and in the fractures of lithologic formations. Approximately one-third of all public supplies and 95% of all rural domestic supplies in the United States use groundwater sources. It is also often drawn for agricultural, municipal, and industrial uses. Correspondingly, understanding and characterization of groundwater resources are critical for strategic improvement of protection policies and security from terrorism.

Groundwater is naturally recharged by infiltration of precipitation, infiltration of stream flows, and leakage from connected aquifers. A unit of rock or an unconsolidated deposit is called an *aquifer* when it can yield a usable quantity of water that meets the water quality standards for the demand or the use for which it is extracted (e.g., municipal potable water demands or agricultural water demands). Under normal hydraulic gradients, or slope of the water table/potentiometric surface, groundwater moves about 10 to 300 feet per year. A special type of aquifer known as a *karst* is found where limestone rock is highly permeated by dissolution channels, voids, and caves. Groundwater in karst aquifers may flow at rates of tens of feet per minute, much like the flow in an open channel. The highest rates of flow and more limited contact between solid aquifer particles and groundwater make karst aquifers highly susceptible to threats from terrorism.

Under natural conditions, groundwater moves from areas of high "hydraulic head" to areas of low hydraulic head, for example, from an elevated recharge area in the mountains to a discharge area along a canyon or a valley floor. Groundwater is recharged and eventually directed to the surface naturally. Natural discharge often takes place at springs, lakes, and seeps and can form oases or wetlands. When groundwater is pumped heavily, the natural flow path or the direction of groundwater flow is frequently interrupted, and groundwater is directed toward the pumping well. Contaminants in groundwater can be captured by the radius of influence of the pumping well.

4.3.1 Limestone Aquifers

A limestone aquifer is a water bearing rock that consists mainly of calcium carbonate and is chiefly produced by deposit of organic remains (Figure 4.1). Limestone consists of fossilized sea shells, shell fragments, calcareous sand, and consolidated limy mud. Its main mineral is calcium carbonate, $CaCO_3$. Dolomite is similar to limestone but has few recognizable fossils; its main minerals are calcium carbonate, $CaCO_3$, and magnesium carbonate, $MgCO_3$ (Hoorman et al. 2009). Both limestone and dolomite are commonly referred to as limestone (Hoorman et al. 2009). Limestone formations usually are a sufficient supply of groundwater because of cracks, faults, and fractures and naturally created solution channels or conduits, which provide water storage capacity for water movement. Groundwater in a limestone aquifer is vulnerable to terrorism using deadly chemicals because the water table is close to the surface and the limestone bedrock is permeable without the advantage of being filtered through soil.

4.3.2 Karst Aquifers

A karst aquifer is limestone or other easily dissolved rock (Figure 4.1) that has been partly dissolved, so that some fractures are enlarged into passages called conduits, which carry the groundwater flow. The karst aquifer is a body of soluble rock that conducts water principally via enhanced (conduit or tertiary) porosity formed by the dissolution of the rock. These aquifers are commonly structured as a branching network of tributary conduits, which connect together to drain a groundwater basin and discharge to a perennial spring (Kentucky Geological Survey 2009). Karst aquifers are very vulnerable to terrorist attacks because contaminated runoff can enter these conduits through sinkholes and swallow holes without the advantage of being filtered through sand and soil.

4.3.3 Aquifer Storage and Recovery Technology

Aquifer storage and recovery (ASR) technology means water can be injected into aquifers and stored there for later use. It has to meet drinking water quality standards, so there is no chance that water already in the ground could be contaminated. Typically, when source water is available, it is injected into a sand aquifer. The same wells are later used to extract the water and distribute it to users. The land overlying the wellfield can also continue to be used for other purposes such as agriculture or grazing (Eckhardt 2009a,b). This technology has been utilized for several years on the east coast, California, Texas, and Florida.

4.3.4 Sandstone Aquifer

Aquifers in sandstone are more widespread than those in all other kinds of consolidated rocks. Although the porosity of well-sorted, unconsolidated sand may be as

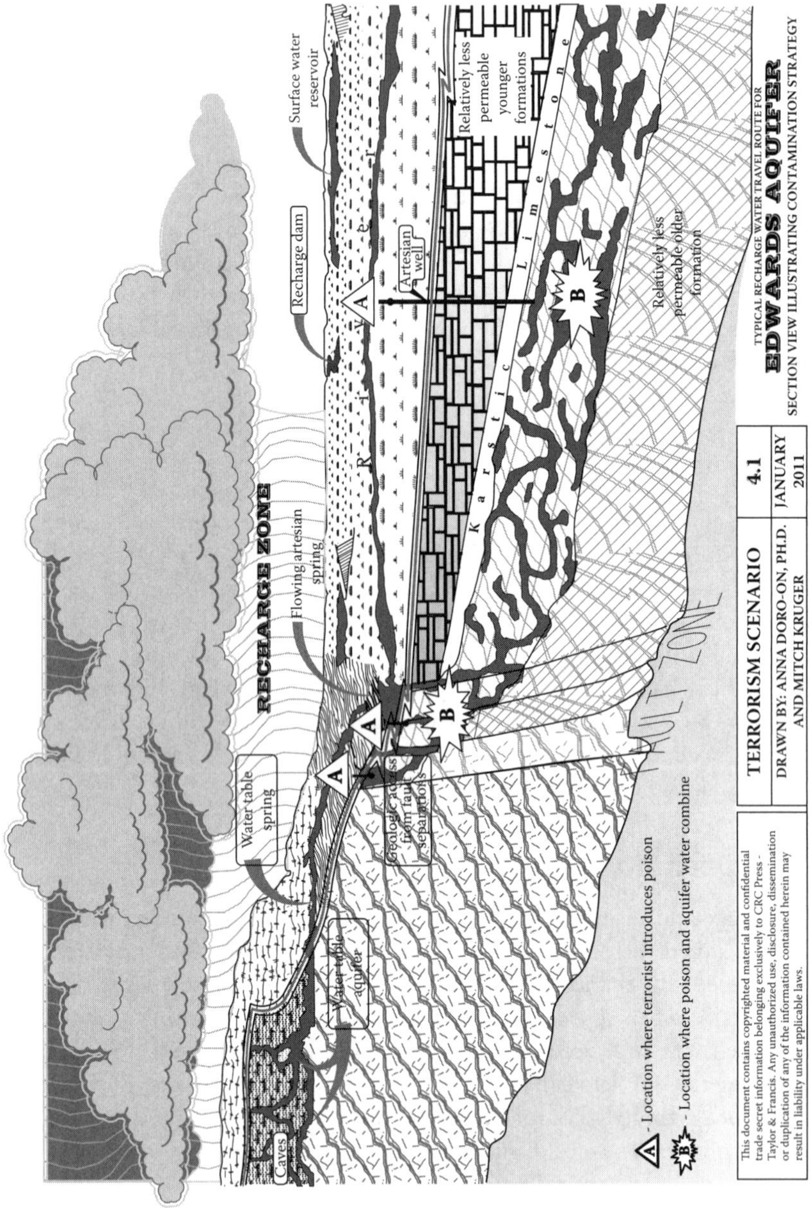

Figure 4.1 Section of karstic limestone aquifer.

high as 50%, the porosity of most sandstones is considerably less. During the process of conversion of sand into sandstone (lithification), compaction by the weight of overlying material reduces not only the volume of pore space as the sand grains become rearranged and more tightly packed but also the interconnection between pores (permeability) (USGS 2009). Sandstone retains some primary porosity unless cementation has filled all the pores, but most of the porosity in these consolidated rocks consists of secondary openings such as joints, fractures, and bedding planes (USGS 2009). Groundwater movement in sandstone aquifers is primarily along bedding planes, but junctions and fractures cut across bedding and provide pathways that allow water to move vertically between bedding planes. Sandstone aquifers commonly grade laterally into fine-grained, low-permeability rocks such as shale or siltstone. Folding and faulting of sandstones following lithification can greatly complicate the movement of water through these rocks (USGS 2009). Sandstone aquifers are largely productive in different places and generate large quantities of water supply.

4.3.5 Terrorism against Groundwater Resources

After the 9/11 attacks, the United States put airports on guard, while terrorist leaders continue to pursue their ultimate mission of launching other massive attacks. Attacking groundwater reserves, one of the most defenseless types of water infrastructure, could breed catastrophe. For instance, the Edwards Aquifer is designated as a "sole source" drinking water supply for the 1.7 million people of San Antonio and the Austin–San Antonio, Texas, corridor (Figure 4.2). Any of the terrorism activity scenarios shown in Figures 4.3 through 4.5 could likely happen.

4.4 Desalination Treatment Facilities

Desalination is the process of converting salt water to fresh water to provide fresh water for human use in regions where the availability of fresh water is, or is becoming, limited. The world's largest desalination plant is the Jebel Ali Desalination Plant located in the United Arab Emirates. This facility uses multistage flash distillation and has the capacity to produce 300 million cubic meters per year. The largest desalination plant in the United States is located in Florida, and operated by Tampa Bay Water, which started desalinating 34.7 million cubic meters of water per year in 2007.

Brackish groundwater has been treated at an El Paso, Texas, plant since 2004. Producing 27.5 million gallons (104,000 m^3) of water day by reverse osmosis, it is a crucial contribution to water supplies in this water-stressed city (EPWU 2010). "Desalinated water may provide a solution for water-stressed regions, but definitely not for economically distressed areas or at high elevation." Unfortunately, that includes some of the places with major water problems. Thus, it may be more

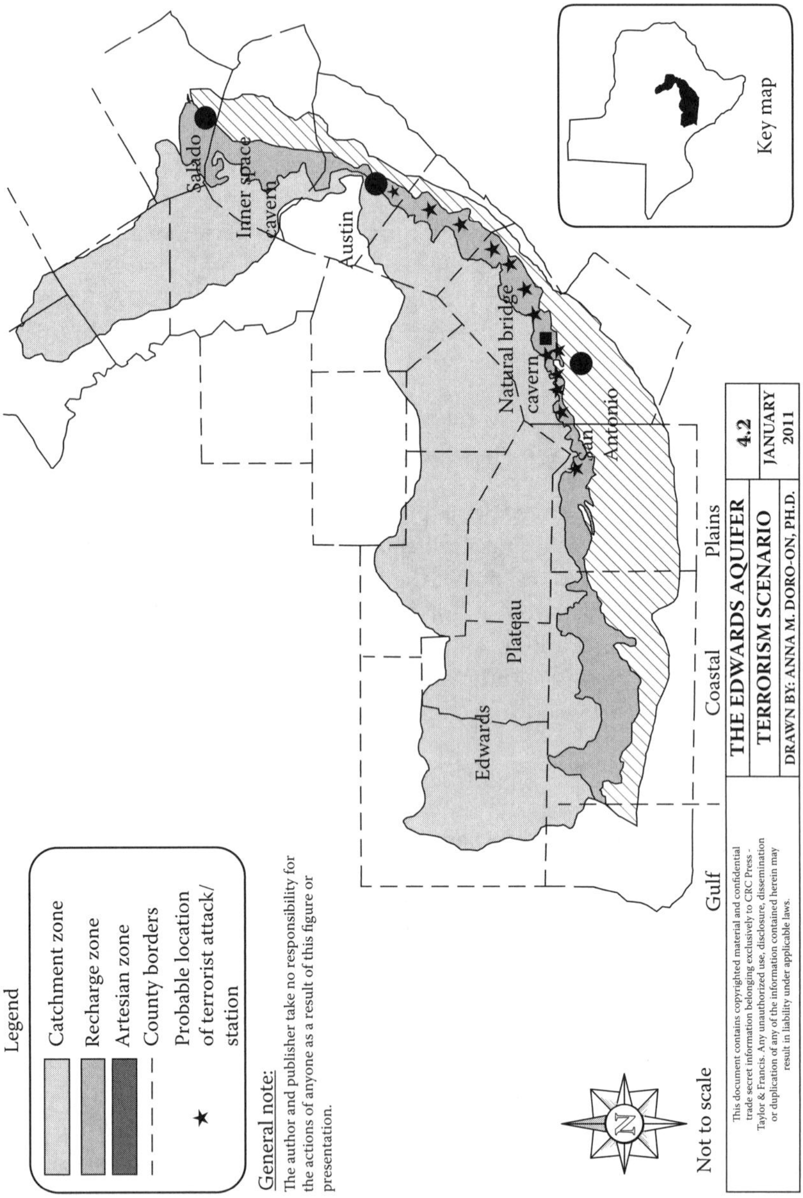

Figure 4.2 Map of possible locations of terrorist attacks against a sole source water supply.

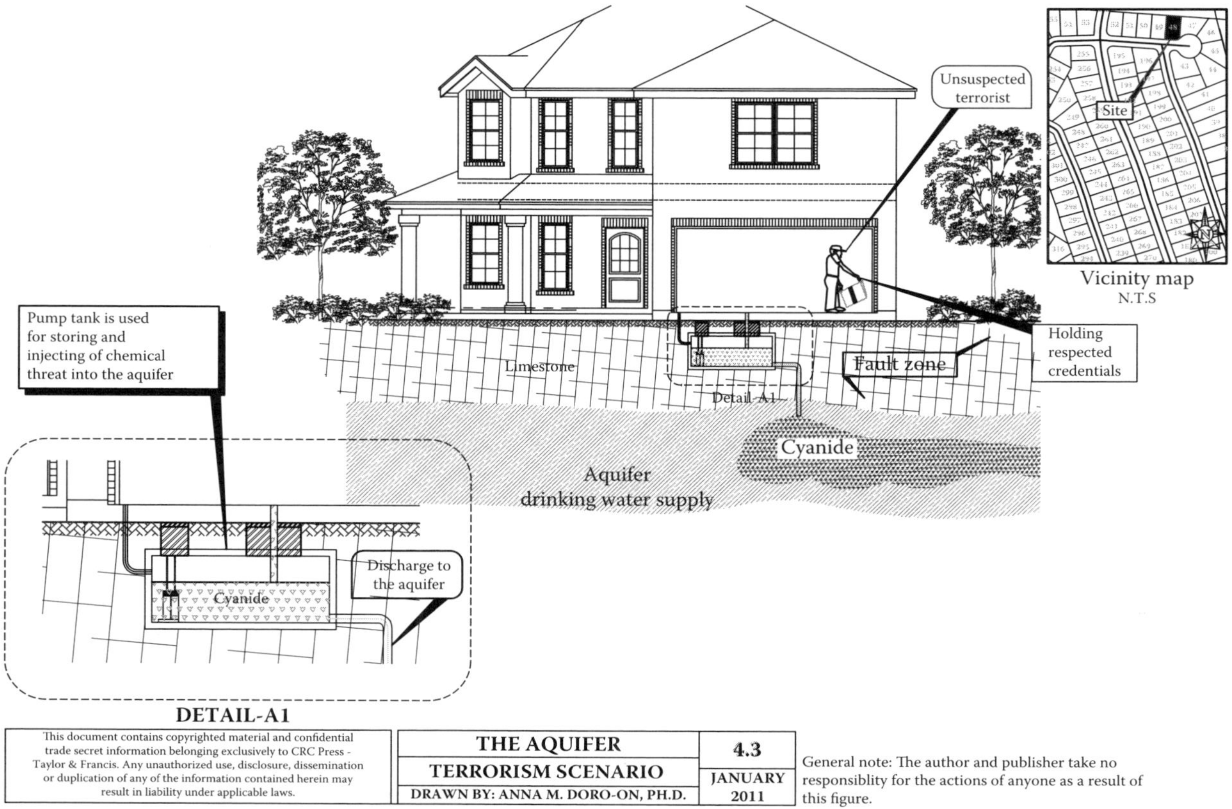

Figure 4.3 Aquifer contaminations beneath a development.

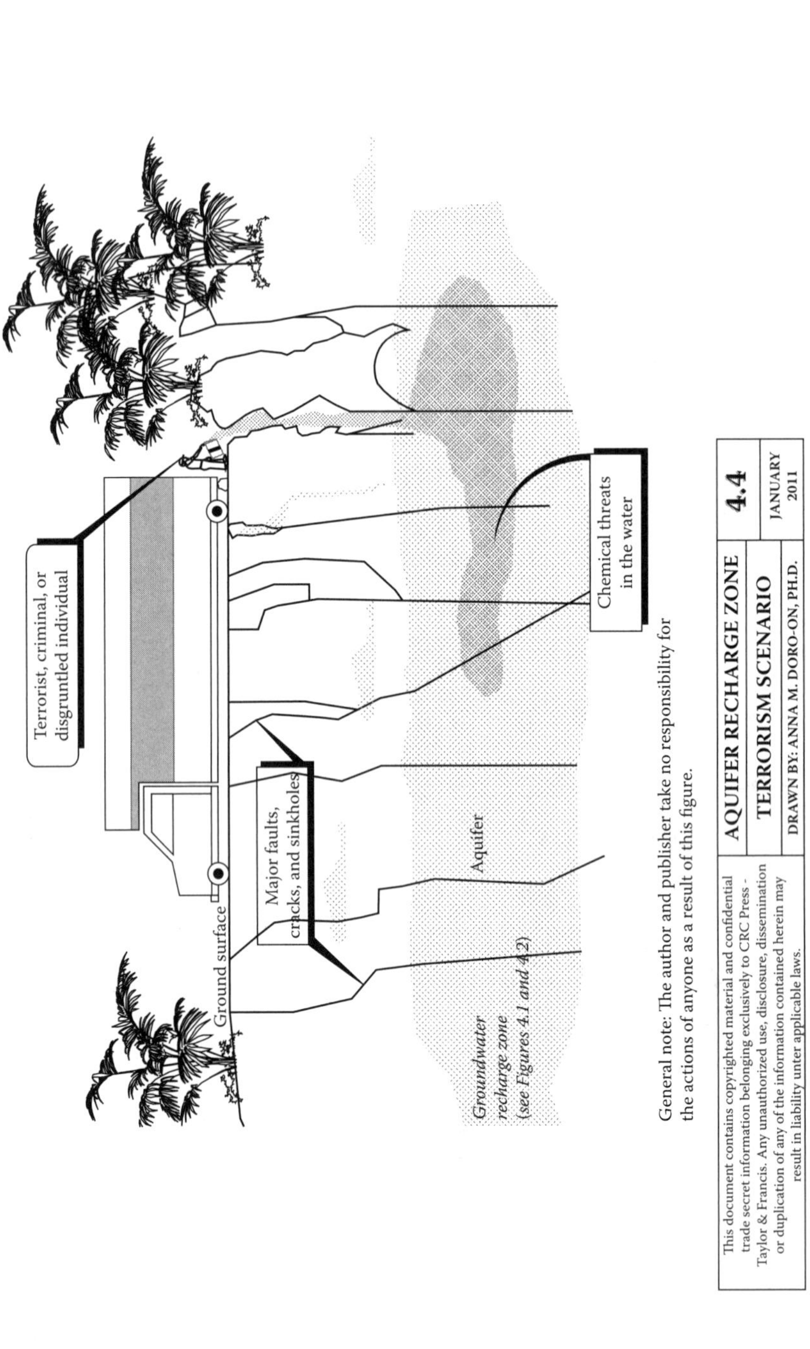

Figure 4.4 Aquifer contaminations through sinkholes, faults, and cracks.

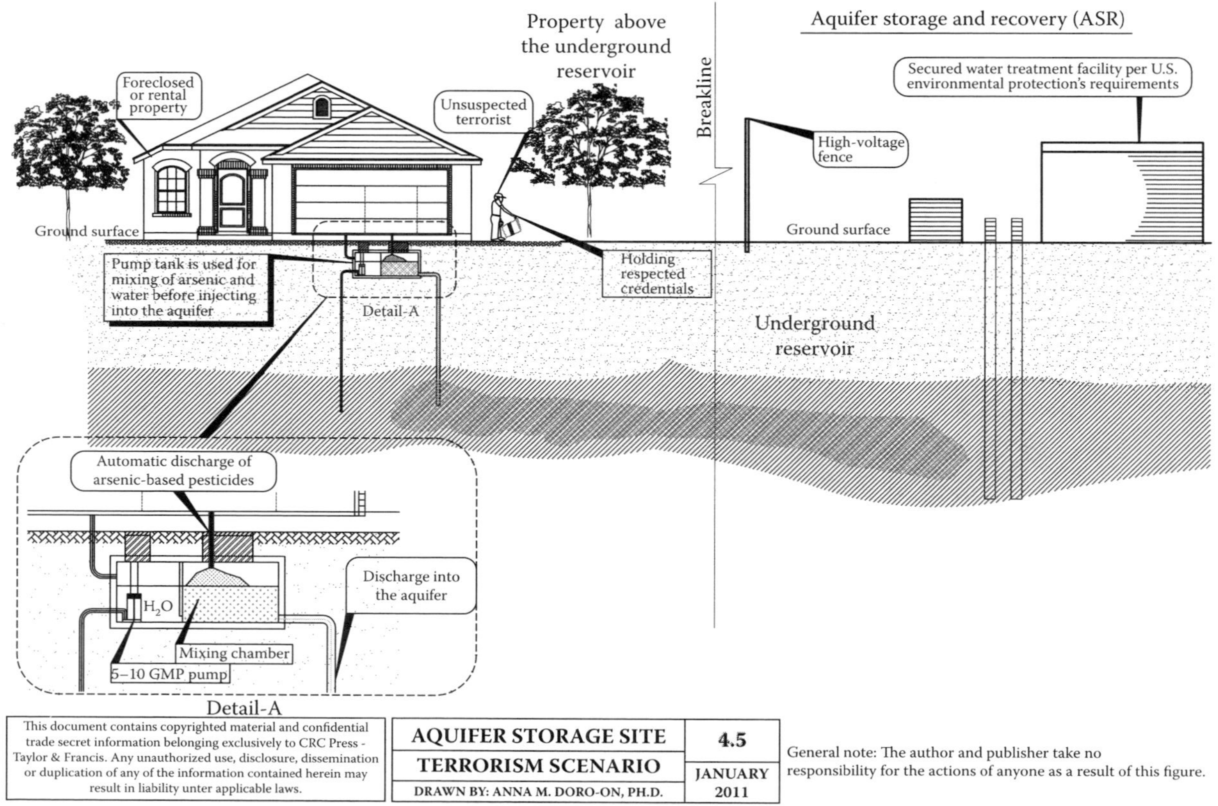

Figure 4.5 Aquifer contaminations through injection of deadly chemicals adjacent to aquifer storage and recovery.

cost effective to transport water from one place to another than to desalinate it. An adversary can create economic distress to water-stressed cities by destroying desalination plants. Protection against any attack should be implemented to avoid economic damages (short or long term).

4.5 Water Tanks

Water tanks usually store water for human consumption. A water tank provides water for drinking, agricultural irrigation, fire suppression, farming livestock, chemical manufacturing, food preparation, and many other possible uses. A ground-based water tank is made of lined carbon steel, and it may receive water from a well or from surface water, allowing a large volume of water to be placed in inventory to be used during peak demand cycles. Elevated water tanks, also known as *water towers*, create a pressure at the ground-level tank outlet of 1 pound per square inch (psi) per 2.31 feet of elevation; thus, a tank elevated to 70 feet creates about 30 psi of discharge pressure. The discharge pressure of 30 psi is sufficient for most domestic and industrial requirements. Unfortunately, water tanks can be very susceptible to vandalism and terrorism using deadly or hazardous chemicals if no chemical detection technology is installed as shown in Figure 4.6. Preventative measures against attacks on water tanks are presented in Chapter 10.

4.6 Reservoirs

A reservoir is an artificial lake normally used to collect and store water as illustrated in Figure 4.7a. Reservoirs may be developed between river valleys by building a dam or may be constructed by excavation in the ground. The storage capacity is divided into three zones: exclusive, multiple-purpose, and inactive, as shown in Figure 4.7b.

4.6.1 *Exclusive Capacity*

The exclusive space is reserved for use by a single purpose. Usually, this serves flood control, although navigation and hydroelectric power have exclusive space in some reservoirs (USACE 1997).

4.6.2 *Multiple-Purpose Capacity*

Multiple-purpose capacity serves seasonal flood control storage, navigation, hydroelectric power, water supply, irrigation, wetland, groundwater supply, recreation, and water quality (USACE 1997).

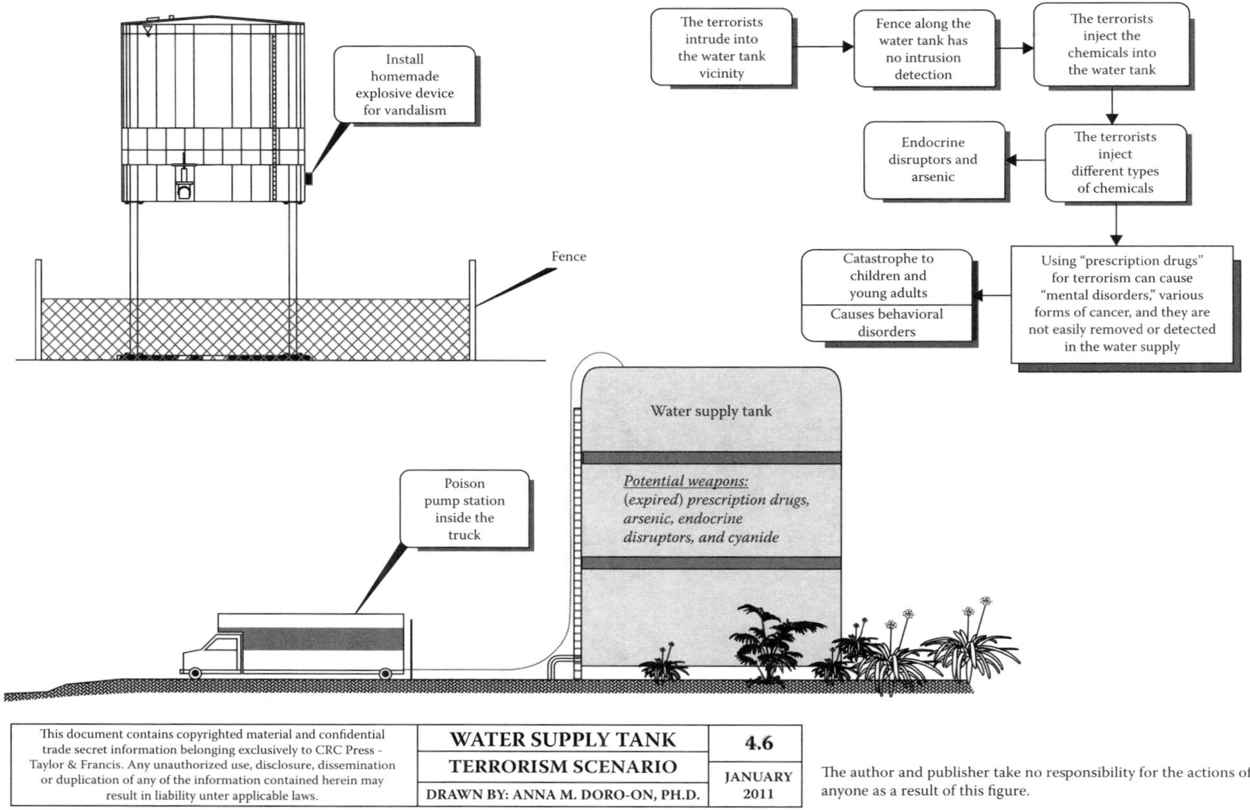

The author and publisher take no responsibility for the actions of anyone as a result of this figure.

Figure 4.6 Water supply tank contaminations.

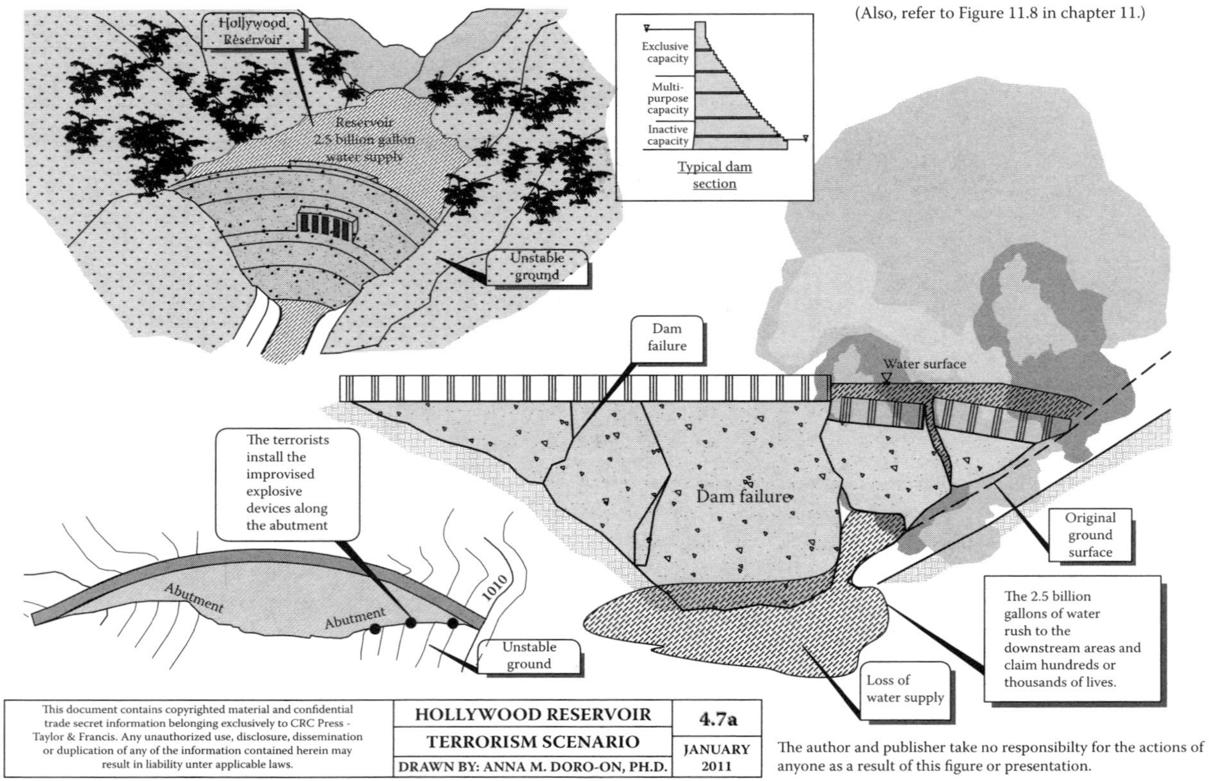

Figure 4.7a Terrorist attack against the Hollywood Reservoir/Mulholland Dam.

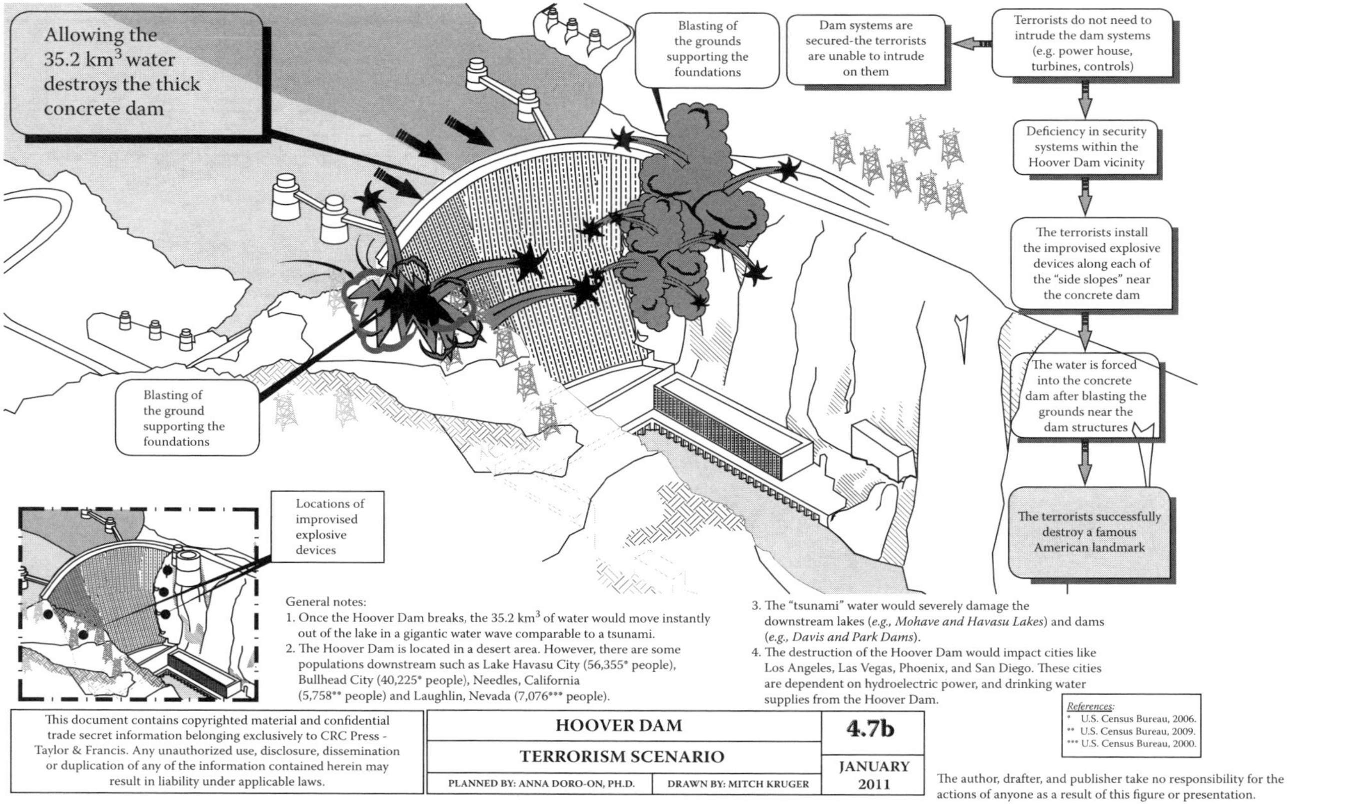

Figure 4.7b Terrorist attack against the Hoover Dam.

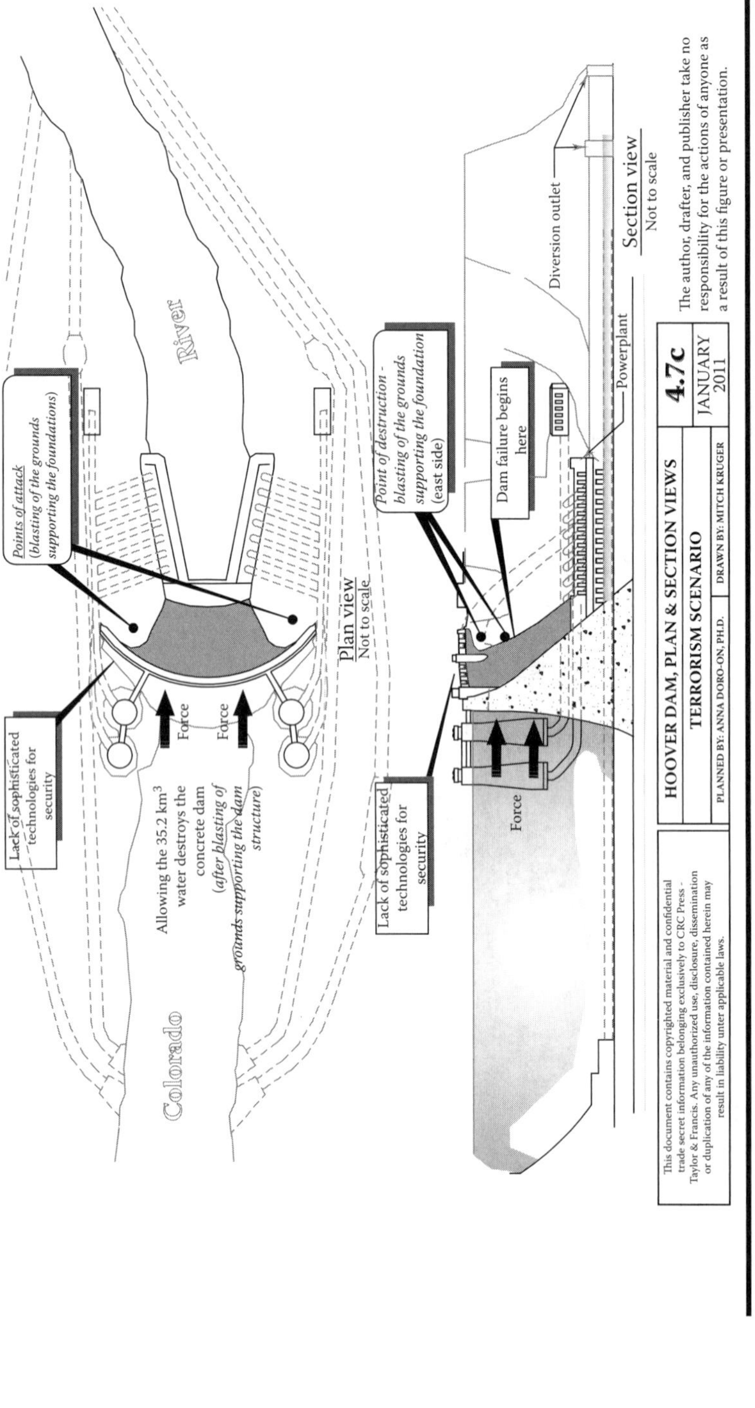

Figure 4.7c Terrorist attack against the Hoover Dam, shown in Plan and Section Views.

4.6.3 Inactive Space

Sediment storage may affect all levels of reservoir storage. Inactive space can be used during drought when it can provide limited but important storage for water supply, irrigation, recreation, fish and wildlife, and water quality (USACE 1997).

4.6.4 Terrorism against Reservoirs

The terrorists might thoroughly consider a different approach in yielding severe damage against the United States by contaminating reservoirs with deadly agents or blasting the dam structures that hold the reservoirs as shown in Figures 4.7a, 4.7b, and 4.7c. Los Angeles is one of the largest metropolitan areas in the United States that has faced a major economic crisis in the recent years. Los Angeles could be a prime choice for terrorism because of its economic vulnerability. Attacking one of the Los Angeles reservoirs such as the Hollywood Reservoir may cause catastrophic destruction comparable to the collapse of the St. Francis Dam in 1928. The Hollywood Reservoir is situated in one of the most famous cities in the whole world, Hollywood. It currently holds 2.5 billion gallons of water and provides the majority of water to Los Angeles. An example of terrorism against the Hollywood reservoir is presented in Figure 4.7a.

4.7 Dams

Dams are barriers that retain water or underground streams, whereas other structures such as levees are used to manage or prevent water flow into specific areas. Dams are generally categorized based on the material used in the structure and the type of design. Concrete gravity dams are structurally designed according to their weight for their stability, hydrostatic forces to their abutment by arch action, and based on the soil characteristic of the site. Whereas, with concrete buttress dams, the hydrostatic force is supported by a slab that spreads the weight or types of loadings to buttresses perpendicular to the axis of the dam. Most dams are constructed within a narrow part of a deep river valley; the side slopes of the valley can then act as natural supports for a dam and its foundations. The primary function of the dam's structure is to fill the space in the natural reservoir boundary left by the stream channel.

Dam failures are usually catastrophic, comparable or worse than the wars in Iraq or Afghanistan when the structure is breached or damaged, particularly those dams located upstream of several communities such as the Kensico Dam in New York. One of the examples of dam catastrophic failure is the St. Francis Dam. According to the Association of the Engineering Geologists Special Publication commemorating the 50th anniversary of the failure of the St. Francis Dam, dated March 1978, it was a concrete gravity–arch dam, designed to create a reservoir as a storage point of the Los Angeles Aqueduct. It was located forty miles northwest of Los Angeles, California, near the present city of Santa Clarita. It was built between 1924 and

1926 under the supervision of William Mulholland, chief engineer and general manager of the Los Angeles Department of Water and Power. Three minutes before midnight on March 12, 1928, the dam catastrophically failed, and the resulting flood killed more than 500 people (AEGS 1978). The collapse is the second greatest loss of life in California's history, after the 1906 San Francisco Earthquake and Fire. After the disaster, the City of Los Angeles immediately reinforced another dam almost structurally identical in shape and design, the Mulholland Dam, which creates the Hollywood Reservoir, also designed and built by Mulholland.

4.7.1 Terrorism against Dams

The terrorists could consider blasting the Hollywood Reservoir, also known as the Mulholland Dam. It has similar engineering design as the St. Francis Dam that collapsed in 1928, which generated catastrophe and over five hundred deaths. Popular dams that large major urban areas are dependent upon for tourism, hydroelectric power, flood control, and water supply are attractive to terrorists for attacks. These include such dams as the Hoover Dam (Figures 4.7b and 4.7c) in the Black Canyon of the Colorado River, the Kensico Dam in New York, and the Glen Canyon Dam on the Colorado River. Figures 4.7a, 4.7b, and 4.7c illustrate terrorist attack scenarios against the Hollywood Reservoir (the Mulholland Dam) and the Hoover Dam.

4.8 Aqueducts

In modern engineering, the term *aqueduct* is used for any system of pipes, ditches, canals, tunnels, and other structures used for the purpose of carrying water. The term aqueduct also applies to any bridge or viaduct that transports water across a gap. The largest aqueducts of all have been built in the United States to supply the country's biggest cities. The Catskill Aqueduct carries water to New York City over a distance of 190 km but is dwarfed by aqueducts in the far west of the country, most notably the Colorado River Aqueduct as shown in Figure 4.8, which supplies water to the Los Angeles area from the Colorado River nearly 400 km to the east, and the 714.5-km California Aqueduct, which runs from the Sacramento–San Joaquin River Delta to Lake Perris. In addition, the Central Arizona Project is the largest and most expensive aqueduct constructed in the United States. It stretches 540.7 km from its source near Parker, Arizona, to the metropolitan areas of Phoenix and Tucson.

4.8.1 Terrorism against Aqueducts

Attacking the most defenseless water infrastructures in the United States' largest and overpopulated cities (e.g., Los Angeles and New York) can easily breed catastrophe.

Terrorists could easily consider both blasting and contaminating the aqueducts as illustrated in Figures 4.9a and 4.9b.

Figure 4.8 California water source map.

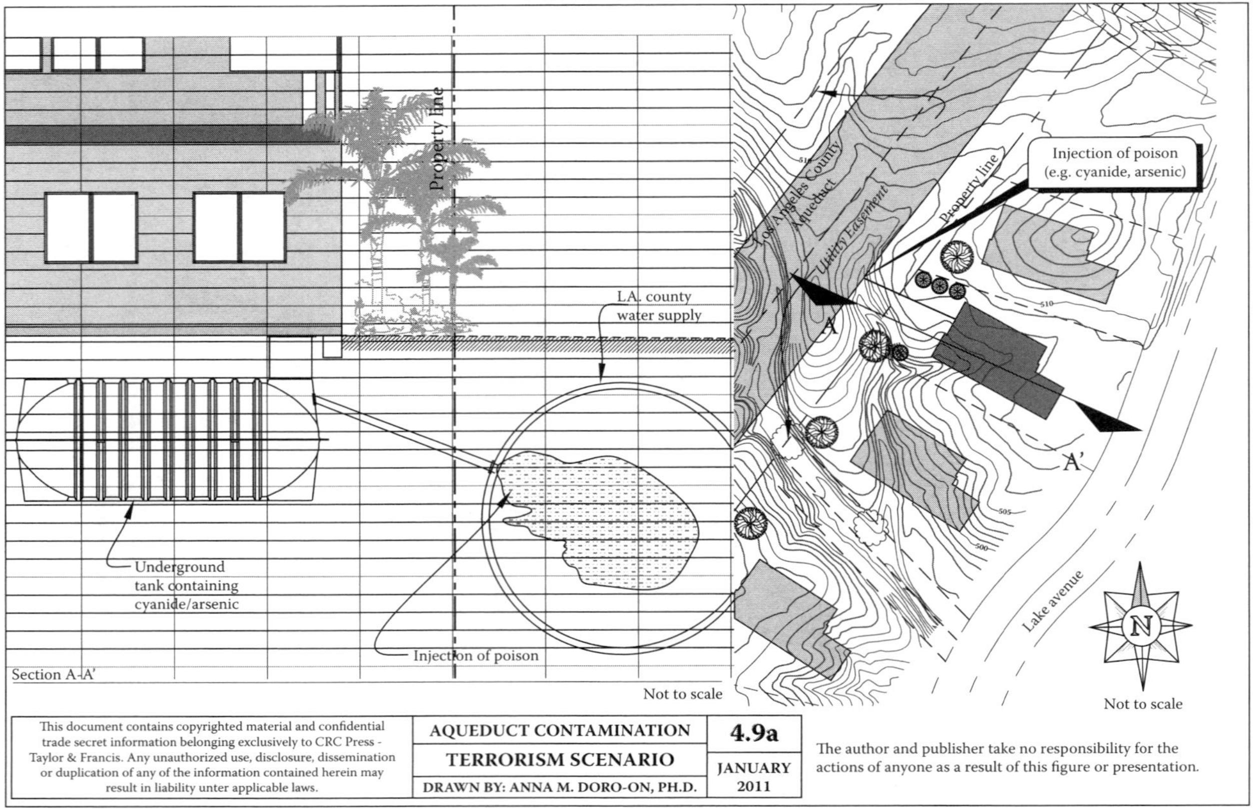

Figure 4.9a **Aqueduct pipeline terrorism.**

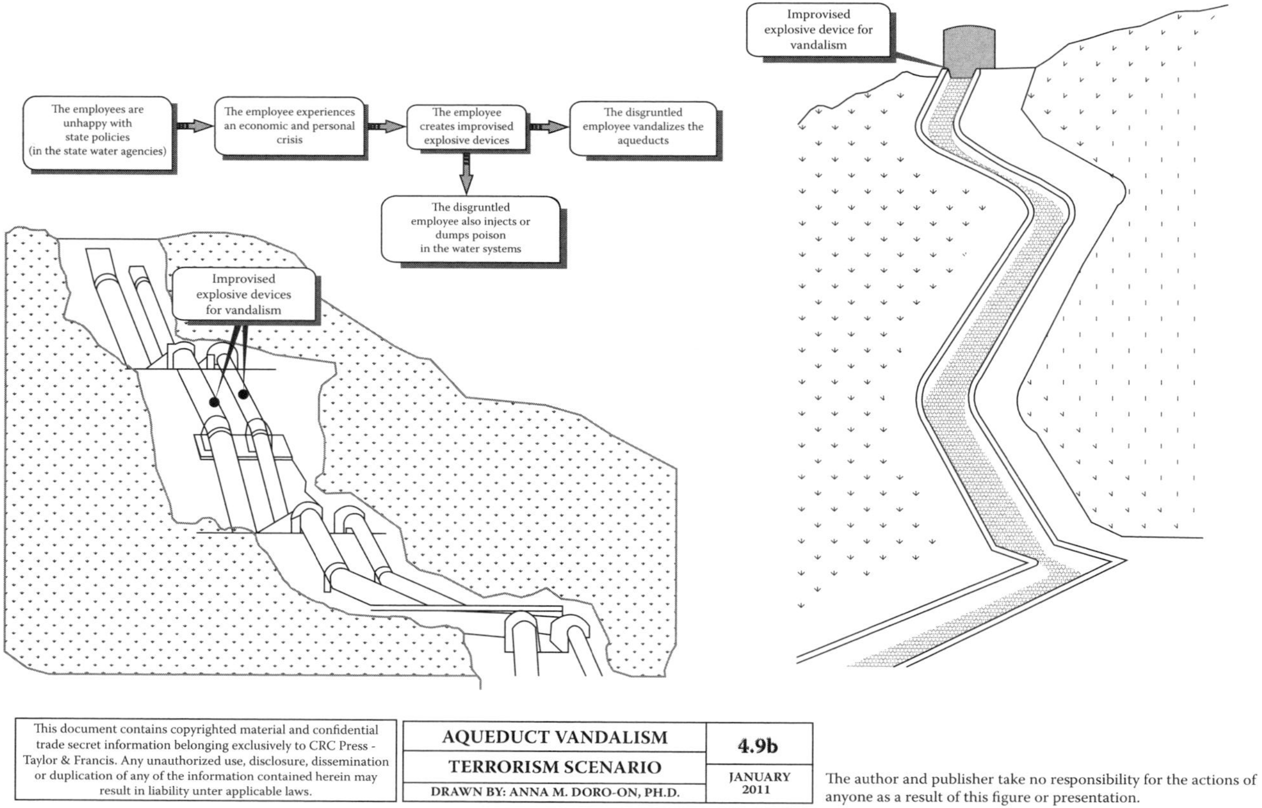

Figure 4.9b **Aqueduct pipeline terrorism.**

4.9 Surface Water

Surface water is water accumulated in a river, lake, stream, or ocean. Surface water naturally recharges through precipitation and is naturally lost by evaporation and subsurface seepage into the groundwater. Meanwhile, terrorists can indirectly attack surface water through the explosion of chemical plants or petroleum refineries with results comparable to the Deepwater Horizon accident that occurred on April 20, 2010. Figure 4.10 illustrates an example of a terrorist attack that can devastate the surface water, groundwater, water mains, and sewer pipelines.

4.10 Municipal Water Treatment Plants

The objective of municipal water treatment plants (MWTPs) is to provide a potable water supply, one that is chemically and bacteriologically safe for human consumption (Hammer 1975). Also, the goal of all MWTPs is to produce potable water that is aesthetically acceptable, which is free from apparent turbidity, color, odor, and objectionable taste. Common water sources for MWTPs are aquifers, rivers, natural lakes, and reservoirs. Disinfection and fluoridation are the simplest treatment processes. The combination of typical treatment processes is presented in Figure 4.11. Chlorine is used to disinfect the water supplies and provides residual protection. Fluoride is added to reduce the incidence of dental caries. Dissolved iron and manganese in groundwater oxidize when contacted with air, forming tiny rust particles that discolor the water. Removal of rust particles is performed by oxidizing the iron and manganese with chlorine or potassium permanganate and removing the precipitates by filtration. Lime and soda ash are mixed with raw water, and settleable precipitate is removed. Carbon dioxide is applied to stabilize the water prior to final filtration. Aeration is the common first step in the treatment of most groundwater to strip out dissolved gases and add oxygen. The primary process in surface water treatment is chemical clarification by coagulation, sedimentation, and filtration, as shown in Figure 4.11. Lake and reservoir water has a more uniform quality year-round and requires a lesser degree of treatment than river water. The challenge in waterworks operation is to process these waters to a safe, potable product acceptable for domestic use.

4.10.1 Terrorism against Municipal Water Treatment Plants

After the 9/11 series of attacks, the EPA maintained that the MWTPs were secure and safe. It is unlikely the major terrorists would exert a great amount of effort on attacking MWTPs, where there is no benefit of creating mass casualties and no optimization of fear by just blasting the MWTPs. However, governing agencies should keep MWTPs secured and safe from vandalism and from the attack of amateur terrorists, whose aim is to create short-term media attention

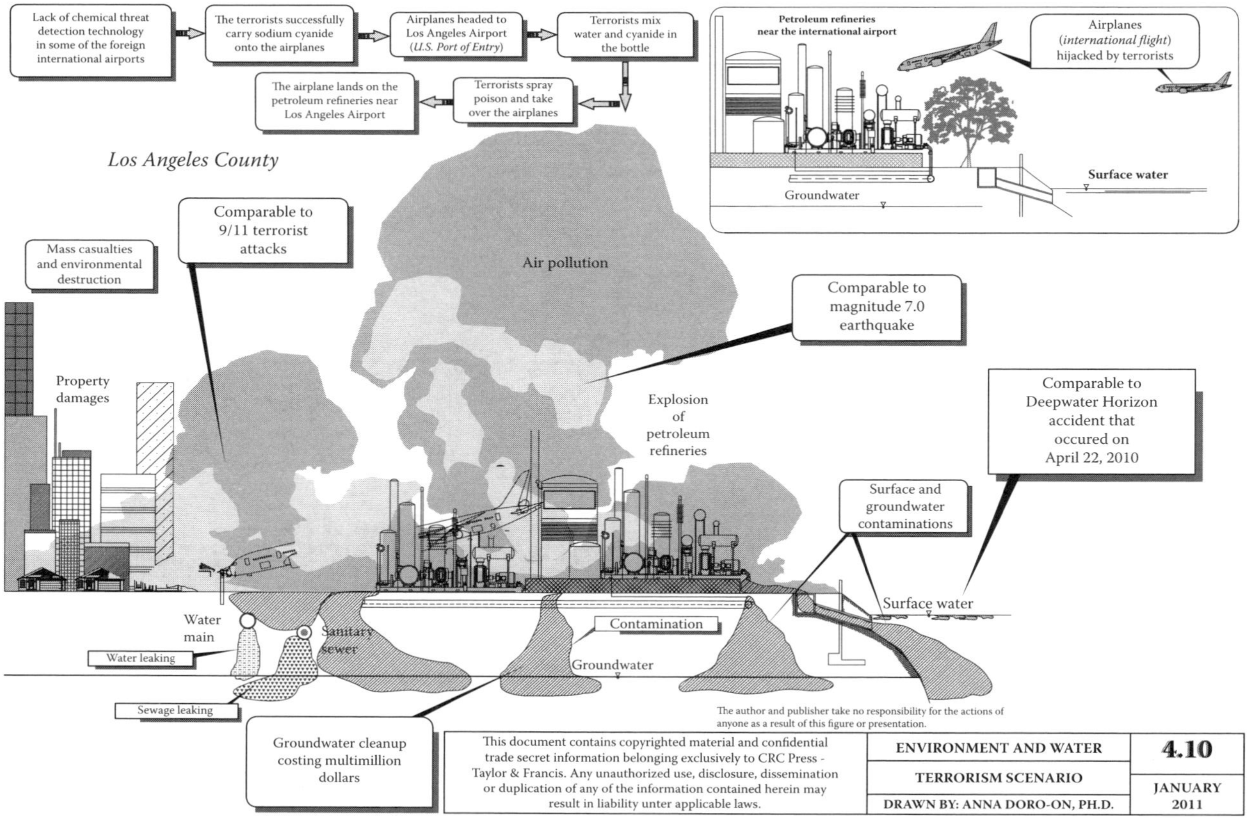

Figure 4.10 Explosion of petroleum refineries to create massive water body contamination.

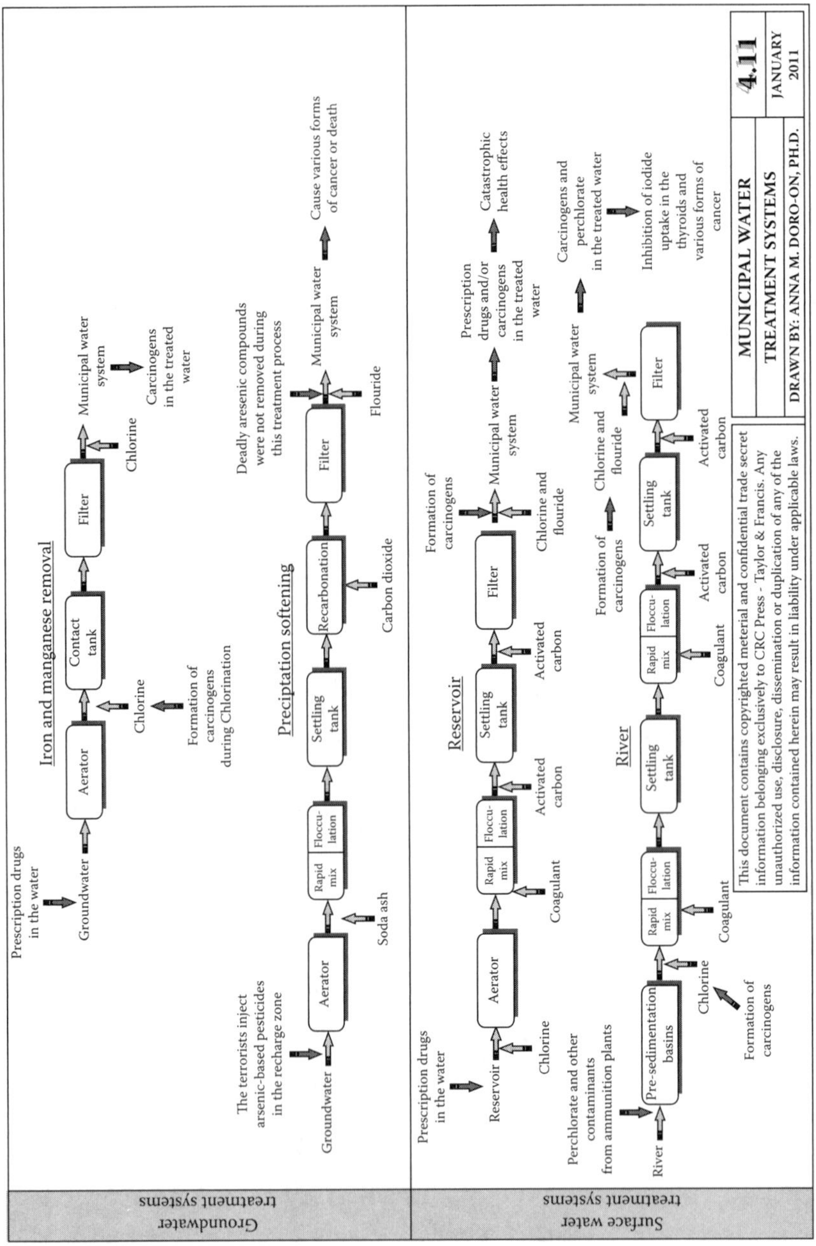

Figure 4.11 Schematic flow diagrams of the typical municipal water treatment system.

and temporary service disruption. Contaminating the original water source with deadly agents such as cyanide and arsenic compounds is the most likely way of attacking MWTPs because once the highly contaminated water is disinfected with chlorine in the plants, the water can be more hazardous and it could easily disrupt MWTP operation. An example of a preventive measure is presented in Chapter 10.

4.11 Municipal Wastewater Treatment Plants

Conventional wastewater treatment may involve three stages, called primary, secondary, and tertiary treatments. Primary treatment consists of temporarily holding the sewage in a quiescent basin where heavy solids can settle to the bottom while oil, grease, and lighter solids float to the surface. The settled and floating materials are removed, and the remaining liquid may be discharged or subjected to secondary treatment. Secondary treatment removes dissolved and suspended biological matter. Secondary treatment is typically performed by indigenous, waterborne microorganisms in a managed habitat. Secondary treatment may require a separation process to remove the microorganisms from the treated water prior to discharge or tertiary treatment. Tertiary treatment is sometimes defined as anything more than primary and secondary treatment. Treated water is sometimes disinfected chemically or physically prior to discharge into a stream, river, bay, lagoon, or wetland, or it can be used for the irrigation of a golf course, green way, or park. If it is sufficiently clean, it can also be used for groundwater recharge or agricultural purposes.

4.11.1 Terrorism against Municipal Wastewater Treatment Plants

Currently, DHS and the EPA have municipal wastewater treatment plants (MWWTPs) secured. It is recognized that major terrorists and their leaders (Al Qaeda) could plan another series of terrorist attacks; however, it is unlikely they would exert a great amount of effort on MWWTPs, as there is no gain of generating mass casualties by blasting the MWWTPs. The terrorists want Americans dead. Thus, governing agencies should keep MWWTPs secure from vandalism and from the attack of amateur terrorists, whose main goal is to create short-term media attention and temporary service disruption. The most vulnerable components of MWWTPs are the major inlet sanitary sewers that carry wastewater to the treatment plants. In addition, terrorists can easily intrude MWWTPs through the *outfall* area as shown in Figure 4.12. If any of those main sewers are compromised, it can injure the public health, environment, and water resources. Figure 4.12 shows a typical municipal wastewater treatment facility.

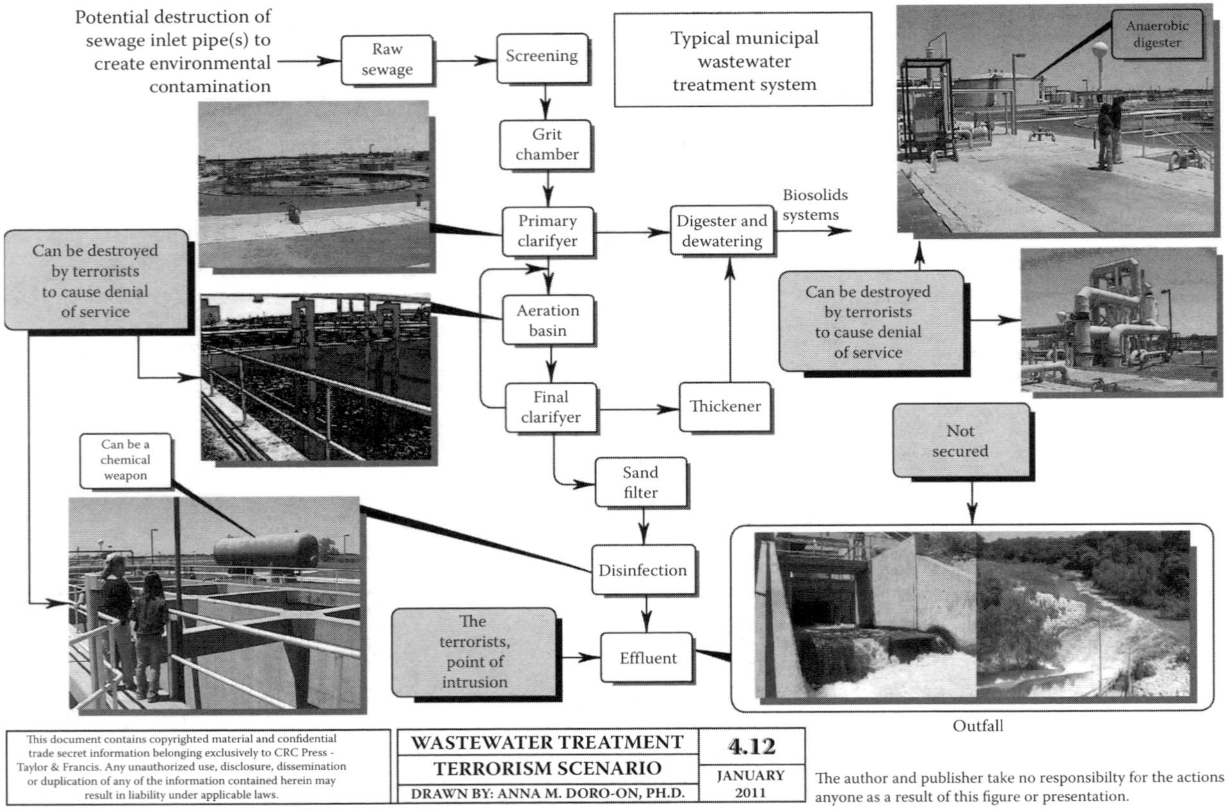

Figure 4.12 Schematic flow diagrams of the typical municipal wastewater treatment facility.

4.11.2 Major Sewer Pipelines and Manholes

Currently, sanitary sewer pipelines and manholes near major assets are not secured. Terrorists could install improvised explosive devices (IEDs) with highly explosive materials presented in Chapter 3 and the scenario illustrated in Chapter 11. If any of those main sewers are compromised and utilized as accessories for attacks, the public health, environment, and water resources could be injured easily.

4.12 Impacts

Water infrastructure vandalism and terrorism could impact the public in the following ways: create catastrophe, produce long-term damage to safe drinking water, injure public morale and confidence, optimize fear, generate economic chaos, cause water shortages or outages, cause irreversible damage to water resources, and injure the environment. These impacts could have serious consequences, which the United States should be concerned with to keep water infrastructure safe and secure. Hence, risk assessment and risk acceptability analysis for water infrastructure are critically needed in identifying the probabilities of prospective events of terrorism. The identification of these terrorism risks may lead to strategic improvements of the security of water infrastructure in the United States, and make it more difficult for attacks to succeed and lessen the impact of attacks that may occur.

References

Association of Engineering Geologist. 1978. *Failure of the St. Francis Dam*. San Francisquito Canyon, Near Saugus, California: Association of Engineering Geologist-Southern California Section Publication.

Association of State Drinking Water Administrators. *Public Health Protection Threatened by Inadequate Resources for State Drinking Water Programs: An Analysis of State Drinking Water Programs Resources, Needs, and Barriers*. Washington, DC: Association of State Drinking.

Haddix. 2003. *Cost of Illness in the 1993 Waterborne Cryptosporidium Outbreak, Milwaukee, WI*. Centers for Disease Control. 426–31.

Court Young, H. 2005. *Understanding Water and Terrorism*. Denver: BurgYoung Publishing, LLC.

Eckhardt, G. 2009. Introduction to Edwards Aquifer. http://www.edwardsaquifer.net/intro.html (accessed May 20, 2009).

Eckhardt, G. 2009a. Aquifer Storage and Recovery. http://edwardsaquifer.net/asr.html (accessed May 23, 2009).

Eckhardt, G. 2009b. Introduction to Edwards Aquifer. http://edwardsaquifer.net/intro.html (accessed May 20, 2009).

El Paso Water Utilities. 2010. Water Setting the Stage for the Future. http://epwu.org/water/desal_info.html (accessed June 3, 2010).

El Paso Water Utilities. 2010. Water Setting the Stage for the Future. http://www.epwu.org/water/desal_info.html (accessed June 3, 2010).

Hoorman, J. J., J. M. Raab, K. M. Boone, and L. C. Brown. 2009. Defiance County Groundwater Resources. http://ohioline.osu.edu/aex-fact/0490_20.html (accessed July 1, 2009).

Kentucky Geological Survey, 2009. Introduction to Karst Groundwater. http://www.uky.edu/KGS/water/general/karst/karstintro.htm (accessed June 10, 2009).

U.S. Army Corps of Engineers. 1997. Engineering and Design, Hydrologic Engineering Requirements for Reservoir, Engineer Manual 1110-2-1420. http://artikel-software.com/file/Hydrologic%20Engineering%20Requirements%20for%20Reservoirs.pdf.

U.S. Census Bureau. 2000. *2000 Census of Population and Housing*. http://www.census.gov/prod/cen2000/phc-3-30.pdf (accessed September 24, 2010).

U.S. Census Bureau. 2006. *State and County American Fact Finder*-Population of Bullhead City. http://quickfacts.census.gov/qfd/states/04/0408220.html (accessed September 24, 2010).

U.S. Census Bureau. 2006. *State and County American Fact Finder*-Population of Lake Havasu City. http://quickfacts.census.gov/qfd/states/04/0439370.html (accessed September 24, 2010).

U.S. Census Bureau. 2009. *State and County American Fact Finder*-Population of Needles, CA. http://factfinder.census.gov/servlet/STTable?-geo_id=06000US0607192080&-qr_name=ACS_2009_5YR_G00_S1201&-ds_name=ACS_2009_5YR_G00_ (accessed July 25, 2010).

U.S. Department of the Interior-Bureau of Reclamation. 2010. Hoover Dam. http://usbr.gov/lc/hooverdam/ (accessed October 1, 2010).

U.S. Geological Survey. 2009. *Groundwater Atlas of the United States*. http://pubs.usgs.gov/ha/ha730/ch_a/A-text4.html (accessed March 15, 2009).

Chapter 5

Regulatory Policies for the Protection of Water Infrastructure

The U.S. water supply was designated as one of the eight national infrastructures vital to the security of the United States through the issuance of Executive Order (EO) 13010. EO 13010 established the President's Commission on Critical Infrastructure Protection, which concluded in 1997 that there was inadequate protection against chemical or biological contamination of water supplies and insufficient technology for the detection, identification, and measurement of contaminants (Nuzzo 2006). In response to the Commission's findings, President Clinton issued Presidential Decision Directive 63 (PDD 63) in May 1998, which designated the U.S. Environmental Protection Agency (EPA) as the lead federal agency responsible for protecting the U.S. water supply from intentional physical, chemical, and biological attacks (Nuzzo 2006).

According to the EPA, Title IV of the Bioterrorism Act of 2002 pertains to drinking water security and safety requiring vulnerability assessments and emergency response plans for large- and medium-sized water systems. Consequently, as of 2006, the EPA has the lead for developing surveillance, monitoring systems of water contamination events, and implementing emergency preparedness/response plans per Homeland Security Presidential Directive 9 (HSPD 9). Based on the 2006 EPA report, HSPD 9 directs the agency to develop a network of integrated federal and state water testing laboratories.

The potential success of the program and plan presented by the EPA is unclear. According to an EPA Inspector General Report in 2003, the EPA has not issued standards for water infrastructure and has not obtained or analyzed data to develop a baseline for water security (USEPA 2003; Grosskruger 2006). Without established standards and benchmarks, the water industry and the government have no idea on what exactly constitutes vulnerability (Grosskruger 2006). Grosskruger pointed out that the EPA simply focused on complying with the 2002 Bioterrorism Act, which required completing vulnerability assessments instead of developing standards. Moreover, the EPA has proceeded with a heavy emphasis on issuing water infrastructure guidance, developing systems for information sharing, and partnering approaches via a heavy-handed approach. According to the U.S. Department of Homeland Security (DHS), protection for water infrastructure against chemical, biological, and radiological threats is inadequate and technology for the detection, identification, and measurement of contaminants is insufficient. Furthermore, it is important to perform a comprehensive review on U.S. regulatory policies for groundwater, water supply systems, dams, and reservoirs.

5.1 U.S. Regulatory Policies for Groundwater and Water Supply System Protection

The regulatory policies and standards for water supply are presented in the following sections.

5.1.1 Safe Drinking Water Act

The Safe Drinking Water Act (SDWA) allows states to develop a Comprehensive State Groundwater Protection Program to protect underground water reserves (e.g., aquifers, underground reservoirs, or aquifer storage and recover facilities) from contamination. Under this program, a state can require an agricultural establishment or other agribusiness to use designated Best Management Practices (BMPs) to help prevent contamination of groundwater by nitrates, phosphates, pesticides, microorganisms, or petroleum products; however, these apply only to agricultural operations that are subject to public water system supervision (USEPA 2010b). The Act does not cover private wells based upon U.S. Code 42 U.S.C.§300f(4)(A). Likewise, bottled water is regulated by the Food and Drug Administration (FDA) under the federal Food, Drug, and Cosmetic Act.

5.1.1.1 Title 40 of the Code of Federal Regulations

Maximum contaminant levels (MCLs) are standards that are set by the EPA in Title 40 of the Code of Federal Regulations. An MCL is the legal threshold limit on the amount of a hazardous substance that is allowed in drinking water under the SWDA. The MCL standards and treatment technique (TT) are shown in Tables 5.1 to 5.6. Thus, National Secondary Drinking Water Regulations are nonenforceable guidelines regarding contaminants that may cause cosmetic or aesthetic effects on drinking water. The EPA recommends secondary standards for water systems but does not require systems to comply (Table 5.7).

Table 5.1 Microorganism MCL Standards

Contaminant	*MCLG*[a] *(mg/L)*[b]	*MCL or TT*[a] *(mg/L)*[b]
Cryptosporidium	zero	TT[c]
Giardia lamblia	zero	TT[c]
Heterotrophic plate count	n/a	TT[c]
Legionella	zero	TT[c]
Total coliforms (including fecal coliform and *E. coli*)[d]	zero	5.0%[e]
Turbidity	n/a	TT[c]
Viruses (enteric)	zero	TT[c]

Source: Data from U.S. Environmental Protection Agency, EPA 816-F-09-004, 2010.

[a] Definitions:

Maximum contaminant level goal (MCLG)—The level of a contaminant in drinking water below which there is no known or expected risk to health. MCLGs allow for a margin of safety and are nonenforceable public health goals.

MCL—The highest level of a contaminant that is allowed in drinking water. MCLs are set as close to MCLGs as feasible using the best available treatment technology and taking cost into consideration. MCLs are enforceable standards.

Maximum residual disinfectant level goal (MRDLG)—The level of a drinking water disinfectant below which there is no known or expected risk to health. MRDLGs do not reflect the benefits of the use of disinfectants to control microbial contaminants.

TT—A required process intended to reduce the level of a contaminant in drinking water.

Maximum residual disinfectant level (MRDL)—The highest level of a disinfectant allowed in drinking water. There is convincing evidence that addition of a disinfectant is necessary for control of microbial contaminants.

[b] Units are in milligrams per liter (mg/L) unless otherwise noted. Milligrams per liter are equivalent to parts per million.

[c] EPA's surface water treatment rules require systems using surface water or groundwater under the direct influence of surface water to (1) disinfect their water, and (2) filter their water or meet criteria for avoiding filtration.

[d] Fecal coliform and E. coli are bacteria whose presence indicates that the water may be contaminated with human or animal wastes.

[e] No more than 5.0% samples total coliform-positive in a month.

Table 5.2 Disinfectant MCL Standards

Contaminant	*MCLG*[a] *(mg/L)*[b]	*MCL or TT*[a] *(mg/L)*[b]
Bromate	zero	0.010
Chlorite	0.8	1.0
Haloacetic acids (HAA5)	n/a[c]	0.060
Total trihalomethanes (TTHMs)	n/a[c]	0.080

Source: Data from U.S. Environmental Protection Agency, EPA 816-F-09-004, 2010.

[a] Definitions:
MCLG—The level of a contaminant in drinking water below which there is no known or expected risk to health. MCLGs allow for a margin of safety and are nonenforceable public health goals.
MCL—The highest level of a contaminant that is allowed in drinking water. MCLs are set as close to MCLGs as feasible using the best available treatment technology and taking cost into consideration. MCLs are enforceable standards.
MRDLG—The level of a drinking water disinfectant below which there is no known or expected risk to health. MRDLGs do not reflect the benefits of the use of disinfectants to control microbial contaminants.
TT—A required process intended to reduce the level of a contaminant in drinking water.
MRDL—The highest level of a disinfectant allowed in drinking water. There is convincing evidence that addition of a disinfectant is necessary for control of microbial contaminants.

[b] Units are in milligrams per liter (mg/L) unless otherwise noted. Milligrams per liter are equivalent to parts per million.

[c] Although there is no collective MCLG for this contaminant group, there are individual MCLGs for some of the individual contaminants:
(a) Trihalomethanes: bromodichloromethane (zero); bromoform (zero); dibromochloromethane (0.06 mg/L): chloroform (0.07mg/L); (b) Haloacetic acids: dichloroacetic acid (zero); trichloroacetic acid (0.02 mg/L); monochloroacetic acid (0.07 mg/L). Bromoacetic acid and dibromoacetic acid are regulated with this group but have no MCLGs.

5.1.2 Bioterrorism Act: Title IV-Drinking Water Security and Safety

Bioterrorism Act Title IV requires community drinking water systems serving populations of more than 3,300 persons to conduct assessments of their vulnerabilities to terrorist attack or other intentional acts and to defend against adversarial actions that might substantially disrupt the ability of a system to provide

Table 5.3 Disinfection By-product MCL Standards

Contaminant	*MRDLG[a] (mg/L)[b]*	*MRDL[a] (mg/L)[b]*
Chloramines (as Cl_2)	MRDLG = 4[a]	MRDL = 4.0[a]
Chlorine (as Cl_2)	MRDLG = 4[a]	MRDL = 4.0[a]
Chlorine dioxide (as ClO_2)	MRDLG = 0.8[a]	MRDL = 0.8[a]

Source: Data from U.S. Environmental Protection Agency, EPA 816-F-09-004, 2010.

[a] Definitions:

MCLG—The level of a contaminant in drinking water below which there is no known or expected risk to health. MCLGs allow for a margin of safety and are non-enforceable public health goals.

MCL—The highest level of a contaminant that is allowed in drinking water. MCLs are set as close to MCLGs as feasible using the best available treatment technology and taking cost into consideration. MCLs are enforceable standards.

MRDLG—The level of a drinking water disinfectant below which there is no known or expected risk to health. MRDLGs do not reflect the benefits of the use of disinfectants to control microbial contaminants.

TT—A required process intended to reduce the level of a contaminant in drinking water.

MRDL—The highest level of a disinfectant allowed in drinking water. There is convincing evidence that addition of a disinfectant is necessary for control of microbial contaminants.

[b] Units are in milligrams per liter (mg/L) unless otherwise noted. Milligrams per liter are equivalent to parts per million.

a safe and reliable supply of drinking water (USEPA 2010a). The provision also requires the EPA to focus on prevention, detection, preparedness, response, and recovery. Hence, each administrator also focuses on how to best stop and contain contaminated water flow to keep it from reaching the public. They also develop classes and training exercises for Community Water Systems (CWS) employees. The EPA should consider many ways a terrorist might disrupt, destroy, or contaminate a water supply based on simple to bolder terrorism activity scenarios and then take measures to prevent them. When creating these scenarios, the administrator must consider the security of the following among other issues: (1) storage and distribution facilities; (2) water pipes; (3) water collection facilities; (4) pretreatment and treatment plants; (5) electric and computer systems; (6) original water source; (7) large stormwater, chemical, and sewage pipelines adjacent to water supply distribution systems; (8) outfall of wastewater treatment plants (outfall can be a point of intrusion); and (9) explosive and petroleum plants near water bodies or groundwater resources.

Table 5.4 Inorganic Chemical MCL Standards

Contaminant	*MCLG[a] (mg/L)[b]*	*MCL or TT[a] (mg/L)[b]*
Antimony	0.006	0.006
Arsenic	0[c]	0.010 as of 01/23/06
Asbestos (fiber >10 μm)	7 million fibers per liter	7 MFL
Barium	2	2
Beryllium	0.004	0.004
Cadmium	0.005	0.005
Chromium (total)	0.1	0.1
Copper	1.3	TT[d]; action level = 1.3
Cyanide (as free cyanide)	0.2	0.2
Fluoride	4.0	4.0
Lead	zero	TT[d]; action level = 0.015
Mercury (inorganic)	0.002	0.002
Nitrate (measured as nitrogen)	10	10
Nitrite (measured as nitrogen)	1	1
Selenium	0.05	0.05
Thallium	0.0005	0.002

Source: Data from U.S. Environmental Protection Agency, EPA 816-F-09-004, 2010.

[a] Definitions:

MCLG—The level of a contaminant in drinking water below which there is no known or expected risk to health. MCLGs allow for a margin of safety and are non-enforceable public health goals.

MCL—The highest level of a contaminant that is allowed in drinking water. MCLs are set as close to MCLGs as feasible using the best available treatment technology and taking cost into consideration. MCLs are enforceable standards.

MRDLG—The level of a drinking water disinfectant below which there is no known or expected risk to health. MRDLGs do not reflect the benefits of the use of disinfectants to control microbial contaminants.

TT—A required process intended to reduce the level of a contaminant in drinking water.

MRDL—The highest level of a disinfectant allowed in drinking water. There is convincing evidence that addition of a disinfectant is necessary for control of microbial contaminants.

Table 5.4 Inorganic Chemical MCL Standards (*Continued*)

[b] Units are in milligrams per liter (mg/L) unless otherwise noted. Milligrams per liter are equivalent to parts per million.
[c] Lead and copper are regulated by a TT that requires systems to control the corrosiveness of their water.
[d] Each water system must certify, in writing, to the state (using third-party or manufacturer's certification) that when acrylamide and epichlorohydrin are used to treat water, the combination (or product) of dose and monomer level does not exceed the levels specified, as follows:
Acrylamide = 0.05% dosed at 1 mg/L (or equivalent)
Epichlorohydrin = 0.01% dosed at 20 mg/L (or equivalent)

Table 5.5 Organic Chemical MCL Standards

Contaminant	*MCLG[a] (mg/L)[b]*	*MCL or TT[a] (mg/L)[b]*
Acrylamide	zero	TT[c]
Alachlor	zero	0.002
Atrazine	0.003	0.003
Benzene	zero	0.005
Benzo(a)pyrene (PAHs)	zero	0.0002
Carbofuran	0.04	0.04
Carbontetrachloride	zero	0.005
Chlordane	zero	0.002
Chlorobenzene	0.1	0.1
2,4-D	0.07	0.07
Dalapon	0.2	0.2
1,2-Dibromo-3-chloropropane (DBCP)	zero	0.0002
o-Dichlorobenzene	0.6	0.6
p-Dichlorobenzene	0.075	0.075
1,2-Dichloroethane	zero	0.005
1,1-Dichloroethylene	0.007	0.007
cis-1,2-Dichloroethylene	0.07	0.07
trans-1,2-Dichloroethylene	0.1	0.1

(Continued)

Table 5.5 Organic Chemical MCL Standards (*Continued*)

Contaminant	*MCLG[a] (mg/L)[b]*	*MCL or TT[a] (mg/L)[b]*
Dichloromethane	zero	0.005
1,2-Dichloropropane	zero	0.005
Di(2-ethylhexyl) adipate	0.4	0.4
Di(2-ethylhexyl) phthalate	zero	0.006
Dinoseb	0.007	0.007
Dioxin (2,3,7,8-TCDD)	zero	0.00000003
Diquat	0.02	0.02
Endothall	0.1	0.1
Endrin	0.002	0.002
Epichlorohydrin	zero	TT[c]
Ethylbenzene	0.7	0.7
Ethylene dibromide	zero	0.00005
Glyphosate	0.7	0.7
Heptachlor	zero	0.0004
Heptachlor epoxide	zero	0.0002
Hexachlorobenzene	zero	0.001
Hexachlorocyclopentadiene	0.05	0.05
Lindane	0.0002	0.0002
Methoxychlor	0.04	0.04
Oxamyl (Vydate)	0.2	0.2
Polychlorinated biphenyls (PCBs)	zero	0.0005
Pentachlorophenol	zero	0.001
Picloram	0.5	0.5
Simazine	0.004	0.004
Styrene	0.1	0.1

Table 5.5 Organic Chemical MCL Standards (*Continued*)

Contaminant	*MCLG[a] (mg/L)[b]*	*MCL or TT[a] (mg/L)[b]*
Tetrachloroethylene	zero	0.005
Toluene	1	1
Toxaphene	zero	0.003
2,4,5-TP (Silvex)	0.05	0.05
1,2,4-Trichlorobenzene	0.07	0.07
1,1,1-Trichloroethane	0.20	0.2
1,1,2-Trichloroethane	0.003	0.005
Trichloroethylene	zero	0.005
Vinyl chloride	zero	0.002
Xylenes (total)	10	10

Source: Data from U.S. Environmental Protection Agency, EPA 816-F-09-004, 2010.

[a] Definitions:

MCLG—The level of a contaminant in drinking water below which there is no known or expected risk to health. MCLGs allow for a margin of safety and are non-enforceable public health goals.

MCL—The highest level of a contaminant that is allowed in drinking water. MCLs are set as close to MCLGs as feasible using the best available treatment technology and taking cost into consideration. MCLs are enforceable standards.

MRDLG—The level of a drinking water disinfectant below which there is no known or expected risk to health. MRDLGs do not reflect the benefits of the use of disinfectants to control microbial contaminants.

TT—A required process intended to reduce the level of a contaminant in drinking water.

MRDL—The highest level of a disinfectant allowed in drinking water. There is convincing evidence that addition of a disinfectant is necessary for control of microbial contaminants.

[b] Units are in milligrams per liter (mg/L) unless otherwise noted. Milligrams per liter are equivalent to parts per million.

[c] Each water system must certify, in writing, to the state (using third-party or manufacturer's certification) that when acrylamide and epichlorohydrin are used to treat water, the combination (or product) of dose and monomer level does not exceed the levels specified, as follows:

Acrylamide = 0.05% dosed at 1 mg/L (or equivalent)

Epichlorohydrin = 0.01% dosed at 20 mg/L (or equivalent)

Table 5.6 Radionuclide MCL Standards

Contaminant	*MCLG*[a] *(mg/L)*[b]	*MCL or TT*[a] *(mg/L)*[b]
Alpha particles	none[c] ---------- zero	15 picocuries per liter (pCi/L)
Beta particles and photon emitters	none[c] ---------- zero	4 millirems per year
Radium 226 and radium 228 (combined)	none[c] ---------- zero	5 pCi/L
Uranium	zero	30 µg/L as of 12/08/03

Source: Data from U.S. Environmental Protection Agency, EPA 816-F-09-004, 2010.

[a] Definitions:

MCLG—The level of a contaminant in drinking water below which there is no known or expected risk to health. MCLGs allow for a margin of safety and are non-enforceable public health goals.

MCL—The highest level of a contaminant that is allowed in drinking water. MCLs are set as close to MCLGs as feasible using the best available treatment technology and taking cost into consideration. MCLs are enforceable standards.

MRDLG—The level of a drinking water disinfectant below which there is no known or expected risk to health. MRDLGs do not reflect the benefits of the use of disinfectants to control microbial contaminants.

TT—A required process intended to reduce the level of a contaminant in drinking water.

MRDL—The highest level of a disinfectant allowed in drinking water. There is convincing evidence that addition of a disinfectant is necessary for control of microbial contaminants.

[b] Units are in milligrams per liter (mg/L) unless otherwise noted. Milligrams per liter are equivalent to parts per million.

[c] Lead and copper are regulated by a TT that requires systems to control the corrosiveness of their water.

5.2 Funding for Protection Research

The EPA provides grants to organizations that provide training, technical assistance, and tool development for water security (USEPA 2010d). For example, beginning in 2002 the EPA provided counterterrorism grants to ensure that drinking water

Table 5.7 Secondary Drinking Water Standards

Contaminant	*Secondary Standard*
Aluminum	0.05 to 0.2 mg/L
Chloride	250 mg/L
Color	15 (color units)
Copper	1.0 mg/L
Corrosivity	noncorrosive
Fluoride	2.0 mg/L
Foaming agents	0.5 mg/L
Iron	0.3 mg/L
Manganese	0.05 mg/L
Odor	3 threshold odor number
pH	6.5–8.5
Silver	0.10 mg/L
Sulfate	250 mg/L
Total dissolved solids	500 mg/L
Zinc	5 mg/L

Source: Data from U.S. Environmental Protection Agency, EPA 816-F-09-004, 2010.

utilities receive technical assistance and training on homeland security issues, including vulnerability assessments, and emergency response plans.

5.3 Enforcement of Regulations

States and the EPA maintain a formal enforcement program to ensure that violations related to water supply systems are promptly addressed and that public health is protected from hazard, however, the success of this program is currently ambiguous when there are notorious contaminants (e.g., prescription drugs, perchlorate, or other toxic contaminants generated by ammunition/explosive plants) that are not entirely regulated and monitored. The 1996 SDWA amendments also require that primacy states have administrative penalty authority.

5.4 Agencies Involved in Protection Policies

Consequently, the EPA has evolved to be in charge of developing surveillance and monitoring systems to provide early detection and awareness of water contamination events per HSPD 9. The EPA works with other federal agencies such as the Centers for Disease Control and Prevention (CDCP), the Federal Bureau of Investigation (FBI), the Department of Defense (DOD), and water organizations such as the Water Environment Research Foundation (WERF) to improve information on water security technologies and conduct research for water security.

5.5 Federal Regulations for Dams, Reservoirs, and Other Water Systems

The regulations involved in protecting dam, reservoir, and other water systems are presented in the following sections.

5.5.1 Water Resources Development Act

Most people recognize and acknowledge the Clean Air Act, the Clean Water Act, and the Coastal Zone Management Act, but are unfamiliar with the Water Resources Development Act (WRDA). However, it is a very important piece of legislation that can have a dramatic impact on your favorite beach or surf spot because WRDA funds the U.S. Army Corps of Engineers (USACE). Almost every coastal community in America is somehow affected by a USACE project. It is responsible for many coastal armoring projects, beach fill projects, channel dredging for navigation, construction of dams and flood control projects, and wetlands developments. According to USACE (2006), U.S. Fish and Wildlife Service and National Wildlife Federation (2007), in 1986, Congress passed a landmark WRDA bill containing 300 new projects, which, in fact, requires all local sponsors pay a portion of project costs.

5.5.2 Dam Safety and Security Act

The Dam Safety and Security Act of 2002, signed into law on December 2, 2002, reauthorized the National Dam Safety Program (NDSP) for four more years and added enhancements to the 1996 Act that are designed to safeguard dams against terrorist attacks (FEMA 2010a).

5.5.3 River and Harbors Act of 1899

The River and Harbors Act of 1899 through USACE prohibits the unauthorized discharging or dumping of nonliquid waste in navigable waterways.

5.5.4 The Federal Water Power Act of 1920

The Federal Water Power Act of 1920 controls hydroelectric dam construction and improvements on all navigable waterways. This allows the federal government to standardize and require private electric utilities to obtain licenses prior to legal operation. The federal government also constructed hydroelectric dams to provide cost-effective power or electricity and an emergency water supply and to persuade private electric companies to maintain reasonable prices.

5.6 Funding for Protection Research Related to Dams, Reservoirs, and Other Water Systems

Research development is crucial for dam safety and security development.

According to FEMA, the National Dam Safety Review Board recently developed a five-year strategic plan for research needs in dam safety and security. The goal in developing the strategic plan was to ensure that priority would be given to those projects demonstrating a high degree of collaboration and expertise, and the likelihood of producing products that will contribute to the safety of dams in the United States (FEMA 2010b).

In recent years, research funds have been allocated to workshops related to dam safety. The research topics presented at the workshops are utilized for further studies and future research advancements.

5.7 Agencies and Programs Involved in the Protection Policies for Dams, Reservoirs, and Other Water Systems

As lead federal agency for the NDSP, the Federal Emergency Management Agency (FEMA) is responsible for coordinating, managing, and evaluating efforts to secure the safety of dams through the U.S. Congress' passage of the NDSP Act of 2006 that reauthorized the program for five more years, through 2011. The program makes federal funds of approximately \$3.5 million per year available to the states, which are primarily responsible for protecting the public from dam failures and disasters. Also, FEMA is responsible for providing financial aid to states' dam safety programs where eligibility is based on national performance criteria, under the NDSP Act of 2006. The potential success of FEMA's financial assistance and programs for dam safety and security is currently doubtful, because most dams are still unsafe and unsecured. The aid and programs should heavily focus on engineering technology development, and policy and security improvements for dam structure failure and disaster including terrorist attacks.

5.7.1 The National Dam Safety Review Board

The National Dam Safety Review Board advises FEMA on national dam safety priorities, coordinates federal–state activities, and evaluates states' development in meeting national performance criteria defined under the NDSP Act of 2006. Board membership includes representatives from federal and state agencies and the private sector.

5.7.2 The Interagency Committee on Dam Safety

According to FEMA, the Interagency Committee on Dam Safety (ICODS) coordinates federal regulatory policies, guidelines, and efforts related to dam safety. ICODS is chaired by FEMA and includes the following federal agencies, which build, own, operate, or regulate dams: Department of Agriculture (USDA); Department of Energy (DOE); Department of the Interior (DOI); Department of Labor (DOL); Mine Safety and Health Administration (MSHA); Federal Energy Regulatory Commission (FERC); Department of State, International Boundary and Water Commission (U.S. Section); Nuclear Regulatory Commission (NRC); and Tennessee Valley Authority (TVA). ICODS's main goal is to maintain the safety of dams that pose a risk to the public and damage to property.

5.7.3 The Association of State Dam Safety Officials

The Association of State Dam Safety Officials (ASDSO) is a national nonprofit organization of state and federal dam safety regulators, dam owners and operators, science and engineering consultants, manufacturers and suppliers, academia, and contractors. The organization supports programs and policies designed to increase dam safety and security.

5.7.4 The United States Society on Dams

The United States Society on Dams (USSD) is a national professional organization that represents mainly private sector interests with representatives from the private sector and some governmental agencies. The USSD seeks to advance the technology of dam engineering, construction, operation, maintenance, and dam safety; to foster socially and environmentally responsible water resource projects; and to promote awareness of the role of dams in managing the nation's water resources.

References

Federal Emergency Management Agency. 2010a. *About the National Dam Safety Program.* http://www.fema.gov/plan/prevent/damfailure/ndsp.shtm (accessed October 10, 2010).

Federal Emergency Management Agency. 2010b. *Dam Safety Research.* http://www.fema.gov/plan/prevent/damfailure/research.shtm (accessed October 11, 2010).

Grosskruger, P. 2006. *Analysis of U.S. Water Infrastructure from a Security Perspective*. Master's Thesis, The U.S. Army War College, Carlisle Barracks, Pennsylvania. http://www.strategicstudiesinstitute.army.mil/pdffiles/ksil353.pdf (accessed July 4, 2006).

Nuzzo, J. 2006. *The Biological Threat to U.S. Water Supplies: Toward a National Water Security Policy*. Biosecurity and Bioterrorism: Biodefense Strategy, Practice, and Science, v(4) Nov. 4, 2006. Mary Ann Liebert, Inc. http://www.liebertonline.com/doi/pdf/10.1089/bsp.2006.4.147 (accessed May 20, 2008).

U.S. Environmental Protection Agency. 2003. EPA Inspector General Report, "EPA Needs a Better Strategy to Measure Changes in the Security of the Nation's Water Infrastructure," Sept. 11, 2003, cover letter. http://www.epa.gov/oig/reports/2003/HomelandSecurityReport2003M00016.pdf. (accessed July 4, 2006).

U.S. Environmental Protection Agency. 2010a. Requirement of the Public Health Security and Bioterrorism Preparedness and Response Act 2002. http://water.epa.gov/infrastructure/watersecurity/lawsregs/bioterrorismact.cfm (accessed July 25, 2010).

U.S. Environmental Protection Agency. 2010b. Comprehensive State Groundwater Protection Program. http://www.epa.gov/oecaagct/tdri.html (accessed July 4, 2010).

U.S. Environmental Protection Agency. 2010c. National Primary & Secondary Drinking Water Regulations, EPA 816-F-09-004.

U.S. Environmental Protection Agency. 2010d. Grants and Funding. http://water.epa.gov/infrastructure/watersecurity/funding/index.cfm (accessed May 5, 2010).

Chapter 6

Introduction to Risk and Vulnerability Assessment

6.1 Introduction

This chapter is an introduction to the qualitative and quantitative risk and vulnerability assessment, operational formulations, and models for homeland critical infrastructure protection, especially for water supply systems, recommended by prominent authors, the U.S. Department of Homeland Security (DHS), U.S. Environmental Protection Agency (EPA), Central Intelligence Agency, Federal Emergency Management Agency (FEMA), National Counterterrorism Center, U.S. Department of Defense, National Security Council, Public Health Service, National Commission on Terrorism, and other government and state agencies, together with the integrated approach to risk assessment with *cumulative prospect theory* for water infrastructure security and safety.

6.2 Standard Risk and Vulnerability Strategies and Models

6.2.1 Basic Homeland Security Risk Assessments

According to the DHS, risk is the projected (or expected) loss from a future sequence of events with an unwanted outcome. The total system risk (R) is the summation of the risks from all possible events. The events arising from one or many threats and initiating actions may lead to different risk scenarios, which in turn have many possible outcomes. The basic DHS risk (R) assessment formula is as follows: the *threat* (T)

to a target/area multiplied by *vulnerability* (V) of the target/area multiplied by *consequence* (C) of an attack on that target/area or

$$R = T \times V \times C \tag{6.1}$$

6.2.2 Model-Based Vulnerability Analysis

According to critical infrastructure authors and experts, scalable or model-based vulnerability analysis (MBVA) is an extensive tool of analysis that combines risk analysis, fault tree method, event tree method, and network analysis based on the principles of probability and cost minimization. The MBVA was mostly used by computer science, network science, and information technology professionals and later on was adopted for homeland critical infrastructure risk/vulnerability analysis. In MBVA, hubs are identified, hub vulnerabilities are organized and quantified using fault tree, all possible events are organized as an event tree, and an optimal investment strategy is computed that minimizes risk (Lewis 2006). The primary procedure of MBVA includes the network analysis, which is basically rooted in the scale-free network theory, proposed by Derek J. de Solla Price in 1965, called *cumulative advantage* or *preferential attachment*. Albert-Laszlo Barabasi rediscovered and popularized the network theory in 1999. In this theory, critical infrastructure is modeled as a network, with nodes and links conceptually representing areas, power lines, power generators, cyber technologies, or sector assets and relationships among those assets. As this approach only produces a specific subset of networks, scientific communities are considering alternative techniques. A further discussion is presented in Chapter 7.

6.2.3 Water/Wastewater Vulnerability Self-Assessment Tools

The vulnerability self-assessment tool (VSAT) is a qualitative risk assessment method recommended by the EPA and American Water Works Association, which provides a structured approach to assess the vulnerabilities of utilities. This tool analyzes the risk of utility assets and identifies critical assets and potential single points of failure in water infrastructure utilities. Examples of the assessment using VSAT are presented in Chapter 7.

6.2.4 Security Vulnerability Self-Assessment Guide for Small Drinking Water Systems

The EPA recommended the security vulnerability self-assessment to assist small water systems serving populations of at least 3,300 and fewer than 50,000 to determine possible vulnerable components and identify security measures that should be considered in order to protect the system and the public health from hazard and denial of water service. This guide was designed for use by water system personnel and was developed by the Association of State Drinking Water Administrators and the National Rural Water Association in association with the EPA. More details about completing the security vulnerability self-assessment are provided in Chapter 7.

6.2.5 *Automated Security Survey and Evaluation Tool*

The Automated Security Survey and Evaluation Tool is a self-guided software program designed to help small and medium-sized drinking water systems to complete a vulnerability assessment and to improve their security and their responsiveness to a range of threats (USEPA 2010a,b,c,d).

6.2.6 *Risk Analysis and Management for Critical Asset Protection* Plus

The Risk Analysis and Management for Critical Asset Protection (RAMCAP) model has been developed and used for risk analysis of critical national infrastructures. It is a quantitative method that estimates values of risk and resilience, and benefits of improving risk and resilience based on evaluations of vulnerability, potential threat, and consequence. The RAMCAP *Plus* model defines risk as the product of potential threat, vulnerability, and consequence:

$$\begin{aligned} \text{risk} &= \text{threat} \times \text{vulnerability} \times \text{consequence} \\ R &= T \times V \times C \end{aligned} \tag{6.2}$$

where R is risk, the potential for loss or harm, calculated as the combination of the probability and consequence of an adverse event; T is threat, any event that can create potential damage to an asset or population; in terms of terrorism risk, threat is based on the analysis of the capability of a terrorist to launch actions injurious to an infrastructure or public health; V is vulnerability, any weakness in an infrastructure's design, implementation, or operation that can be attacked by a terrorist to generate disaster; in this risk analysis, vulnerabilities that can cause potential damage or catastrophe are usually determined and identified; C is consequence or the outcome of an event occurrence.

In this method, resilience is the central purpose and not part of the risk formulation. Resilience is broadly defined as the ability to operate or survive through an attack or disaster and its ability to return to its full function.

For the asset owner, the level of resilience for a particular threat is expressed as (ASME-ITI, 2009)

$$\text{resilience}_{(\text{owner})} = \text{lost revenue} \times \text{vulnerability} \times \text{threat} \tag{6.3a}$$

For the community, the level of resilience for a particular threat is expressed as

$$\text{resilience}_{(\text{community})} = \text{lost economic activity in the community} \times \text{vulnerability} \times \text{threat} \tag{6.3b}$$

The lost revenue is the product of the duration of service (in days), the extent of service denial (in units of service denied per day), and the price per unit (in dollars, estimated at present levels), which are all essential parts of estimating the owner's financial loss:

$$\text{loss revenue} = \text{duration of denial} \times \text{severity of denial} \times \text{price per unit} \tag{6.3c}$$

The lost economic activity in the community is the level of decrease in output to direct customers and the indirect losses (multiplier effect) throughout the economy of a given region due to denial of service. Practical examples and assessment of this method are presented in Chapter 7.

6.2.7 CARVER Matrix

CARVER (acronym for *criticality, accessibility, recuperability, vulnerability, effect,* and *recognizability*) is a decision tool used for rating the relative desirability of potential targets and for properly allocating attack resources. For every potential target, a lowest value of 1 to a highest value of 5 is assigned for each CARVER factor, thereby creating a CARVER matrix. Then, by totaling the six CARVER values, a total score for each target can be calculated, and the scores represent the targets' relative prioritization. Moreover, the higher the CARVER score is, the more significant a target becomes. The CARVER matrix presented in this book is designed and refined for water infrastructure (including dams, aqueducts, and reservoirs) qualitative risk and vulnerability assessment based on multiple experts and government agencies.

6.2.7.1 Criticality

Criticality is a target value and is the main aspect in targeting. A target is considered *critical* when the magnitude of the destruction of the target has a potential effect on political or economic operations, or any operations of safety and security. The value of a target will change depending on the condition, requiring the use of time-sensitive methods to respond to changing conditions. For instance, when an area has

Table 6.1 Assigning Criticality Values

Criteria	*Scale*
Immediate termination in outcome; target cannot function without it	9–10
Loss would reduce mission performance considerably, or two-thirds reduction in outcome	7–8
Loss would reduce mission performance, or one-third reduction in outcome	5–6
Loss may reduce mission performance, or 10% reduction in outcome	3–4
No significant effect on outcome	1–2

few water supply systems (e.g., reservoirs and aqueducts), water supply tanks may be less critical as targets; however, safeguarding water supply tanks may be critical to maneuvering conventional forces, which requires the use of the water supply in the tanks (e.g., emergency response fire protection). Criticality depends on several factors: (1) time, which is crucial in evaluating the rapidness of the impact on operations of the destruction of a target; (2) the magnitude of outcome due to target destruction; (3) the presence of substitutes for the outcome product; and (4) perspective or relativity, which is important in determining the number of targets and in evaluating their conditions. Table 6.1 shows how criticality values are assigned on CARVER matrixes.

6.2.7.2 Accessibility

A target is accessible when terrorists can physically intrude upon the target or if the target can be hit by direct or indirect methods. Accessibility varies with the intrusion/exit, survival and escape potential of the target zone, the security situation, and the need for barrier penetration. The four basic steps identifying accessibility are as follows: (1) intrusion from the staging base to the target zone; (2) movement from the point of entry to the target; (3) mobility to the target's critical object; and (4) the ability of the terrorist to escape. The use of obstacle equipment and methods should always be considered when evaluating accessibility. The ability of the terrorists to survive is not usually associated with a target's accessibility. The factors considered when evaluating accessibility include, but are not limited to, the following: (1) advance warning systems; (2) detection devices; (3) defense capabilities within the target zone; (4) transportation systems; (5) terrain and location; (6) concealment; (7) population density; (8) barriers; (9) weather conditions; and (10) roadways.

It is crucial to measure the time it could take for the terrorists to penetrate barriers along each way based on the relative ease/difficulty of movement, and the likelihood of detection. Hence, the use of standoff weapons should be incorporated in the assessment. Table 6.2 shows how accessibility values are assigned on CARVER matrixes.

Table 6.2 Assigning Accessibility Values

Criteria	*Scale*
Easily accessible, standoff weapons can be used, or away from security	9–10
Inside a perimeter fence but outdoors or easily accessible outside	7–8
Easily accessible, inside a building or a structure but on ground level floor	5–6
Difficult to gain access, inside a building or a structure but on the top floor or in the basement; climbing or crawling required	3–4
Not accessible (very difficult to gain access)	1–2

Table 6.3 Assigning Recuperability Values

Criteria	*Scale*
Extremely difficult to replace or recovery requires 1 year or more	9–10
Difficult to replace or recover in less than a year (<1 year)	7–8
Can be replaced or recovered in a relatively short time (months)	5–6
Easily replaced or recovered in a short time (weeks)	3–4
Easily replaced or recovered in a short time (days)	1–2

6.2.7.3 Recuperability

Recuperability is the period needed to recover or circumvent the destruction inflicted on the target. It varies with the sources and type of targeted components and the availability of spare parts. Table 6.3 shows how recuperability values are assigned on CARVER matrixes.

6.2.7.4 Vulnerability

Vulnerability is a measure of the ability of the terrorists to destroy the target object, and the scale of the critical component needs to be compared with the ability of the terrorists to destroy. Primarily, the terrorists may tend to choose particular targets and cause permanent damage and maximize effects through the use of weapons, resulting in destruction of the targets. Table 6.4 shows how vulnerability values are assigned on CARVER matrixes.

6.2.7.5 Effect

The effect of an asset attack is a measure of possible security, political, environmental, and sociological impacts. The type and intensity of consequences will help

Table 6.4 Assigning Vulnerability Values

Criteria	*Scale*
Vulnerable to long-range target designation; special operations forces definitely have the knowledge and expertise to attack	9–10
Vulnerable to light weapons; special operations forces probably have the knowledge and expertise to attack	7–8
Vulnerable to medium weapons; special operations forces may have the knowledge and expertise to attack	5–6
Vulnerable to special weapons; special operations forces probably have no impact	3–4
Invulnerable to all but the most extreme targeting measures; special operations forces do not have much capability to attack	1–2

Table 6.5 Assigning Effect Values

Criteria	*Scale*
Overwhelming positive effects; no significant negative effects, favorable sociological impact	9–10
Moderate positive effects; few significant negative effects	7–8
No significant effects; neutral; some adverse impact on public and environment	5–6
Moderate negative effects; few significant positive effects; adverse impact on public and environment	3–4
Overwhelming negative effects; no significant positive effects; assured adverse impact on public and environment	1–2

analysts and decision makers select targets and target components for attack. Usually, the effect of a target attack includes the effect on the local population, but now effects also include (1) the triggering of countermeasures, (2) economic aftershocks, (3) national panic and chaos, and (4) collateral damage to other targets. Table 6.5 shows how effect values are assigned on CARVER matrixes.

6.2.7.6 Recognizability

An asset's recognizability is the level to which it can be identified and perceived by intelligence, survey, and exploration.

Table 6.6 Assigning Recognizability Values

Criteria	*Scale*
The asset is clearly recognizable under all conditions and from a distance; it requires very minimal or no training to attack.	9–10
The asset is easily recognizable, which requires a very minimal amount of training for recognition.	7–8
The asset is difficult to recognize at night or in bad weather conditions, or might be confused with other assets or target components.	5–6
The asset is difficult to recognize at night or in bad weather conditions; it is easily confused with other assets or asset components; it requires extensive training for recognition, as it is hard to recognize, with the potential for confusion.	3–4
The asset cannot be recognized under any conditions, except by experts.	1–2

Other factors that influence recognizability include the size, popularity, and complexity of the asset; the existence of distinctive asset signatures; and the technical sophistication and training of the terrorists (e.g., technical expertise to destroy Hoover Dam or Glen Canyon Dam). Table 6.6 shows how recognizability values are assigned on CARVER matrixes.

An individual target is evaluated for each CARVER factor by entering the appropriate value into the matrix. Once all the potential targets are evaluated, values for each potential target are added. The summation represents the relative desirability of each potential target, constituting a prioritized list of targets. Those targets with the highest totals are attacked first.

6.2.8 CARVER Plus Shock

The CARVER plus Shock method assesses the vulnerabilities in the food sector (USDA 2010) and can be used in water infrastructure. It helps a decision maker to think like a terrorist and identify the most attractive targets for attacks as detailed in Section 6.2.7. Terrorists attempt to achieve strong emotional responses and cognitive reactions from their target audience. CARVER has six attributes, as discussed in Section 6.2.7, for evaluating the attractiveness of a target. The modified CARVER tool assesses a seventh attribute, the combined health, environmental, economic, and cognitive reactions, and psychological impact, or the Shock attributes of a target. Table 6.7 shows how Shock values are assigned on CARVER plus Shock.

Once the ranking on each of the attribute scales has been quantified for a given node, the ranking on all the scales is summed to give an overall value for that node.

Table 6.7 Assigning Shock Values

Criteria	*Scale*
Asset has major historical, cultural, religious, or other symbolic importance; loss of over 5,000 lives; national economic impact of more than $100 billion	9–10
Asset has high historical, cultural, religious, or other symbolic importance; loss of between 500 and 5,000 lives; national economic impact between $50 and $100 billion	7–8
Asset has moderate historical, cultural, religious, or other symbolic importance; loss of between 150 and 500 lives; national economic impact between $5 and $50 billion	5–6
Asset has little historical, cultural, religious, or other symbolic importance; loss of less than 150 lives; national economic impact between $50 million and $5 billion	3–4
Asset has no historical, cultural, religious, or other symbolic importance; loss of less than 50 lives; national economic impact less than $50 million	1–2

Source: Derived from U.S. Department of Agriculture, http://fda.gov/Food/FoodDefense/CARVER/default.htm. 2010.

The nodes with the highest overall rating have the highest probable risk and should be the focus for the protective measures. Practical examples and assessment of this tool are presented in Chapter 7.

6.2.9 Freight Assessment System

The Freight Assessment System (FAS) is designed to minimize the risks of potential terror threats to the nation and is associated with the estimated 50 billion pounds of domestic cargo transported annually by air carriers. Congress directed the Transportation Security Administration (TSA) in 2004 to develop a system to identify and target increased-risk cargo, and recently mandated the TSA to implement security plans to support a requirement of 50% screening of all cargo by February 2009, and 100% screening of all cargo by August 2010. This new policy is called the *Certified Cargo Screening Program*, which is designed to connect via portals to TSA legacy and future systems to enable data transfer and receive risk-based data. Additionally, FAS will eventually share crucial information with other modes of transportation: rail, highway, and motor carriers.

System-Based Risk Management Asset Assessment is used in the FAS, utilizing an analytical approach that seeks to develop technology and policy for preventive measures to reduce the risks to those assets that are critical to the sector's strategic risk objectives. Assessment is similar to other risk assessments in that it estimates the chances of a specific set of events occurring and/or their potential consequences. Risk assessments carry a range of interpretations that vary within industries. Also, the fundamental understanding of what properly constitutes the risk assessment process can vary. In the context of homeland security, risk assessments typically focus on threats, vulnerabilities, and consequences (TVC).

$$\text{relative risk} = f(\text{threat, vulnerability, consequence}) \tag{6.4}$$
$$\text{likelihood of a successful attack}$$

Separate analyses are associated with each term (e.g., threat analysis and vulnerability analysis). A set of activities represent the TVC analyses and are input into a resulting risk assessment model. The output of a risk assessment model provides a relative scoring, either qualitative or quantitative, for the assets under study. This risk assessment enables the development of outcome-focused countermeasures designed to reduce the overall risk to the assets under study.

6.2.10 Federal Emergency Management Agency HAZUS-MH

HAZUS (HAZards United States) is a geographic information system–based natural hazard loss assessment model created by FEMA. The current version is HAZUS-MH MR4, where MH stands for *multi-hazards*. Currently, HAZUS can model floods, hurricanes, and earthquakes. Also, it is used for mitigation and recovery as well as preparedness and response (FEMA 2010). Moreover, this model can also be applied in terrorism risk assessment under special conditions. First, it quantitatively defines the exposure of a selected area; second, it systematically characterizes the magnitude of the hazard affecting the exposed area; and third, it exploits and utilizes the exposed area and the hazard to quantify the potential losses.

6.2.11 Chemical Security Assessment Tool

The DHS Chemical Security Assessment Tool (CSAT) was created to identify facilities that have a high level of risk. After the initial registration for the process, a series of questions, called "Top Screen," is used to determine if the facility should be regulated by the Chemical Facility Anti-Terrorism Standards. Top Screen results define and qualify the risk level and appropriate responsibilities. Screening threshold quantities (STQs—quantities that trigger Top Screen requirements) for chemicals in these categories tend to be quite high (5,000 pounds or greater), with the exception of chemicals such as phosgene, whose threshold is 500 pounds. Facilities in this category may already be covered under the EPA's Risk Management Program.

6.2.11.1 Chemical Weapons/Chemical Weapon Precursors

According to DHS, certain Chemical Weapon Convention (CWC) toxic chemicals and precursors (e.g., chlorosarin and methylphosphonyldifluoride) have a very low threshold of 100 grams, cumulative. Other CWC hazardous chemicals and precursors have thresholds between 2.2 and 220 pounds.

6.2.11.2 Chemicals That Qualify as a Weapon of Mass Effect

This category includes certain chemicals of interest classified as DOT Division 2.3 (poisonous gases), Zone A or Zone B. The thresholds for Zone A and B chemicals are 15 and 45 pounds, respectively.

6.2.11.3 Chemicals That Qualify as an Improvised Explosive Device

This includes certain chemicals that may be used as or developed into an improvised explosive device. This category includes nitric acid (typically used for nitration) with a threshold of 400 pounds and hydrogen peroxide with a threshold of 400 pounds.

6.2.11.4 Sabotage or Contamination of Chemicals

The sabotage or contamination of chemicals, if mixed with other chemicals, has the potential to create potential catastrophic consequences. This category includes chemicals that can produce a poisonous gas when mixed with water (DOT Division 4.3, water-reactive materials).

6.2.11.5 Mission-Critical Chemicals

The term *mission-critical chemicals* applies to facilities that supply 20% or more of the domestic production of any chemical to infrastructure sectors.

6.2.12 Automated Targeting System

The Automated Targeting System (ATS) is a DHS-computerized system that, for every person and cargo that crosses U.S. borders, scrutinizes a large volume of data related to the designated individual and assigns a rating based on which the person may be considered in a risk group of terrorists or other criminals. Currently, ATS consists of six modules that focus on exports, imports, passengers and crew (airline passengers and crew on international flights, passengers and crew on sea carriers), private vehicles crossing land borders, and import trends over time (DHS 2006, 2007). ATS is consistent in its evaluation of risk associated with individuals and is used to support the overall Custom and Border Protection (CBP) law enforcement mission.

6.2.12.1 ATS-Inbound

ATS-Inbound is the primary decision support tool for inbound targeting of cargo and is available to CBP officers at all major ports (air/land/sea/rail) throughout the United States; it assists CBP personnel in the Container Security Initiative decision-making process (DHS 2006, 2007).

6.2.12.2 ATS-Outbound

ATS-Outbound is the outbound cargo–targeting of exports that pose a high risk of containing hazardous chemicals (e.g., narcotics, illegal goods, or possible materials used for WMD) and Federal Aviation Administration violations. It utilizes Export Declaration data from CBP's automated export system.

6.2.12.3 ATS-Passenger

ATS-Passenger is used at all U.S. airports and seaports receiving international flights and ships to evaluate and screen passengers and crew that could be a potential risk for creating danger to homeland or violation of U.S. law.

6.2.12.4 ATS-Land

ATS-Land (ATS-L) provides risk assessment of private passenger vehicles crossing the United States borders for security and screening by inspection of license plate numbers of vehicles. ATS-L permits CBP officers to compare information evaluate the assessment, and cross-reference the treasury enforcement communications system (TECS) crossing data, TECS seizure data, and state Department of Motor Vehicle data while using weighted rule sets to provide risk scores.

6.2.12.5 ATS-International

ATS-International provides foreign customs authorities with controlled access to automated cargo targeting capabilities (DHS 2006) and provides critical collaboration to other countries to enhance the security of international supply chains and increase protection to avoid disruption by terrorists.

6.2.12.6 ATS-Trend Analysis and Analytical Selectivity

ATS-Trend Analysis and Analytical Selectivity allow CBP to examine, trace, identify and target action violators of U.S. laws regarding international trade. The trend analysis functions thoroughly review historical statistics that provide an overview of trade activity and can support in determining illegal trade activity.

6.2.13 Risk Lexicon

The DHS Risk Lexicon provides a set of terms for use by the homeland security risk community and represents an important milestone in building a unified approach to homeland security risk management and enabling integrated risk management for the department (DHS 2008a). It is developed for the following reasons among others: (1) the promulgation of a common language to create uniformity and unity in communications between the DHS and its partners, (2) support for exchanging data and information essential to interoperability among risk practitioners, and (3) the improvement of relationships and credibility by providing consistency in the usage of terms by the risk community across DHS and its components.

6.2.14 Microbial Risk Assessment Framework

The EPA Microbial Risk Assessment (MRA) framework has two methods of estimating risk: chemical risk assessment, and ecological risk assessment.

6.2.14.1 Chemical Risk Assessment

Chemicals in foods were first addressed in a public policy framework in the mid-1800s, when they were viewed as adulterants. In the early twentieth century, chemical food safety and toxicity were major issues; these chemicals were later regulated as if a "safe threshold" existed (Hutt 1997; NRC 1983). As animal testing produced more data, the method of chemical risk assessment led to intense debates and interrogation.

Chemical risk assessment relies on static modeling techniques, which cannot represent dynamic processes such as disease transmission, and on the assumption that each exposure is an independent event (USEPA 2008b). Infection is a function of dose and, rather than mortality, may be the health outcome of concern (USEPA 2008b). When secondary transmission is involved, dynamic modeling and population scale measures of risk are crucial (Eisenberg et al. 2006).

Some of the problems that researchers have identified as not addressed by the Red Book paradigm for microorganisms include the following:

1. Microorganisms can propagate, evolve into different life stages, and die off.
2. Virulence varies during a pathogen's life cycle and between different pathogen strains.
3. Pathogens behave differently under particular conditions in terms of temperature, environment, and time.
4. Microbial pathogens are not evenly distributed in the environment, which presents very asymmetrical likelihood of exposure.
5. Person-to-person transmission occurs in many infectious diseases.
6. Attack rates and infection rates differ.

Although the MRA steps are often the same as those for chemical risk assessments, the emphasis, elements, and conduct are different in various stages of MRA due to the dynamic nature of the agents and population (ECFS 1997). In particular, changes in the population's immunity and susceptibility status are not considered in the traditional chemical risk assessments.

6.2.14.2 Ecological Risk Assessment

During the 1980s and early 1990s, ecological risk assessment evolved at the EPA (USEPA 1992). Although ecologic analysts were informed by the chemical risk assessment approach, they recognized that as a paradigm, it did not entirely fit their needs. Ecological risk assessment demands an extensive, systems-oriented context, which requires analysts with different backgrounds and experiences. As a result, lack of a clearly defined problem statement makes ecological risk assessments difficult and inefficient. Because the Red Book did not include a problem formulation step, ecological risk assessors created one and explained how to implement it (USEPA 1992).

6.2.14.3 MRA for Drinking Water

Numerous studies on drinking water exposure to pathogens have been conducted (e.g., Eisenberg et al. 2005; Haas et al. 1993). The typical sequence of MRA exposure assessment includes the following: (1) source and cause of contamination, (2) source of water, (3) treatment technology, (4) storage and distribution system, (5) customer's piping system, plumbing and point-of-use devices, (6) tap water, and (7) consumption/exposure. Uniquely, this MRA separates exposure and effects assessment into two distinct processes. The exposure assessment is the method of detecting biological threats in the water supply system for determining the recovery efficiency, viability/infectivity and specificity, and monitoring untreated or treated drinking water. The effects assessment is the process of the dose–response analysis based on human application studies; hit theory for infection; pathogen dynamics; host dynamics; and the spectrum of health consequences.

6.2.14.4 MRA for Wastewater

Several MRA have modeled the risks associated with water reuse applications. In these evaluations, exposure assessment has included examination of the (1) source and cause of contamination, (2) fate and transport of contaminants, (3) treatment technology, (4) post-treatment storage, (5) distribution and use, and (6) consumption.

6.2.14.5 A Need to Improve MRA

There are many requirements for new research and technologies to improve MRA precision and accuracy. Data are crucially needed for several components of MRA paradigms; such research is time consuming and requires resources. With comprehensive paradigms to help risk assessors or analysts identify and consider the potential factors involved in biological threat–related illness, MRA will become increasingly successful in providing assessment and information effectively to protect the public from health hazards.

6.2.15 Pareto Principle (80–20 Rule)

The Pareto Principle, commonly known as the 80–20 rule, is used to compare the results with other methods of risk rate determination and to verify whether this principle can be applied to the *event tree analysis*. According to the Pareto Principle, in today's context, 80% of car accidents might be caused by 20% of car drivers, or 80% of the traffic pollution is produced by 20% of the vehicles. The Pareto Principle is widely used in engineering risk assessment and project management, i.e., used in the analysis of running a project within the budget (Doro-on 2009). The Pareto Principle will be utilized to illustrate a comparison of the results of an engineering judgment with an engineering judgment based on the Pareto Principle. The original observation was with regard to income and wealth. Pareto noticed that 80% of Italy's wealth was owned by 20% of the population. He then performed surveys on other countries and discovered that a similar outcome and distribution applied.

6.2.16 Sandia National Laboratories Security Risk Assessment Methods

A risk assessment method has been refined by Sandia National Laboratories to assess risk at various types of facilities and critical infrastructures. The method is based on the risk equation provided by Garcia (2008) in her book, *The Design and Evaluation of Physical Protection Systems*:

$$\text{risk} = P_A \times (1 - P_E) \times C \tag{6.5}$$

where P_A is the likelihood of adversary attack, P_E is security system effectiveness, $1 - P_E$ is adversary success, and C is consequence of loss of the asset.

The primary step in this risk assessment method is the characterization of the facility, which includes identification of the undesired events and the respective critical assets. Guidelines for determining terror threats and for using the definition of the threat to estimate the probability of adversary attack against the facility is included. Hence, relative values of consequence and the effectiveness of the security system against the adversary attack are estimated.

6.2.17 Security Vulnerability Assessment Method

The first step in the process of estimating risks is to identify and analyze the threats and the vulnerabilities facing a facility by an SVA. The SVA is a systematic process that evaluates the likelihood that a threat against a facility will be successful (API-NPRA 2004). The objective of conducting an SVA is to identify security hazards, threats, and vulnerabilities facing a facility and to evaluate the countermeasures for protecting the public, workers, national interests, the environment, and the facility (API-NPRA 2004). The basic approaches to estimate the potential risk are (1) deter, (2) detect, (3) delay, (4) deny, (5) defeat, and (5) respond. Appropriate approaches for managing security vastly depend on the individual characteristics of the facility, including the type of facility and the threats facing the facility. Accordingly, in the SVA process, risk is a function of

1. Consequences of a successful attack against a facility.
2. Likelihood of a successful attack against a facility.

Likelihood is a function of (1) the attractiveness for the potential attack, (2) the magnitude of the consequence, and (3) the degree of vulnerability of the asset. The SVA process does not recommend preventive measures but provides analysis and estimation of vulnerabilities.

6.2.18 ASME RA-S Probabilistic Risk Assessment

ASME RA-S is the American Society of Mechanical Engineers Probabilistic Risk Assessment Standard for non-light water reactor nuclear power plant applications. The Probabilistic Risk Assessment (PRA) is a mainstream regulatory tool that contributes to the decision-making process for plant design, operation, and maintenance. Hence, the principle and standard of PRA can be applied in the PRA process for water infrastructure risk analysis.

In March 1999, the General Accounting Office (GAO) issued GAO/RCED-99-95, "Nuclear Regulation: Strategy Needed to Regulate Safety Using Information on Risk." GAO pointed out that it is important to "develop standards on the scope and detail of risk assessments needed for utilities to determine that changes to their plants' design will not negatively affect safety." The standard establishes requirements for a PRA ranging from a limited-scope to a "full-scope" PRA. The meaning of "full scope" includes but is not limited to (1) sources of radioactive material (or other vital material relating to water infrastructure) both within and outside the reactor core or system; (2) a full set of plant-operating states covering all anticipated operating and shutdown modes; (3) a full set of initiating events, such as fires and floods, seismic events and transportation accidents; (4) a definition of event sequences to a level that is necessary and sufficient to characterize mechanistic source terms and offsite radiological and chemical consequences to public health and safety; and (5) a quantification of the event sequence frequencies, mechanistic source terms, offsite radiological and chemical consequences, risk, and associated uncertainties, and using this information consistent with the scope of PRA. Currently, ASME RA-S does not cover accidents resulting from acts of terrorism. However, the PRA standard procedure can be used as a guideline in risk assessment for water infrastructure protection from terror threats because the degree of consequences generated from a catastrophic nuclear power plant accident is almost comparable to the magnitude of consequences from a terrorist attack (e.g. the psychological response created by a nuclear power plant accident is almost the same as terrorism).

6.2.19 Checkup Program for Small Systems

Checkup Program for Small Systems (CUPSS) is an infrastructure management tool for small drinking water and wastewater utilities. CUPSS provides a straightforward systematic approach based on EPA's Simple Tools for Effective Performance Guide series. The EPA's Office of Groundwater and Drinking Water created CUPSS with assistance and support from state agencies, some technical groups, EPA regional offices, and small wastewater and water supply utilities. With this collaborative approach, the EPA was able to develop tools for implementation of an asset management program and develop effective asset management plans (USEPA 2010a,d).

6.2.20 Water Health and Economic Analysis Tool

WHEAT was developed by the EPA to be compatible with water sector risk assessment methods such as RAMCAP (USEPA 2010c,d). According to the EPA, the tool is intended for drinking water utility owners and operators to quantitatively and qualitatively assesses public health impacts, utility financial costs, and regional economic impacts of an adverse event such as terrorist attack.

6.2.21 Water Contaminant Information Tool

The Water Contaminant Information Tool (WCIT) helps in protecting drinking water systems and wastewater systems from the effects of intentional or accidental contamination (USEPA 2010b). It also provides information and guidelines for water treatment utilities, public health officials, and agencies responsible for the protection of water supplies from contamination. Introduced in late 2005, WCIT is a password-protected online database containing information on 93 contaminants of concern: chemical, biological, and radiological substances that pose a serious threat if introduced into drinking water systems or wastewater systems (USEPA 2010b,d). Unlike other databases, WCIT supports water-specific data and covers both regulated and non-regulated contaminants, including data on other chemical threats used for terrorism.

6.3 Historical Perspective of Prospect Theory

Prospect theory is a theory of decision making under risk conditions. Decisions are based on judgments under conditions of uncertainty, when it is difficult to *foresee* or predict the consequences of events with clarity. Also, prospect theory directly addresses how these preferences are framed and assessed in the decision-making development. Daniel Bernoulli was the first to introduce the concept of systematic bias in decision making based on a "psychophysical" model. Bernoulli used a coin toss game known as *St. Petersburg's Paradox* to demonstrate the limitations of expected value as a normative decision rule, which led him to analyze *utility function* to explain people's choice of behavior (Kahneman and Tversky 1979). He assumed that people tried to maximize their utility and not their expected value. Bernoulli's function proposed that utility was not merely a linear function of wealth but rather a subjective evaluation of outcome (McDermott 1998). The concave shape of the function introduced the idea of declining minor utility, through which changes away from the starting point have less impact than those that are closer. For instance, Bernoulli's utility function argues that $1 is a lot compared with nothing; people will, therefore, be reluctant to part with this dollar. However, to most people, $101 is not significantly different from $100. Because Bernoulli's concave utility function assumed that increments in utility decreased with increasing wealth, the expected utility model implicitly assumed risk aversion (McDermott 1998). Thus, prospect theory is based on psychophysical models, such as those that originally inspired Bernoulli's expected value proposition. Tversky and Kahneman (1979) applied psychophysical principles to investigate and examine judgment and decision analysis. People are not conscious of how the brain interprets vision into prospect. People make decisions based on how their brains comprehend facts or information and not exclusively on the basic utility that a certain choice obtains for a decision maker.

6.3.1 Expected Utility Theory

According to Kahneman and Tversky (1979), decision making under risk can be observed as a preference between prospects or gambles. A prospect $(x_1, p_1; \dots; x_n, p_n)$ is a contract that yields outcome x_i with probability p_i, where $p_1+p_2+\dots+p_n = 1$. To simplify notation, we omit null outcomes and use (x, p) to denote the prospect $(x, p; 0, 1-p)$ that yields x with probability p, and 0 with probability $1-p$. The (riskless) prospect that yields x with certainty is denoted by (x) (Kahneman and Tversky 1979).

6.3.2 Classical Prospect Theory

According to Kahneman and Tversky (1979), the *classical prospect theory* distinguishes two phases in the choice process: *framing* and *valuation*. In the *framing* stage, the decision maker develops a representation of the consequences that are crucial to the decision. In the *valuation* stage, the decision maker assesses the value of each prospect and chooses systematically. Then, the decision maker is considered to evaluate each of the refined prospects, and to select the prospect of highest value. The overall value of a refined prospect, designated as V, is expressed in terms of two scales, π and v. The first scale, π, associates a decision weight $\pi(p)$ with each probability, p, which reveals the influence of p on the overall value of the prospect.

6.4 Cumulative Prospect Theory

Prospect theory was modified by Kahneman and Tversky (2000) into cumulative prospect theory. The five foremost *phenomena of choice*, as detailed in Sections 6.4.1 to 6.4.5, which violated the standard model and presented a basic challenge, must be met by any adequate descriptive theory of choice as pointed out by Kahneman and Tversky.

6.4.1 Framing Effects

The rational theory of choice assumes description invariance: equivalent formulations of a choice problem should give rise to the same preference order (Arrow 1982). Contrary to this assumption, there is much evidence that variations in the framing of options yield systematically different preferences (Tversky and Kahneman 1986).

6.4.2 Nonlinear Preferences

According to the expectation principle, the utility of a risky prospect is linear in outcome probabilities. Allais's (1953) famous example challenged this principle by showing that the difference between probabilities of 0.99 and 1.00 has more impact on preferences than the difference between 0.10 and 0.11. More recent studies observed nonlinear preferences in choices that do not involve sure things (Camerer and Ho 1991).

6.4.3 Source Dependence

People are eager to bet on an uncertain event based on the magnitude of uncertainty and on its source or cause. More recent evidence indicates that people often prefer to bet on an event, although the former probability is vague and the latter is clear (Heath and Tversky 1991).

6.4.4 Risk Seeking

Risk aversion is primarily assumed in economic analysis of decisions under uncertainty.

6.4.5 Loss Aversion

One of the basic phenomena of choice under both risk and uncertainty is that losses loom larger than gains (Kahneman and Tversky 1991).

6.5 Advances in Prospect Theory

The new theory explains loss aversion, risk seeking, and nonlinear preferences in terms of the value and the weighting functions (Kahneman and Tversky 2000).

In cumulative prospect theory as characterized by Kahneman and Tversky (2000), S is a definite set of states of nature (subsets of S are called "events"). X is a set of consequences or "outcomes." Assume X is a neutral outcome, denoted as 0. All other elements of X are gains (+) or losses (–). A prospect f is then represented as a sequence of pairs (x_i, A_i), which yields x_i if A_i occurs, where $x_i > x_j$ iff $i > j$ and (A_i) is a partition of S. The positive part of f, denoted as f^+, is obtained by

$$f^+(s) = f(s) \quad \text{if} \quad f(s) > 0$$

$$f^+(s) = 0 \quad \text{if} \quad f(s) \leq 0$$

The negative part of f, denoted as f^-, is defined similarly.

Based on Kahneman and Tversky (2000), there exists an increasing value function, $v: X \rightarrow \text{Re}$, satisfying $v(x_0) = v(0) = 0$, and capacities W^+ and W^-, such that for $f = (x_i, A_i), -m \leq i \leq n$,

$$V(f) = V(f^+) + V(f^-) \tag{6.6}$$

$$V(f^+) = \sum_{i=0}^{n} \left(\pi_i^+ v(x_i)\right) \tag{6.6a}$$

$$V(f^-) = \sum_{i=m}^{n} \left(\pi_i^- v(x_i)\right) \tag{6.6b}$$

where the decision weights $\pi^+(f^+) = (\pi_0^+,\ldots,\pi_n^+)$ and $\pi^-(f^-) = (\pi_{-m}^-,\ldots,\pi_0^+)$ are defined by

$$\pi_n^+ = W^+(A_n),\ \pi_{-m}^- = W^-(A_{-m})$$

$$\pi_i^+ = W^+(A_i \cup \ldots \cup A_n), - W^+(A_{i+1} \cup \ldots \cup A_n)$$

Letting $\pi_i = \pi_i^+$ if $i \geq 0$ and $\pi_i = \pi_i^-$ if $i < 0$, Equations 6.6a and 6.6b reduce to

$$V(f^+) = \sum_{i=-m}^{n} \left(\pi_i^+ v(x_i)\right) \tag{6.6c}$$

The decision weight π_i^+, in relation to a positive outcome, is the difference between the capacities of the events: "the outcome is at least as good as x_i" and "the outcome is at least as bad as x_i" (Kahneman and Tversky 2000). The decision weight π_i^-, associated with negative outcome, is at least as bad as x_i and is strictly worse than x_i (Kahneman and Tversky 2000). Thus, the decision weight associated with an outcome can be deduced as W^+ and W^-. It follows readily from the definitions of π and W that for both positive and negative prospects, the decision weights add to 1 and for mixed prospects, the sum can be either lesser or greater than 1 as defined by individual weights (Kahneman and Tversky 2000):

$$\begin{aligned}
&\pi_n^+ = \omega^+(p_n),\ \ \pi_{-m}^- = \omega^-(p_m) \\
&\pi_i^+ = \omega^+(p_n + \ldots + p_n) - \omega^+(p_{i+1} + \ldots + p_n), \quad 0 \leq i \leq n-1 \\
&\pi_i^- = \omega^-(p_{-m} + \ldots + p_i) - \omega^-(p_{-m} + \ldots + p_{i-1}), \quad 1 - m \leq i \leq 0
\end{aligned}$$

where ω^+ and ω^- are increasing functions from the unit interval, $\omega^+(0) = \omega^-(0) = 0$, and $\omega^+(1) = \omega^-(1) = 1$.

According to Kahneman and Tversky (2000), cumulative prospect theory broadens the original theory as follows:

1. It applies to any controlled or limited prospect and it can be extended to continuous distributions.
2. It applies to both probabilistic and uncertain prospects and can coordinate some form of source dependence.
3. The enhanced theory allows different decision weights for gains and losses thereby simplifying the initial version that assumes $\omega^+ = \omega^-$.
4. Consequently, the cumulative prospect theory presented herein will be integrated into risk acceptability analysis as detailed in Chapters 9 and 11.

6.6 A Need for Risk Acceptability Analysis

This chapter adequately presented the introductory information about the qualitative/quantitative methods and tools for risk and vulnerability analysis recommended by renowned authors, experts, and governmental agencies. Some of the most standardized models will be further discussed and presented in detail in Chapter 7 with illustrative examples. All the models presented in this chapter are adequate in qualitatively and quantitatively identifying the risk and vulnerability for the components of a specific sector such as the water sector. Most of these risk and vulnerability models concluded that the *terror risk* against U.S. infrastructure is *unacceptable* to society. Chapters 8, 9, and 11 prove to the readers that *risk acceptability* can be achieved.

References

Allais, M. 1953. Le comportement de l'homme rationnel devant le risqué, critique des postulats et axioms de l'econle Amercaine. *Econometrica* 21:503–46.

American Society of Mechanical Engineers Technologies, Committee on Nuclear Risk Management. 2008. Standard for Probabilistic Risk Assessment for Advanced Non-LWR Nuclear Power Plant Applications. http://cstools.asme.org/csconnect/pdf/CommitteeFiles/27523.pdf (accessed June 2, 2010).

American Society of Mechanical Engineers Technologies Institute, LLC. 2009. *All-Hazards Risk and Resilience Prioritizing Critical Infrastructure Using the RAMCAP Plus Approach.* New York: ASME.

American Petroleum Institute-National Petrochemical and Refiners Association. 2004. Security Vulnerability Assessment Methodology for the Petroleum and Petrochemical Industries, 2nd Edition. http://www.npra.org/docs/publications/newsletters/SVA_2nd_edition.pdf (accessed August 14, 2010).

Arrow, K. J. 1982. Risk perception in psychology and economics. *Econ Inquiry* 20:1–9.

Camerer, C. F., and T. H. Ho. 1991. *Nonlinear Weighting of Probabilities and Violations of the Betweeness Axiom.* Unpublished Manuscript. The Wharton School, University of Pennsylvania.

Doro-on, A. M. 2009. "Risk Assessment Embedded with Cumulative Prospect Theory for Terrorist Attacks on Aquifer of Karstic Limestone and Water Supply System," PhD diss. University of Texas at San Antonio. San Antonio, Texas.

Eisenberg J. N. S. 2006. Application of a Dynamic Model to Assess Microbial Health Risks Associated with Beneficial Uses of Biosolids: Final Report to the Water Environment Research Foundation. WERF Report 98-REM-1a. London: IWA Publishing.

Eisenberg, J. N., X. Lei, A. H. Hubbard, et al. 2005. The role of disease transmission and conferred immunity in outbreaks: Analysis of the 1993 *Cryptosporidium* outbreak in Milwaukee, Wisconsin. *Am J Epidemiol* 161(1):62–72.

European Commission on Food Safety. 1997. Opinion on principles for the development of risk assessment of microbial hazards under the hygiene of foodstuffs directive 93/43/Eec. Available online at http://europa.eu.int/comm/food/fs/sc/oldcomm7/out7_en.html.

Garcia, M. L. 2008. *The Design and Evaluation of Physical Protection Systems.* 2nd ed. Butterworth-Heinemann.

Haas, C. N., J. B. Rose, C. Gerba, et al. 1993. Risk assessment of virus in drinking water. *Risk Anal* 13:545–52.

Heath, C., and A. Tversky. 1991. Preference and belief: Ambiguity and competence in choice under uncertainty. *J Risk Uncertain* 4:5–28.

Hutt, P. B. 1997. Law and risk assessment in the United States. In: Molak, V., ed. *Fundamentals of risk analysis and risk management.* Boca Raton, Florida: CRC Lewis Publishers.

Kahneman, D., and A. Tversky. 1979. Prospect Theory: An analysis of decision under risk. *Econometrica* 47(2):263–92.

Kahneman, D., and A. Tversky. 2000. *Choices, Values and Frames.* New York: Russell Sage Foundation and Cambridge University Press.

Lewis, T. 2006. *Critical Infrastructure Protection in Homeland Security: Defending a Networked Nation.* John Wiley and Sons, Inc.

McDermott, R. 1998. *Risk Taking in International Politics-Prospect Theory in American Foreign Policy.* University of Michigan Press. http://press.umich.edu/pdf/0472108670.pdf (accessed December 5, 2007).

National Research Council. 1983. Risk assessment in the federal government: managing the process. Washington, DC: National Academy Press.

Tversky, A. 1967. Additivity, utility, and subjective probability. *J Math Psychol* 4:175–201.

Tversky, A., and D. Kahneman. 1974. Judgment under uncertainty: Heuristics and biases. *Science* 185:1124–31.

U.S. Department of Agriculture. 2010. CARVER Plus Shock Method. http://fda.gov/Food/FoodDefense/CARVER/default.htm (accessed October 1, 2010).

U.S. Department of Homeland Security. 2006. Privacy Impact Assessment for the Automated Targeting System. Custom and Borders Protection, ATS, November 22, 2006. http://dhs.gov/xlibrary/assets/privacy/privacy_pia_cbp_ats.pdf (accessed June 3, 2010).

U.S. Department of Homeland Security. 2007. Privacy Impact Assessment for the Automated Targeting System. Custom and Borders Protection, ATS, August 3, 2007. http://dhs.gov/xlibrary/assets/privacy/privacy_pia_cbp_ats.pdf (accessed June 3, 2010).

U.S. Department of Homeland Security. 2008a. *DHS Risk Lexicon, Risk Steering Committee.* Washington, DC.

U.S. Environmental Protection Agency. 2008b. Foundations and Frameworks for Human Microbial Risk Assessment. http://www.epa.gov/raf/files/epa_mra_fw_comparison_report_0609.pdf (accessed August 3, 2010).

U.S. Environmental Protection Agency. 2003. *EPA Needs to Assess the Quality of Vulnerability Assessments Related to the Security of the Nation's Water Supply.* Report No. 2003-M-0013. Washington, DC: September 24, 2003.

U.S. Environmental Protection Agency. 2010a. Checkup Programs for Small Systems. http://water.epa.gov/infrastructure/drinkingwater/pws/cupss/index.cfm (accessed August 22, 2010).

U.S. Environmental Protection Agency. 2010b. Water Contaminant Information Tool. http://water.epa.gov/scitech/datait/databases/wcit/index.cfm (accessed October 23, 2010).

U.S. Environmental Protection Agency. 2010c. Water Health and Economic Analysis Tool. http://yosemite.epa.gov/ow/SReg.nsf/6eb546814ef524b08525779f006a4b79/f85fc2b35fc3797f852577b500655494/$FILE/WHEAT%20User%20Manual.pdf (accessed October 23, 2010).

U.S. Environmental Protection Agency. 2010d. Water security- Tools and Technical Assistance. http://water.epa.gov/infrastructure/watersecurity/techtools/index.cfm (accessed October 22, 2010).

U.S. Federal Emergency Management Agency. 2010. HAZUS. http://fema.gov/plan/prevent/hazus/(accessed October 22, 2010).

Chapter 7

Standard Risk and Vulnerability Assessment

7.1 Introduction

This chapter presents the standard qualitative and quantitative risk and vulnerability assessment, operational formulations, and models for homeland critical infrastructure protection, especially for water supply systems, mostly used and recommended by prominent authors, the U.S. Department of Homeland Security (DHS), the U.S. Environmental Protection Agency (EPA), the Federal Emergency Management Agency (FEMA), the U.S. National Counterterrorism Center (NCTC), Department of Defense, National Security Council, Public Health Service, National Commission on Terrorism (NCT), and other governmental and state agencies. Examples and applications of these standard processes and models will be illustrated in this chapter.

7.2 Standard Homeland Security Risk Assessment and RAMCAP *Plus* Processes

The basic DHS risk assessment method, risk (R) = threat (T) × vulnerability (V) × consequence (C), introduced in Section 6.2.1, is widely used and recommended by experts and is used in conjunction with other vulnerability assessment processes such as Risk Analysis and Management for Critical Asset Protection (RAMCAP Plus; see Section 6.2.6). An illustrative example is provided below. Seven steps are used in the risk estimation: (1) asset classification, (2) threat definition, (3) consequence analysis,

(4) vulnerability analysis, (5) threat evaluation, (6) risk analysis, and (7) value of consequence. These seven steps of risk estimation for terrorist attacks against the aquifer and water supply system of the San Antonio metropolitan area are shown in Tables 7.1 through 7.7.

Consequence scales for fatalities, injuries, financial losses to owners or operators, and economic losses to the regional community have been developed based on the RAMCAP Plus process for the San Antonio, Texas area, which is dependent solely on the vulnerable aquifers of karstic limestone (Doro-on 2009). There are different methods for estimating the consequence scale for specific sectors, events, and situations or conditions for the designated area. It also depends on the political, pollution, environmental, and economic status of the area (e.g., community, city, or nation). The consequence scale is estimated on the basis of available official statistical data. If the community has scarce or limited statistical information, an estimate

Table 7.1 Step 1: Asset Characterization for San Antonio, Texas Water Infrastructure

(a) Edwards Aquifer "Sole source water supply" Recharge zones (sinkholes, faults, cracks, caves, springs, wells and dams) are unprotected.
(b) Twin Oaks Aquifer Storage and Recovery (ASR) Facility San Antonio Water System (SAWS) stores excess Edwards Aquifer drinking water during rainy times in a large-scale underground water storage facility in south Bexar County for use during dry south Texas summers (SAWS 2010). Above ground of ASR can be leased for public use. Adjacent properties are also open for public use.
(c) Dos Rios Wastewater Treatment & Water Recycling Plant Discharge point is not secured.
(d) Medio Creek Water Treatment Plant Secured with surveillance and fence.
(e) Bexar Metropolitan Water Treatment Secured with surveillance and fence.
(f) Twin Oaks Water Treatment Plant Used for treating Edwards before storing into the underground reservoir. Secured with surveillance and fence.
(g) Water tanks Unsecured.

Table 7.2 Step 2: Threat Characterization for San Antonio, Texas Water Infrastructure

(a) Inject poison (arsenic-based pesticide such as "fire ants control") to Edwards Aquifer via sinkholes, cracks, wells, and cave.
(b) Injection of poison (arsenic-based pesticide such as "fire ants control") into the underground reservoir (ASR facility).
(c) Attack security personnel, bomb water, and wastewater treatment facilities. Destroy the building structures within the facilities. Mainly destroy major water and wastewater (inlet/outlet) pipelines.
(d) Inject poison (arsenic-based pesticide such as "fire ants control") to water tanks located in major communities.

can be based on an area with a similar situation that has ample data. In addition, the available aerial photographic, topographic, or geographic information system (GIS) maps provided by state or local agencies and the U.S. Geological Survey can assist in preparing the consequence scale by including information specific damage, fatalities, injuries, people, equipment, and materials. In Table 7.3, the range of each bin increases by a factor of two over the next bin. The use of a scaling factor creates a logarithmic scale for scientific presentation purposes, in this case one at base 2. As will be seen later, the vulnerability scale presented in Table 7.4 uses a scale factor of two, enabling construction of a conditional risk table of consequence and vulnerability scales with the sum of their "bin numbers" being the logarithm of the conditional risk (ASME-ITI 2009). The bin numbers and vulnerability scale values presented in this book are solely designed and modified for water infrastructure based on the RAMCAP Plus process and DHS methods. This will produce a convenient, qualitative display of results since the conditional risk matrix will contain diagonal lines of constant risk.

7.2.1 Fatalities and Serious Injuries

In RAMCAP Plus, human safety and health consequences should be expressed in terms of the number of fatalities and the number of serious injuries that occur immediately as a result of disaster events (e.g., lost work time and disability).

7.2.2 Financial and Economic Impacts

Usually, economic and financial impacts are measurements of consequences in analyzing risks from terrorism and natural disasters. The owners and operators of the water infrastructure are responsible for maintaining the security of their facilities, reliability of their services, and financially sustainable operations. The general public served by the facility is normally represented by public

Table 7.3 Step 3: Consequence Analysis for San Antonio, Texas Water Infrastructure

(a) Financial and economic impacts Replacement costs Business interruption Negative impact on other sectors Loss of business dependent on clean/safe water supply
(b) Human health catastrophe
(c) Environmental impact Injure endangered species Damage other natural resources (e.g., animals, lakes, and springs)
(d) Psychological impacts Damage public confidence and morale
(e) Severe contamination of Edwards aquifer
(f) Negative impacts to Air Force bases in San Antonio
(g) Regional/national security functionality impacts

Consequence Scale—Fatalities									
Number of Fatalities									
Consequence criteria ("bin numbers")	0	1	2	3	4	5	6	7	8
Ranges in number of fatalities	0–20	21–40	41–80	81–160	161–320	321–640	641–1,280	1,281–2,560	2,561–5,120
Consequence criteria ("bin numbers")	9	10	11	12	13	14	15	16	
Ranges in number of fatalities	5,121–10,240	10,241–20,480	20,481–40,960	40,961–81,920	81,921–163,840	163,841–327,682	327,681–655,360	655,361+	
Consequence Scale—Injuries									
Number of Injuries									
Consequence criteria ("bin numbers")	0	1	2	3	4	5	6	7	8
Ranges in number of injuries	0–20	21–40	41–80	81–160	161–320	321–640	641–1,280	1,281–2,560	2,561–5,120
Consequence criteria ("bin numbers")	9	10	11	12	13	14	15	16	
Ranges in number of injuries	5,121–10,240	10,241–20,480	20,481–40,960	40,961–81,920	81,921–163,840	163,841–327,682	327,681–655,360	655,361+	

(*Continued*)

Table 7.3 Step 3: Consequence Analysis for San Antonio, Texas Water Infrastructure (*Continued*)

Consequence Scale—Financial Impacts to the Owner/Operator									
Financial Loss									
Consequence criteria ("bin numbers")	0	1	2	3	4	5	6	7	8
Owner/operator financial loss ($1M)	0–20	21–40	41–80	81–160	161–320	321–640	641–1,280	1,281–2,560	2,561–5,120
Consequence criteria ("bin numbers")	9	10	11	12	13	14	15	16	
Owner/operator financial loss ($1M)	5,121–10,240	10,241–20,480	20,481–40,960	40,961–81,920	81,921–163,840	163,841–327,682	327,681–655,360	655,361+	
Consequence Scale—Economic/Financial Impacts ($1M)									
Number of Fatalities									
Consequence criteria ("bin numbers")	0	1	2	3	4	5	6	7	8
Regional community economic loss ($1M)	0–20	21–40	41–80	81–160	161–320	321–640	641–1,280	1,281–2,560	2,561–5,120
Consequence criteria ("bin numbers")	9	10	11	12	13	14	15	16	
Regional community economic loss ($1M)	5,121–10,240	10,241–20,480	20,481–40,960	40,961–81,920	81,921–163,840	163,841–327,682	327,681–655,360	655,361+	

Consequences		
Edwards Aquifer Recharge Zone Contamination Using Arsenic-/Cyanide-Based Pesticides (Similar Scenario—"Bhopal Disaster")	*Bin Number*	*Remarks*
Fatalities	8	In communities dependent on raw water (from wells)
	6	In cities
	1–3	In recreational areas such as springs, lakes, and rivers
Injuries	8	
Financial impact to owner/operator (e.g., SAWS and BexarMet)	3	It is approximately 80–160 M to construct an advance treatment system to remove arsenic/cyanide for one treatment system facility.
	6–10	Overall total loss could go higher.
Regional community economic impact	6 (or higher)	Business cannot operate due to severe contamination in the area.
Environmental, psychological, and ecological impacts	Not quantified	Not quantified in RAMCAP Plus

(*Continued*)

Table 7.3 Step 3: Consequence Analysis for San Antonio, Texas Water Infrastructure (*Continued*)

Consequences		
Blasting of Water and Wastewater Treatment Plants	*Bin Number*	*Remarks*
Fatalities	1	Attacking Dos Rios Wastewater Treatment Plant
	0–1	Blasting Twin Oaks Treatment Plant
	0–1	Blasting Medio Creek/BexarMET Treatment Plants
Injuries	1	
Financial impacts to owner/operator (e.g., SAWS and BexarMet)	2	Repair cost for Dos Rios Treatment Plant + liability cost to employees
	2	Repair cost for treatment plant + liability cost to employees
Regional community economic impact	6 (or higher)	Cost of remediation, recovery, and import of water supply from other city or state

Injection of Poison (Arsenic-/Cyanide-Based Pesticides) to Water Tanks and to the ASR	*Bin Number*	*Remarks*
Fatalities	2 or 3	Affluent cities (e.g., Alamo Heights, Shavano Park, and Hollywood Park) where home owners have "reverse osmosis" treatment installed in their homes and business
	4 (or higher)	Middle class/lower middle class communities
	2	Rural areas
Injuries	5	Overall
Financial impact to owner/operator	2 or 3	Per individual owner/operator of the water tank in the community
	—	Community (cost of treating water/repair/recovery)
Regional community economic impact	6 (or higher)	Remediation, recovery, and construction of advance treatment

Source: Derived from American Society of Mechanical Engineers Innovative Technologies Institute, LLC, *All-Hazards Risk and Resilience Prioritizing Critical Infrastructure Using the RAMCAP Plus Approach*. New York: ASME, 2009.

authorities and public/private partnerships. As indicated by ASME-ITI (2009), in the RAMCAP Plus process, when quantifying the owner's losses, the principle is that value, whether gain or loss, is the incremental discounted net present value of future cash flows. The elements of the owner's loss include, but are not limited to (1) business interruption costs; (2) environmental remediation; (3) costs involved in repair of equipment/structures; (4) replacement costs; (5) liability costs; and (6) other costs contributed by the attack. In the public perspective of water infrastructure, the major concern is the length of time and quantity of service denied and the economic consequences of service denial on direct suppliers and customers of the critical facilities.

The economic loss can be as much as three orders of magnitude greater than the gross revenue losses of the facility. According to ASME-ITI (2009), in the RAMCAP Plus method, estimating the economic impacts of the community requires a regional simulator or an economic model to fully capture cascading failures and indirect/direct consequences and requires a system model that simulates water infrastructure systems. The conventional input-output models used in estimating consequences of a major disruption or attack can lead to major errors. The economic losses shown in Table 7.3 are estimated by approximation using the RAMCAP Plus process. The regional economic loss estimate can serve as a baseline for the resilience of the region and it includes all the affected elements: the severity and time of service denial, economic consequences, and public health impact (due to severe contamination). Table 7.3 shows the consequences based on the threats shown in Table 7.2.

7.2.3 Vulnerability Analysis

Vulnerability analysis estimates the conditional likelihood that a threat will have the consequences estimated in Step 3. It estimates the probability that the terrorist will be successful in executing a specific attack. Table 7.4a presents Step 4 of RAMCAP Plus.

Table 7.4a Vulnerability Analysis for the San Antonio, Texas Water Infrastructure

Vulnerability Analysis
(a) Easy access to Edwards Aquifer recharge zone. No surveillance or security on major roads and freeway leading to wells, faults, sinkholes, and caves.
(b) Wastewater treatment facilities are secured but not on the discharge point.
(c) Water treatment facilities do not have sophisticated fence and surveillance technology.
(d) Above-ground and underground reservoirs (in the ASR vicinity) are open for public use.
(e) Water tanks do not have sophisticated fence and surveillance technology.

Table 7.4b Vulnerability Scale

Bin (Category)		*Probabilities in Decimal Description*
5	a	.85–1.00
	b	.65–.84
	c	.40–.64
4		.20–.398
3		.10–.198
2		.05–.099
1		.025–.049
0		<.024

Source: Derived from American Society of Mechanical Engineers Innovative Technologies Institute, LLC, *All-Hazards Risk and Resilience Prioritizing Critical Infrastructure Using the RAMCAP Plus Approach*. New York: ASME, 2009.

The vulnerability scale is shown in Table 7.4b, and the scale uses the same factor of two between successive categories, as in consequence ranking. This is effective for plotting a resultant risk matrix. Category 5 is further subdivided into three subcategories: a, b, and c. It is feasible for the owner/operator to estimate changes in security and defense level in risk management.

7.2.4 *Threat Assessment*

Threat assessment estimates the probability of each initiating event. In RAMCAP Plus, risk assessment for terrorism consists of weighing available evidence about an adversary and the asset in question. An example of threat assessment is presented in Table 7.5.

7.2.5 *Risk and Resilience Assessment*

The risk and resilience assessment creates the foundation for selecting strategies and tactics to defend against disabling attacks and events by establishing priorities based on the level of risk. The risk imposed by each threat to each asset is calculated from the risk relationship: $R = T \times V \times C$ (Equations 6.1 and 6.2). For the asset owner, the level of resilience for a particular threat is expressed as the product of lost revenue, vulnerability, and threat (see Equation 6.3a).

For the community, the level of resilience for a particular threat is the product of lost economic activity in the community, vulnerability, and threat (Equation 6.3b).

Lost revenue is the product of the duration of service denial, the extent of service denial, and the price per unit (Equation 6.3c). Lost economic activity in the community is the amount of decrease in the loss of output to direct customers and the indirect losses throughout the economy of a given region due to denial of service and its extent.

Table 7.5 Threat Assessment for the San Antonio, Texas Water Infrastructure

Threat Assessment (Three Methods)
(a) Numerical ratio method: *This estimate can be based on historical data, intelligence information, or various assumptions.* 1. Let T = number of attacks attempted in the United States. 2. Assume the probability of attacking water infrastructure is equal to all other 18 sectors. 3. Probability (P) = $T/18$ for terrorist attacks. 4. Assume based on the data, there are W facilities. 5. Let W = 10,000 facilities; $P = (T/18)/10{,}000 \sim .0000056T$. 6. Assume the particular target being evaluated has 15 major assets: $P = (T/18)(W \times 15)$. 7. Let T = 15 attacks. 8. $P = (15/18)/(50 \times 15) = 5.6 \times 10^{-6}$ events/year.
(b) Comparison of risk tolerance with natural hazard risk uses the idea of risk tolerance and a natural hazard risk to compare with a terrorist risk to deduce a threat probability equating the two risks. 1. Consider the Standard Homeland Security risk equation, $R = C \times V \times T$. 2. Transpose to $T = R/(C \times V)$. 3. Assume water and wastewater treatment plants recovery/reconstruction cost = \$2B. 4. Net cash flow after taxes = \$6B. 5. Tornado risk = 150–250 mph (South Texas). 6. Assume tornado risk for the plants = \$30M. 7. Assume a total owner's loss of \$1B after shutting down for 9 months and for reconstruction. 8. Vulnerability (V) = 0.75 (treatment plants with security system at entrance/exit). 9. $T = (30{,}000{,}000)/(2{,}000{,}000{,}000 \times 0.75) = 0.02$ events/year. 10. Frequency = 1/0.02 = 50 years.

Table 7.5 Threat Assessment for the San Antonio, Texas Water Infrastructure (*Continued*)

Threat Assessment (Three Methods)
(c) Investment breakeven *assumes the decision maker's choices are simple on individual options.* $R = T \times V \times C$, DHS standard risk equation/RAMCAP Plus equation. Minimum benefits to justify the option's cost: (baseline risk-option risk $[R])/\text{Cost}_{\text{option}} > 1.0$. Therefore, $\{[(C_{\text{baseline}} \times V_{\text{baseline}}) - (C_{\text{option}} \times V_{\text{option}})] \times T\}/\text{Cost}_{\text{option}} = 1.0$. Continuing the example from (b), a series of countermeasures were delineated at a cost of \$50M. The option was approximated to decrease vulnerability from 0.75 to 0.40 and reduce the consequence to the owner from \$1B to \$0.3B. $C_{\text{baseline}} = \$1\text{B}$; $C_{\text{option}} = \$0.3\text{B}$ $T = \text{Cost}_{\text{option}}/\{[C_{baseline} \times V_{\text{baseline}}] - (C_{\text{option}} \times V_{\text{option}})\}$ $T = \$50\text{M}/\{(\$1\text{B} \times 0.75) - (\$0.3\text{B} \times 0.40)\}$ $T = \$50\text{M}/\$0.63\text{B} = 0.079$ or a reoccurrence of $1/0.079 = 12.6$ years

Source: Derived from American Society of Mechanical Engineers Innovative Technologies Institute, LLC, *All-Hazards Risk and Resilience Prioritizing Critical Infrastructure Using the RAMCAP Plus Approach*. New York: ASME, 2009.

Table 7.6 Risk and Resilience Assessment for the San Antonio, Texas Water Infrastructure

Threat Assessment		
(a) Example from Step 3:		
Edwards Aquifer Recharge Zone Contamination Using Arsenic-/Cyanide-Based Pesticides (Similar Scenario—"Bhopal Disaster")	*Bin Number*	*Remarks*
Fatalities	8	In communities dependent on raw water (from wells)
	6	In the cities
	1–3	In recreational areas such as springs, lakes, and rivers
Injuries	8	

(Continued)

Table 7.6 Risk and Resilience Assessment for the San Antonio, Texas Water Infrastructure (*Continued*)

Edwards Aquifer Recharge Zone Contamination Using Arsenic-/Cyanide-Based Pesticides (Similar Scenario—"Bhopal Disaster")	*Bin Number*	*Remarks*
Financial impacts to owner/operators (e.g., SAWS and BexarMet)	3	It is approximately \$2B to construct advance treatment system to remove arsenic/cyanide for one treatment system facility.
	6–10	Overall total loss could go higher.
Regional community economic impacts	6 (or higher)	Business cannot operate due to severe contamination in the area.
(b) Consequences are summarized as follows: (Similar scenario—"Bhopal disaster, 1984") Fatalities = 5,000+ Acute injuries = 2,560 Financial Impact to the owners = \$20B Losses to the regional economy = \$100B+ *Consequences of damages to the fishery—environmental, psychological impact is not quantified in RAMCAP Plus.*		
(c) Vulnerability: (Recharge zone is unprotected—no need of any expertise to intrude) 0.95		
(d) Threat: The probability of having an attack per Step 4 is 5.6×10^{-6}.		
(e) Risk: 1. Fatalities: $R_f = 5000 \times 0.95 \times (5.6 \times 10^{-6}) = 0.0266$ lives/year 2. Injuries: $R_i = 2560 \times 0.95 \times (5.6 \times 10^{-6}) = 0.0136$ lives/year 3. Financial impacts to the owner: $R_{O\$} = \$20B \times 0.95 \times (5.6 \times 10^{-6}) = \$106,400$ 4. Economic losses to regional economy: $R_{REGION\$} = \$100B \times 0.95 \times (5.6 \times 10^{-6}) = \$532,000$		

Source: Derived from American Society of Mechanical Engineers Innovative Technologies Institute, LLC, *All-Hazards Risk and Resilience Prioritizing Critical Infrastructure Using the RAMCAP Plus Approach*. New York: ASME, 2009.

7.2.6 Risk and Resilience Management

According to ASME-ITI (2009), in RAMCAP Plus risk and resilience management is the deliberate course of deciding and implementing options (e.g., improving preventive measures, mitigation tactics, building in redundancy, creating emergency response plans, exercise business casualties) and achieving an acceptable level of risk and resilience at an acceptable cost to the organization and the community. Risk and resilience management based on RAMCAP Plus will not be presented in detail in this book. Evaluation of the applicability of this model to water infrastructures are presented in Section 7.3.

7.3 CARVER Matrix

The CARVER Matrix is a tool that evaluates the priority ranking of a given set of targets. The matrix also evaluates the strengths and weaknesses inherent in each target. The CARVER Matrix for terrorism aimed on water infrastructures is presented in Tables 7.7 and 7.8.

7.4 CARVER + Shock

CARVER + Shock is a prioritization tool that can be used to assess the vulnerabilities within an infrastructure as detailed in Section 7.3, with a seventh attribute, *Shock*, added to the original six to assess the combined health, environment, economic, and cognitive reactions and psychological impacts of an attack. The process evaluation for using this tool for water infrastructure is shown in Tables 7.9 and 7.10.

7.5 Model-Based Vulnerability Analysis

According to Ted G. (2006), Lewis (2006) vulnerability is not the same as risk; vulnerability is a probability of a risk event, whereas risk is measured in terms of financial risk, casualty risk, and equipment risk. Risk is the product of vulnerability V (probability ranging from 0 to 1.0) and cost D (an estimate of damages). Lewis pointed out that it is important to distinguish the calculation of vulnerability from that of risk, because vulnerability reduction achieves a different goal than risk reduction. Hence, Lewis is one of the authors in the field of homeland security who popularized the model-based vulnerability analysis (MBVA). Generally, MBVA used network analysis with fault tree modeling to derive vulnerability, risk, and resource allocation strategies. Lewis (2006), in his book, *Critical Infrastructure Protection in Homeland Security*, provided a detailed presentation of the MBVA model for water infrastructure using the Hetch Hetchy Water Supply System

Table 7.7 Strategic CARVER Matrix Application for Water Infrastructure

Strategic CARVER Matrix							
Target Systems	*C*	*A*	*R*	*V*	*E*	*R*	*Total*
Contamination							
1. Edwards Aquifer (as sole source for water supply in a large urban area like the San Antonio metropolitan area)	10	10	10	10	10	10	60[a]
2. Aquifers	9	10	10	10	10	10	59[a]
3. ASR facilities (the underground reservoir)	9	9	10	9	10	8	56[a]
4. Reservoirs	9	8	10	9	10	9	55[a]
5. Aqueducts (open channel type)	9	10	9	10	10	10	58[a]
6. Aqueduct pipeline systems	8	10	9	8	10	9	54[a]
7. Water tanks	5	10	4	10	5	10	44
8. Hydrants	2	4	1	2	3	5	17
9. Surface water (e.g., oceans, rivers, lakes, and springs)	8	10	10	10	8	10	56[a]
Blasting/Explosion							
1. Municipal water treatment plants	7	4	10	5	10	8	44
2. Small community water treatment plants	7	4	8	5	10	8	42
3. Municipal wastewater treatment plants	7	4	8	5	10	8	42
4. Aquifer storage and recovery, water treatment plants	7	4	8	5	10	8	42
5. Desalination plants	10	4	10	5	10	5	44
6. Aqueducts	10	10	6	10	10	10	56[a]
7. Major sewer and water pipelines	5	10	3	10	5	4	37
8. Dams	10	8	10	9	10	10	57[a]

[a] Indicates target system suitable for attack.

Table 7.8 Operational CARVER Matrix Application for Water Infrastructure

Operational CARVER Matrix							
Target Subsystems	*C*	*A*	*R*	*V*	*E*	*R*	*Total*
1. SCADA/cyber components	10	6	6	4	10	3	39[a]
2. Controls	10	3	6	4	5	3	30
3. Major inlet/outlet distribution lines	9	10	4	10	5	10	48[a]
4. Turbines	9	3	5	4	5	3	29
5. Pump stations	6	7	3	6	5	9	36
6. Generators/power lines	8	9	2	6	4	9	38[a]
7. Settling chambers	9	7	3	4	4	9	36
8. Chemicals/feed systems	10	7	3	4	4	9	37
9. Primary treatment chambers	10	7	3	4	4	9	37
10. Secondary/tertiary treatment chambers	10	7	3	4	4	9	37
11. Switching stations	8	3	3	4	4	3	25
12. Water quality monitoring systems	7	7	1	6	4	4	29

[a] Indicates target system suitable for attack.

Table 7.9 Strategic CARVER + Shock Application for Water Infrastructure

Strategic CARVER + Shock								
Target Systems	*C*	*A*	*R*	*V*	*E*	*R*	*Shock*	*Total*
Contamination								
1. Edwards Aquifer (as sole source for water supply in a large urban area like the San Antonio metropolitan area)	10	10	10	10	10	10	9	69[a]
2. Aquifers	9	10	10	10	10	10	7	66[a]

(*Continued*)

Table 7.9 Strategic CARVER + Shock Application for Water Infrastructure (*Continued*)

Strategic CARVER + Shock								
Target Systems	*C*	*A*	*R*	*V*	*E*	*R*	*Shock*	*Total*
Contamination								
3. ASR facilities (underground reservoirs)	9	9	10	9	10	8	6	62[a]
4. Reservoirs	9	8	10	9	10	9	7	62[a]
5. Aqueducts (open channel type)	9	10	9	10	10	10	8	66[a]
6. Aqueduct pipeline systems	8	10	9	8	10	9	9	63[a]
7. Water tanks	5	10	4	10	5	10	2	46
8. Hydrants	2	4	1	2	3	5	0	17
9. Surface water (e.g., oceans, rivers, lakes, and springs)	8	10	10	10	8	10	3	59[a]
Blasting/Explosion								
1. Municipal water treatment plants	7	4	10	5	10	8	4	48
2. Small community water treatment plants	7	4	8	5	10	8	4	46
3. Municipal wastewater treatment plants	7	4	8	5	10	8	4	46
4. Aquifer storage and recovery, water treatment plants	7	4	8	5	10	8	4	44
5. Desalination plants	10	4	10	5	10	5	4	48
6. Aqueducts	10	10	6	10	10	10	1	57[a]
7. Major sewer and water pipelines	5	10	3	10	5	4	1	38
8. Dams	10	8	10	9	10	10	7	57[a]

[a] Indicates target system suitable for attack.

Table 7.10 Operational CARVER + Shock Application for Water Infrastructure

Operational CARVER + Shock								
Target Subsystems (Using Explosives)	*C*	*A*	*R*	*V*	*E*	*R*	*Shock*	*Total*
1. SCADA/cyber components	10	6	6	4	10	3	5	44[a]
2. Controls	10	3	6	4	5	3	1	30
3. Major inlet/outlet distribution lines	9	10	4	10	5	10	2	48[a]
4. Turbines	9	3	5	4	5	3	5	29
5. Pump stations	6	7	3	6	5	9	1	36
6. Generators/power lines	8	9	2	6	4	9	1	38[a]
7. Settling chambers	9	7	3	4	4	9	2	36
8. Chemicals/feed systems	10	7	3	4	4	9	2	37
9. Primary treatment chambers	10	7	3	4	4	9	2	37
10. Secondary/tertiary treatment chambers	10	7	3	4	4	9	2	37
11. Switching stations	8	3	3	4	4	3	1	25
12. Water quality monitoring systems	7	7	1	6	4	4	1	29

[a] Indicates target system suitable for attack.

(HHWSS) of San Francisco, California. His model determines the *site or facility-specific* probable risk and vulnerability of an asset.

7.6 Vulnerability Self-Assessment Tool

The vulnerability self-assessment tool (VSAT) is a qualitative risk assessment method highly recommended by the EPA for water and wastewater treatment facilities. It examines utility assets such as the physical plant, people, knowledge base, information technology, and customers. This is an appropriate tool to provide utility managers a general knowledge of their system's vulnerabilities, preparing for extreme events and business recovery activities.

7.7 Security Vulnerability Self-Assessment Guide for Small Drinking Water Systems

The Security Vulnerability Self-Assessment (SVSA) Guide for small water systems is designed for use by water system personnel. The primary steps in conducting SVSA are presented in Table 7.11.

Table 7.11 Security Vulnerability Self-Assessment Guide

Steps for Security Vulnerability Self-Assessment Guide
1. The personnel should provide basic information (e.g., name, address, identification number).
2. Inventory of the critical components of small water systems should be made.
3. Answer the general questions for the entire water system. The questions are designed to apply to all components of small water systems (e.g., wellhead, surface water intake, pumps or structures within the area).
4. Once the questions are completed, then the personnel can identify the areas where the system has vulnerability concerns.
5. The personnel should prepare an emergency contact list. The names and telephone numbers of emergency responders should be enumerated.
6. A local notification list should be prepared (e.g., police department, fire department, hospital, health department).
7. A service notification list should be prepared (e.g., electrician, plumber, pump specialist, telephone utility).
8. A state notification list should be prepared (e.g., hazmat hotline, emergency management agency, drinking water primacy agency).
9. A media notification list should be prepared (e.g., radio, designated water system spokesperson, television).
10. The personnel should identify threats through the SVSA checklist (e.g., biological, chemical).
11. The personnel should observe suspicious activity and report any suspicious activity.
12. The personnel will report to the state drinking primacy that the assessment has been conducted.

Source: Derived from U.S. Environmental Protection Agency. http://water.epa.gov/infrastructure/watersecurity/techtools/index.cfm, 2010.

7.8 Automated Security Survey and Evaluation Tool (ASSET)

The Automated Security Survey and Evaluation Tool (ASSET) provides facility-specific qualitative vulnerability assessment for small and medium-sized drinking water systems.

7.9 Security Vulnerability Assessment

The risk of a security event is assessed qualitatively by the Security Vulnerability Assessment (SVA) by the National Petrochemical and Refiners Association (NPRA) and American Petroleum Institute (API). The security objectives derived from SVA are to use six basic strategies in the analysis herein to help minimize the risk: (1) deter, (2) detect, (3) delay, (4) deny, (5) defeat, and (6) respond. It has a two-step screening process to focus attention on higher risk events. An example of the general SVA step screening process for water infrastructure is illustrated in Table 7.12, and the SVA ranking levels are presented in Table 7.13.

Table 7.12 General Steps of Security Vulnerability Assessment Screening Process

Screening Process
Step 1: Security risk definition based on the consequences and likelihood of a successful attack against an asset. Some examples of significant consequences in a SVA include the following:
Public health injuries Irreversible damage to water supply systems Public panic and chaos Loss of business viability Water shortage (short term and long term) Economic stress due to remediation cost Mass casualties Disruption of the downstream industry Long-term health effects Damage to the environment
Step 2: Likelihood (probability) definition based on the attractiveness to the adversary of the asset, the degree of threat from terrorism, and the degree of vulnerability.
(a) Asset attractiveness: Effect Potential for causing mass casualties

(Continued)

Table 7.12 General Steps of Security Vulnerability Assessment Screening Process (*Continued*)

Screening Process
Potential for damaging the environment Potential for creating public chaos Potential for damaging the regional or national economy Potential for massive media attention Potential for creating water shortage Target Chemical (and explosive) weapons Iconic targets Usefulness of the process material as a weapon Proximity to a national landmark
(b) Threat: Amateur terrorists/vandals Disgruntled individuals Terrorists Criminals Activists
(c) Vulnerability: Unsecured recharge zone areas (e.g., sinkholes, faults and cracks, wells) Weak economy that needed to lease the land above underground reservoirs (ASR) Unsecured dam and reservoirs (no sophisticated security system) Unsecured aqueducts Weakness or poor relation between employees and management that causes a disgruntled individual within working facilities Deficiencies in the protection policies for the water reserves such as aquifers

Source: Derived from American Petroleum Institute-National Petrochemical and Refiners Association, *Security Vulnerability Assessment Methodology for the Petroleum and Petrochemical Industries,* 2nd Edition, http://www.npra.org/docs/publications/newsletters/SVA_2nd_edition.pdf, 2004.

Table 7.13 SVA Ranking Levels

Ranking Levels	*Threat*	*Attractiveness*	*Vulnerability*
1—Very low	There is no plausible evidence of actual threats against the water infrastructure or other critical infrastructures.	No interest in attacking water infrastructure	There are multiple layers of effective protective measures to the threat.
2—Low	There is a low threat against the water infrastructure.	Some interest	There are effective protective measures in place within or surrounding the water infrastructure, however, at least one weakness exists where an adversary would be capable of defeating the countermeasure.
3—Medium	There is a possible threat to the water infrastructure based on the adversary's desire to compromise similar sectors.	Moderate interest in attacking the water infrastructure	Although there are some effective protective measures, there is not a complete and effective application of these security strategies to protect water infrastructure.
4—High	A potential threat exists against the water infrastructure based on the capability of the adversary and the intent to attack.	High-level interest in attacking the water infrastructure	There are some protective measures to deter, detect, delay, defeat, or respond to the water infrastructure threat but not a complete or effective application of these security strategies.
5—Very high	A plausible threat exists against the water infrastructure and the adversary is capable of launching an attack.	Very high level interest in the water infrastructure	There are no effective protective measures currently in place to deter, detect, delay, defeat, and respond to the threat against water infrastructure.

Source: Derived from American Petroleum Institute-National Petrochemical and Refiners Association, *Security Vulnerability Assessment Methodology for the Petroleum and Petrochemical Industries,* 2nd Edition, http://www.npra.org/docs/publications/newsletters/SVA_2nd_edition.pdf, 2004.

7.10 Requirement of Incremental Risk Acceptability Analysis

This chapter presents the assessment of the qualitative and quantitative methods and tools for risk and vulnerability analysis mostly used by renowned authors, experts, and governmental agencies. All the models presented herein including the tools and methods presented in Chapter 6 are effective in identifying and estimating the *site or facility-specific* risk and vulnerability of an asset. But, these models and methods do not reflect the overall causative risk events surrounding the specific risk conditions and do not entirely prescribe consequence values. Consequently, most of these risk and vulnerability models immediately concluded that the *terrorism risk* is *unacceptable* to society. Probabilistic risk assessment can be an important tool in the policy formulation process for homeland critical infrastructure protection. However, it is important to recognize that perceived risk levels may have far more to do with the feasibility and acceptability of a protection policy and preventive measures than the actual risk levels themselves. While developing policy formulation and preventive measures viewpoint of those who are exposed (or feel they are exposed) to risks need to be considered. It is important to unequivocally involve those who are affected by the policy and program choices in the policy formulation and to effectively convey the information on actual risks to which they are exposed. This is particularly important when the exposure to the risk is perceived as involuntary such as with a *terrorist attack*: the straightforward process of making a decision somewhat "participatory" increases the nature of the risk being voluntary. In these situations, it is also important to build an effective breakdown of those affected by a decision in which risk is inherent and to deal with these communities individually; the effects on each community may be different. Also, if the results of risk assessments are to be really useful in policy and counterterrorism, it is important to maintain various components of risk separately. Groups have different sensitivities to issues like property damage, acute injuries, or fatalities. Chapters 8, 9, and 11 of this book will provide the readers the probabilistic risk assessment model and *risk acceptability* analysis of terrorist attacks against water infrastructures.

References

American Petroleum Institute-National Petrochemical and Refiners Association. 2004. *Security vulnerability assessment methodology for the petroleum and petrochemical industries*, 2nd Edition. http://www.npra.org/docs/publications/newsletters/SVA_2nd_edition.pdf (accessed August 14, 2010).

American Society of Mechanical Engineers Innovative Technologies Institute, LLC. 2009. *All-Hazards Risk and Resilience Prioritizing Critical Infrastructure Using the RAMCAP Plus Approach*. New York: ASME.

Doro-on, A. M. 2009. "Risk assessment embedded with cumulative prospect theory for terrorist attacks on aquifer of karstic limestone and water supply system." PhD diss., University of Texas at San Antonio.

Lewis, T. 2006. *Critical Infrastructure Protection in Homeland Security: Defending a Networked Nation*. Hoboken, NJ: John Wiley and Sons.

San Antonio Water System. 2010. Aquifer Storage and Recovery. http://saws.org/our_water/WaterResources/projects/asr.shtml (accessed July 15, 2008).

U.S. Environmental Protection Agency. 2010. *Security Vulnerability Self-Assessment Guide for Small Water Drinking Systems*. http://water.epa.gov/infrastructure/watersecurity/techtools/index.cfm (accessed October 22, 2010).

Chapter 8

Quantitative Risk Estimation Model

Mathematically, risk can be defined as a function of the probability of occurrence of a negative consequence and the value of that consequence. Events are subsequently aggregated at the terminal end until all risk pathways are described.

Risk has two major components: (1) the existence of a possible unwanted consequence or loss, and (2) an uncertainty in the occurrence of that consequence, which can be expressed in the form of a probability of occurrence (Rowe 1977). The consequence implies a negative value to a risk taker. According to William Rowe (1977) in his book *Anatomy of Risk*, risk is the potential for realizing an unwanted, negative consequence of an event. The concepts connected with the evaluation of consequences are covered in this chapter, including a comprehensive *risk estimation model* for water infrastructure protection against terrorism.

The integrated risk assessment methodology discussed in this chapter is a systematic approach for the analysis and evaluation of alternative policies concerning the protection of water infrastructure including dams, reservoirs, and aqueducts. Also, a comprehensive risk estimation model is developed based on *fault tree analysis*, *event tree analysis*, and the *probabilistic model*. A methodical example involving terrorist attacks using chemical threats that are difficult to destroy, particularly cyanide, arsenic, and prescription drugs (PDs), and including biological threats to water resources, is also discussed. Moroever, we look at their impact on water supplies for San Antonio, Texas, one of the fastest growing large cities, and for

Los Angeles, California, one of the largest overpopulated metropolitan areas in the United States, to illustrate this integrated risk assessment methodology.

8.1 Elements of Risk Assessment

The array of risks covers a wide variety of human experiences involving risks, personal or societal, man-made or natural, with consequences ranging from financial involvement to premature death (Rowe 1977). According to Robert Kates (1976), there are three major analytical steps when risk assessment is applied:

1. Risk identification involves reduction of descriptive uncertainty. Whereas, risk reduction means risk is reduced to some acceptable level.
2. Risk estimation is based on the systematic evaluation of probabilities associated with events having negative consequences.
3. Risk acceptability analysis is based on the quantitative revealed preferences method.

8.1.1 Risk Estimation Process for Terrorist Attacks against Water Infrastructure

Based on the literature research, terrorists can easily access, intrude, and attack the water infrastructure of the United States. San Antonio and Los Angeles are used as examples for terrorist attacks on the aquifer recharge zone, aqueducts, and their urban water supply systems. The following are the reasons why terrorists can easily attack the water supply systems in San Antonio and Los Angeles:

1. No sophisticated technology, policy, or strategy available for securing U.S. borders, which have established an easy inflow of deadly chemicals like arsenic, cyanide, illegal PDs, and endocrine disruptors (EDs) through underground tunnels.
2. No thorough investigation conducted on individuals purchasing large quantity of arsenic and cyanide compounds (e.g., pesticides, herbicides, and arsenic trioxide) in local department stores.
3. No policy requiring governing agencies to conduct thorough background investigations on an individual or a group of individuals purchasing and leasing real estate properties near or in the major aquifer recharge zone, aquifer storage and recovery (ASR) facilities, aqueducts, and reservoirs.
4. No security surveillance on the vicinity of any of the aquifer recharge zones (e.g., dams, wells, lakes, springs, large sinkholes).

5. No regular or thorough inspections conducted on any construction over recharge areas.
6. No regular inspection of underground tanks (e.g., septic tanks, underground storage tanks) in recharge zones.
7. No sophisticated technology for detection of high concentrations of chemical threats (e.g., cyanide, arsenic, and EDs) in underground tanks, wells, major recharge areas, water supply tanks, and other water facilities.
8. Some of the water supply agencies like San Antonio Water System (SAWS) allow land above ASR to be leased or used for public purposes. Also, there is no policy requiring governing agencies to perform thorough investigations on individuals who use the land above or near ASR, as long as these individuals are able to provide financial statements.

There are five significant steps for the risk estimation process for water infrastructure security, which are as follows: (1) causative event, (2) outcome, (3) exposure, (4) probability of consequence, and (5) consequence values. These five steps of the risk estimation process for terrorist attacks on aquifers and water supply systems are shown in Tables 8.1a through e.

Table 8.1a Process of Risk Estimation: Step 1—Causative Events

Step 1—Causative Events	
(a)	Terrorist intrusion to aquifer recharge area (e.g., *dams, reservoirs, sinkholes, cracks, faults, and caves*)
(b)	Terrorist intrusion to water supply system/facilities (e.g., *water supply storage, treatment facilities, aqueducts, and wells*)
(c)	Terrorists purchase and lease homes (or any real estate properties) above the aquifer recharge zone
(d)	Terrorists lease or purchase homes (or any real estate properties) adjacent to aqueduct easements
(e)	Terrorists lease or purchase agricultural properties adjacent to or above the ASR area (e.g., *Twin Oaks ASR Facility; according to SAWS (2010), most land directly above the underground reservoir can continue its prior use and land can be leased*)
(f)	Terrorist intrusion to future water supply projects (e.g., *Carrizo Aquifer, brackish groundwater, Lower Colorado River Authority [LCRA]: Highland Lakes*)

Table 8.1b Process of Risk Estimation: Step 2—Outcome

	Step 2—Outcome
(a)	Dumping or injection of chemical threats (e.g., cyanide, arsenic, EDs, and PDs) into the aquifer recharge area such as sinkholes, wells, faults, cracks, caves, and dams
(b)	Injection of cyanide-/arsenic-based pesticides, expired PDs, and EDs in the water supply tanks and other water storage facilities (e.g., ASR, Winwood tank station, Oliver/Bulverde Sneckner Ranch)
(c)	Injection of biological threats to lakes, reservoirs, aqueducts, and wells
(d)	Injection of cyanide-/arsenic-based pesticides, expired PDs, and EDs into the aqueduct pipelines or major water mains
(e)	Destruction of water treatment facilities and stealing of stored chemicals for contamination
(f)	Injection or dumping of cyanide, arsenic, illegal or expired PDs, and EDs into the water resources on U.S. borders.

Table 8.1c Process of Risk Estimation: Step 3—Exposure

	Step 3—Exposure
(a)	Chemical threats (cyanide and arsenic) in the aquifer recharge area: i. Chemicals mixed in water in aquifer. ii. Chemicals mixed with chlorine during treatment process to form hazardous compounds or trihalomethane (carcinogen) or both. iii. Chemicals will not be oxidized by chlorine once they go to the water treatment plant. iv. Chemicals mixed with other nutrients or chemical compounds present in the water. v. Chemicals diluted in some areas.
(b)	Chemical threats (cyanide- and arsenic-based compounds) in the treated water supply system and storage facilities: i. Chemicals mixed in the water supply. ii. Chemicals mixed with chlorine during treatment process to form hazardous compounds. iii. Chemicals will not be oxidized by chlorine or by any other traditional treatment system except reverse osmosis.
(c)	Chemical threats (a mixture of hazardous chemicals and EDs) in the water supply system: i. A combination of different hazardous chemicals, pesticides, and EDs mix in the water system.

Table 8.1c Process of Risk Estimation: Step 3—Exposure (*Continued*)

	ii. A mixture of hazardous chemicals, pesticides, and EDs are difficult to remove by traditional treatment methods except reverse osmosis. An increase of production of hazardous compounds in water. iii. Hazardous chemicals, pesticides, and EDs are diluted in some areas.
(d)	Chemical threats (arsenic- and cyanide-based compounds) in the raw or untreated water in lakes, rivers, aqueducts, and reservoirs: i. Chemicals contaminate the water system. ii. Chemicals react with other chemical compounds present in the water system. Production of other hazardous compounds in the water.
(e)	Biological threats in raw and untreated water such as that in lakes, rivers, aqueducts, and reservoirs: i. Biological threats mixed in the water system. ii. Biological threats acclimate and propagate in the water system. iii. Biological threats are oxidized by traditional disinfection systems (e.g., ultraviolet [UV] treatment, ozonation, and chlorination). iv. Some of the biological threats will not survive in the presence of light.
(f)	PDs and EDs in the aqueducts and reservoirs: i. PD and ED mixed in the water system. ii. PD and ED mixed with other chemical compounds present in the raw water supply. iii. PD and ED will not be oxidized by chlorine or other traditional primary and tertiary treatment systems. (An advanced process shall be used for treatment.) Potential production of hazardous compounds in the water.
(g)	PDs and EDs in the aquifer recharge zone and ASR: i. PD and ED mixed in water system. ii. PD and ED mixed with other nutrients and chemical compounds present in the water. iii. PD and ED will not be oxidized by chlorine or other disinfection system.
(h)	PDs and EDs in the water supply tanks: i. PD and ED mixed in the water system. ii. PD and ED mixed with other chemical compounds present in the water. A possible formation of hazardous chemicals in the water.

Table 8.1d Process of Risk Estimation: Step 4—Consequence

Step 4—Consequence	
(a)	Catastrophic health effects Short-term effect: Death Long-term effects: i. Cyanide: Damage to nervous system and other diseases. Can immediately cause death. ii. Arsenic: Slowly causes death and cancer. iii. PDs and EDs: Slowly cause death, mental illness, behavior problems, adverse effect on reproductive system, impaired immune functions, and various cancers. iv. Biological threats: Various physical illnesses.
(b)	Disrupt downstream commercial, agriculture, and industry infrastructure i. Contaminate the livestock. ii. Contaminate agricultural products. iii. Contamination and destruction of food supply. iv. Contaminate water for commercial use: restaurants, fast food, supermarkets, and other businesses.
(c)	Injury to animals and aquatic organisms (including endangered species) dependent on clean water resources.
(d)	May create irreversible damage to Edwards Aquifer and other water supply systems (e.g., ASR Twin Oaks, storage tanks, and future water supply projects like brackish groundwater, Highland Lakes, Owen Valley Aqueducts, Los Angeles reservoirs). May also create temporary denial to water supply service.
(e)	Result in economic distress due to the tremendous need for groundwater reserves remediation.
(f)	Cause damage to public morale and confidence.
(g)	No other immediate water supply available after the attack, which can create public panic and chaos.
(h)	Mass casualties.
(i)	Pollute the environment and cause destruction of natural resources dependent on clean water.

Table 8.1d Process of Risk Estimation: Step 4—Consequence (*Continued*)

Step 4—Consequence	
(j)	Result in economic distress due to the tremendous need to construct and use an advanced treatment system for cyanide, arsenic, PD, and ED removal from drinking water.
(k)	Result in economic distress due to the need for emergency response and recovery.
(l)	Pollution to water parks (lakes and rivers).

Table 8.1e Process of Risk Estimation: Step 5—Consequence Values

Step 5—Consequence Values	
(a)	Protection and security policy revision
(b)	Provide detection, intrusion, and surveillance technology
(c)	Improve intelligence
(d)	Provide funding for research on improving technology and policy
(e)	Integration of the Department of Homeland Security and educators for national security improvements

8.2 Risk Estimated by Event Tree Analysis

Event tree analysis provides a systematic logical tracing of sequential events resulting in consequential outcomes (Vesely, Goldberg, Roberts, and Hassl 1981). This method dictates that integral component events and decision nodes in any complex model stem from an initial event, resulting in multiple terminal outcomes, which are clearly documented from cause to result (Shih and Riojas 1990). Figure 8.1 presents an example event tree in which only two level-1 components or events are shown. In Figure 8.1, A might be permitting the highway, where A0 would then be regarded if the highway were not constructed, resulting in a terminal node with the end result of "no contamination as a result of highway permitting." The A1 branch would then represent the implementation of highway construction plans. The B event node might then represent incorporation of the highway as a conduit of passage in a hazardous freight route, options presented by this transitional node being no B0 or yes B1, and so on. Events are subsequently aggregated at the terminal end until all risk pathways are described and detailed.

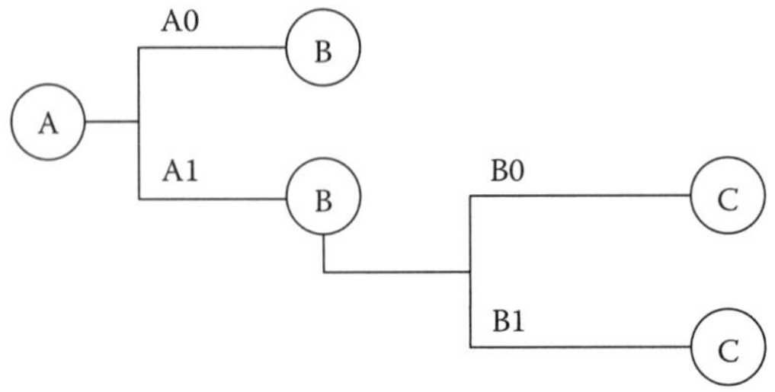

Figure 8.1 An example of an event tree. (Data from Doro-on, A. M., "Risk assessment embedded with cumulative prospect theory for terrorist attacks on aquifer of karstic limestone and water supply system," PhD diss., University of Texas at San Antonio, 2009.)

In development of the aquifer protection model, the following top-level land use policies are analyzed:

- A0: No terrorism worry (recharge zone is designated as a protected watershed)
- A1a: Complete protection of the entire recharge zone with electrified fence and surveillance
- A1d: Partial protection for major recharge facilities
- A1e: Business as usual, no protection against terrorism

Numerous possible events were considered in the development of the fault tree model, related to the transitional events presented in Table 8.2.

Table 8.2 Fault Tree Transitional Events

A	Policy alternatives
B	Facility design
C	Construction practice
D	Personnel hazmat training
E	Facility maintenance practices
F	Facility operational practices
G	Natural phenomena
H	Malicious acts
I	Chemical contamination
J	Abatement action
K	Contaminant/aquifer interaction

Table 8.2 Fault Tree Transitional Events (*Continued*)

L	Plume migration
M	Pollutant mitigation
N	Potable water treatment
O	Potable water delivery

Accordingly, in the aquifer protection example contamination of the potable water distribution system (O1) is the terminal node of significance. Using the event tree as a building block, the fault tree is readily developed. Contamination of the aquifer becomes the top event of the fault tree, and all failures in the event tree become subevents. Symbols describing the logic of fault tree nodes are unique, but fairly common to engineers and managers (Figure 8.2). The basic symbols in the tree include

- Event symbols—indicating event and status
- Gate symbols—indicating logical relationships between input and output events (AND, NAND, OR, XOR, PRIORITY, NOT, etc.)
- Transfer symbols—mechanisms uniting multiple sections or pages

Logic describing the top three transitional nodes of the fault tree is presented graphically in Figures 8.3a through d.

8.3 Estimation of Risk and Risk Factors

There are several methods for illustrating, estimating, and evaluating risk data. At this point, two methods are widely used: (1) computation of risk rates, and (2) computation of losses to life expectancy. Also, there are main approaches of acquiring and comparing risk rates for terrorist attacks on water infrastructure such as risk rates/factors based on fatal transportation accidents or nuclear power plant accidents using the two methods of computing risk rates and losses to life expectancy and risk rates/factors based upon engineering judgment.

8.3.1 Calculation of Risk Rate

For a certain category of events, i, such as transportation accidents, a number of such accidents N_i will occur in a given period of years t_i. The data can be based on information from government agencies such as the U.S. Census Bureau and the U.S. National Highway Traffic Safety Administration. The mean number of accidents per year, $\bar{N}$, is calculated by the formula

$$\bar{N} = \frac{N_i}{t_i} = \text{mean number of accidents or events per year}$$

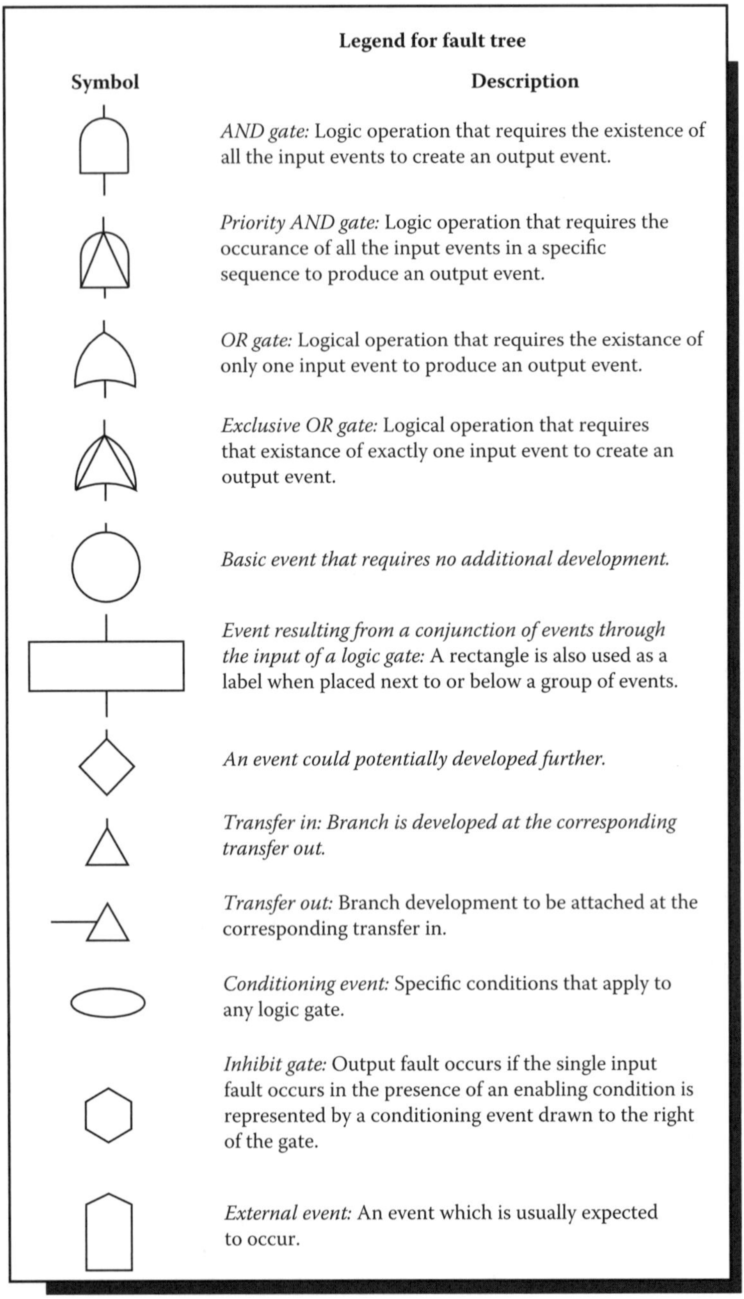
Legend for fault tree
Symbol
Description
AND gate: Logic operation that requires the existence of all the input events to create an output event.
Priority AND gate: Logic operation that requires the occurance of all the input events in a specific sequence to produce an output event.
OR gate: Logical operation that requires the existance of only one input event to produce an output event.
Exclusive OR gate: Logical operation that requires that existance of exactly one input event to create an output event.
Basic event that requires no additional development.
Event resulting from a conjunction of events through the input of a logic gate: A rectangle is also used as a label when placed next to or below a group of events.
An event could potentially developed further.
Transfer in: Branch is developed at the corresponding transfer out.
Transfer out: Branch development to be attached at the corresponding transfer in.
Conditioning event: Specific conditions that apply to any logic gate.
Inhibit gate: Output fault occurs if the single input fault occurs in the presence of an enabling condition is represented by a conditioning event drawn to the right of the gate.
External event: An event which is usually expected to occur.

Figure 8.2 Legend of fault tree.

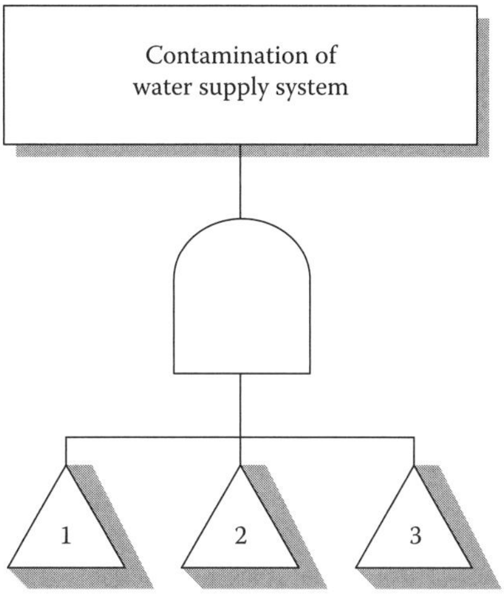

Figure 8.3a The top three transition nodes of the fault tree. This figure shows the contamination of a potable water system. (Data from Doro-on, A. M., "Risk assessment embedded with cumulative prospect theory for terrorist attacks on aquifer of karstic limestone and water supply system," PhD diss., University of Texas at San Antonio, 2009.)

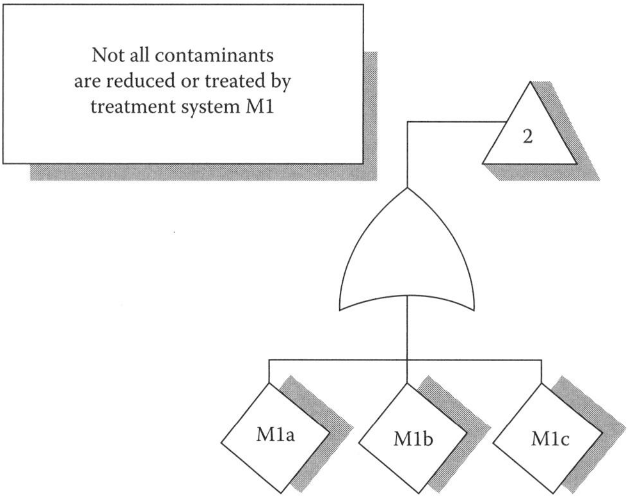

Figure 8.3b The top three transition nodes of the fault tree. This figure shows that not all contaminants are reduced or treated by treatment system M1. (Data from Doro-on, A. M., "Risk assessment embedded with cumulative prospect theory for terrorist attacks on aquifer of karstic limestone and water supply system," PhD diss., University of Texas at San Antonio, 2009.)

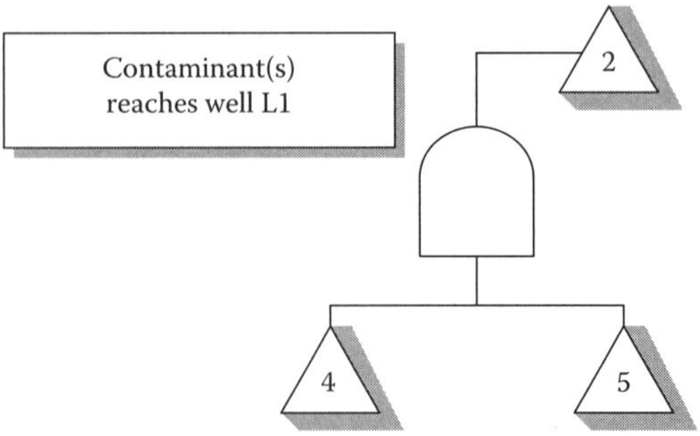

Figure 8.3c The top three transition nodes of the fault tree. The figure shows contaminants spreading and reaching the groundwater well. (Data from Doro-on, A. M., "Risk assessment embedded with cumulative prospect theory for terrorist attacks on aquifer of karstic limestone and water supply system," PhD diss., University of Texas at San Antonio, 2009.)

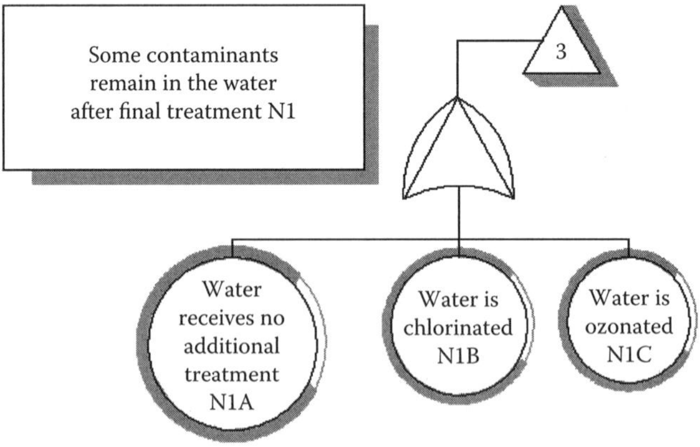

Figure 8.3d The top three transition nodes of the fault tree. The figure shows that some contaminants remain and produce carcinogens and other hazardous compounds in the water system after final treatment N1. (Data from Doro-on, A. M., "Risk assessment embedded with cumulative prospect theory for terrorist attacks on aquifer of karstic limestone and water supply system," PhD diss., University of Texas at San Antonio, 2009.)

For each event, j, of class I, there will be a number of consequence measures for consequences of different nature:

F_{ij} = total fatalities for accidents ij
F_{ij1} = total fatalities under voluntary risk conditions
F_{ij2} = total fatalities under involuntary risk conditions

I_{ij} = total injuries
I_{ij1} = total injuries, voluntary risk
I_{ij2} = total injuries, involuntary risk
D_{ij} = cost of event in dollars

Other consequences such as illness, security, or quality of life factors (e.g., economic, social, natural, political, health, and physical) can all be covered in the same process with clear definitions and detailed classification:

$$F_{ij1} + F_{ij2} = F_{ij}$$
$$I_{ij1} + I_{ij2} = I_{ij}$$

The formulations used for quantification of the average number of fatalities or injuries or cost in terms of voluntary and involuntary risk conditions recommended by Rowe are as follows:

$$\bar{F} = \frac{1}{N_i}\sum_j F_{ij} = \text{mean number of fatalities, per accident of type } i$$
$$\bar{F} = \frac{1}{N_i}\sum_j F_{ij1} = \text{mean number of fatalities (voluntary risk)}$$
$$\bar{F} = \frac{1}{N_i}\sum_j F_{ij2} = \text{mean number of fatalities (involuntary risk)}$$

A similar process is used for quantifying the mean number of injuries $\bar{I}_i$, $\bar{I}_{i1}$, and $\bar{I}_{i2}$ and the mean cost D_i. The populations at risk are designated as follows:

P_i = total population at risk based on statistical data
P_{i1} = population subject to voluntary risks
P_{i2} = population subject to involuntary risks

Then the number of fatalities, injuries, and costs annually for each class of event N_i is of the form

$\bar{N}_i \times \bar{F}_i$ = mean value of fatalities annually
$\bar{N}_i \times \bar{I}_i$ = mean value of injuries annually
$\bar{N}_i \times \bar{D}_i$ = average annual costs

The risk to an individual is

$$\bar{f}_i = \frac{\bar{N}_i \times \bar{F}_i}{P_i} = \text{mean probability of death to an individual at risk annually}$$
$$\bar{k}_i = \frac{\bar{N}_i \times \bar{I}_i}{P_i} = \text{mean probability of injury to an individual at risk annually}$$

The death rate per 100,000 people at risk f_i is

$$f_i = \bar{f}_i \times 10^5 = \frac{N_i \times \bar{F}_i \times 10^5}{\bar{P}_i}$$

And the injury rate per 100,000 people at risk k_i is

$$k_i = \bar{k}_i \times 10^5 = \frac{N_i \times \bar{I}_i \times 10^5}{\bar{P}_i}$$

Therefore, the voluntary and involuntary risk rates can be quantified.

However, not all the members within the population at risk automatically experience the same risk. Thus, the risk rates shall be subdivided into different group exposures. The segregation of exposed and protected groups or populations can also be quantified by using the equations of the *degree of containment index* (CI) recommended by Rowe, as follows:

$$\text{Containment index, CI} = \frac{\bar{f}_i}{\bar{f}_i'}$$

where $\bar{f}_i$ is the risk to an individual in the exposed population and $\bar{f}_i$ is the risk to an individual in the protected population. Since P_i is the exposed population and $T - P_i$ is the protected population, T is the overall population (e.g., the total population of the United States). Alternately, P_i is the protected population if the populations examined are not jointly all-inclusive.

$$\text{CI} = \frac{T - P_i}{P_i} \times \frac{g_i}{g_i'}$$

where

$$g_i = N \times F_i$$
$$g_i' = N_i \times F_i'$$
$$P_i < T$$
$$g_i' \le g_i$$

$$\text{Letting, CI} = \frac{T - P_i}{P_i} \times \log \frac{g_i + 1}{g_i' + 1}$$

8.3.2 Life Expectancy Models

Baldewicz (Baldewicz et al. 1974) developed a model for assessing risk data based on loss of life expectancy. Assuming that all insults (definitions) for a given risk system are linearly independent, the total rate of loss of life, based on 10^6 exposure hours to each stressor, is taken as $\dot{L} = 10^6 \sum_i \omega_i \dot{L}_i = 10^6 \sum_i \omega_i \frac{L_i}{T_i}$ where $\dot{L} = \frac{L_i}{T_i}$ (loss of life expectancy in years per exposure hour for the *i*th insult), L_i = lost years of life expectancy, T_i = time of exposure in hours, and ω_i = coefficient of insult intensity (between 0 and 1).

8.4 Fault Tree Analysis

Integrating the statistical likelihood of component events, the fault tree can then be used to estimate the overall probability of occurrence for a desired end result of a failure sequence (Schreiber 1982). A fault tree analysis can be simply described as an analytical technique, whereby an undesired state of the system is specified (usually a state that is critical from a safety standpoint), and the system is then analyzed in the context of its environment and operation to find all credible ways in which the undesired event can occur (Vesely, Goldberg, Roberts, and Hassl 1981). Vesely et. al. (1981) pointed out that the fault tree is not in itself a quantitative model, but it is rather a qualitative model that can be evaluated quantitatively and often is. The legend and descriptions for the top three transition nodes of the fault tree are presented in Table 8.3. Moreover, Table 8.4 presents the probability equations utilized in describing the top three transition events of the fault tree models illustrated in Figures 8.3a through d.

Table 8.3 Legend and Descriptions for the Top Three Transition Nodes of the Fault Tree

Legend	*Description*
1	Contaminant must reach the well.
2	The efforts to remove the pollutant must be unsuccessful.
3	The treatment that was being used as a final treatment before the aquifer was contaminated must fail to remove all the contaminants that slip past the equipment.
4	Mixing with water supply.
5	Dilution.
M1a	Arsenic.
M1b	Cyanide.
M1c	Arsenic and cyanide.

Source: Data from Doro-on, A. M., "Risk assessment embedded with cumulative prospect theory for terrorist attacks on aquifer of karstic limestone and water supply system," PhD diss., University of Texas at San Antonio, 2009.

Table 8.4 Probability Equations Utilized in Describing the Top Three Transition Events of the Fault Tree Models Illustrated in Figures 8.3a through d

O, P	$= P(1) \cdot P(2) \cdot P(3)$
L1, P(1)	$= P(4) \cdot P(5)$
M1, P(2)	$= P(M1a + M1b + M1c)$ $= P(M1a) + P(M1b) + P(M1c) - [P(M1a \cdot M1b) + P(M1b \cdot M1c)$ $+ P(M1c \cdot M1a)] - P(M1a \cdot M1b \cdot M1c)$
N1, P(3)	$= P(N1a + N1b + N1c)$ $= P(N1a) + P(N1b) + P(N1c) - 2[P(N1a \cdot N1b) \cdot P(N1b \cdot N1c)$ $\cdot P(N1c \cdot N1a)] + 3P(N1a \cdot N1b \cdot N1c)$

Source: Data from Doro-on, A. M., "Risk assessment embedded with cumulative prospect theory for terrorist attacks on aquifer of karstic limestone and water supply system," PhD diss., University of Texas at San Antonio, 2009.

The general form of the event tree analysis for terrorist attacks on groundwater resources is presented in Figure 8.4. Moreover, the probability estimation of successful terrorist attacks based upon Figure 8.4 is very high as presented in Section 8.4.1.

8.4.1 Probability Estimation Based on Probability Model in Figure 8.4

If P denotes an epidemic caused by an aquifer contamination, the assigned risk rates for the general form of an event tree on Figure 8.4 are as follows:

$$P(T) = 0.95$$
$$P(C) = 0.80$$
$$P(M/C) = 0.001$$
$$P(D/MC) = 0.95$$
$$P(O/MCD) = 0.75$$
$$P(Pb/MCDO) = 0.80$$
$$P(E/MCDOH) = [0.95 \times 0.80 \times 0.001 \times 0.95 \times 0.75 \times 0.80] = 4.33 \times 10^{-5}$$
$$= \text{probability of a successful attack against groundwater}$$

Additional detailed presentations of the event tree analysis and risk acceptability analysis for water infrastructure terrorism are provided in Chapter 9.

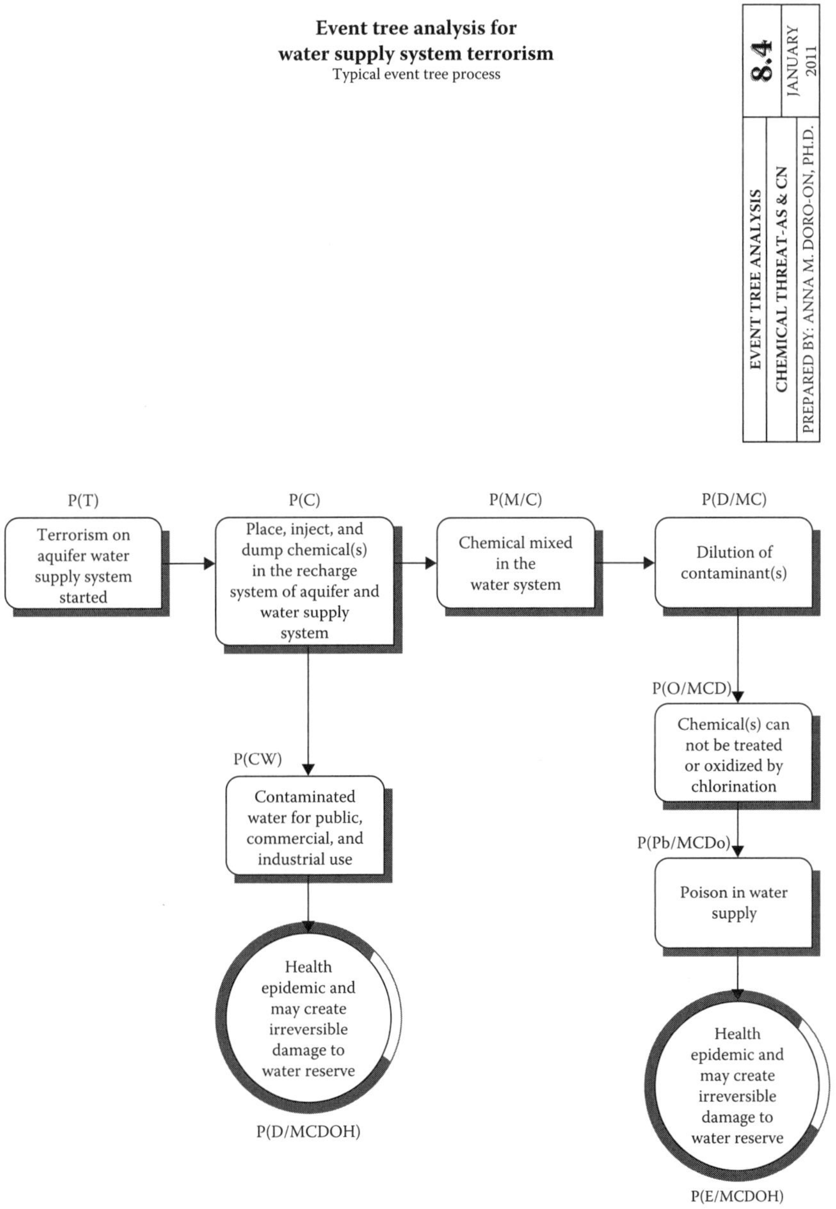

Figure 8.4 General form of an event tree analysis for water supply system terrorism.

References

Baldewicz, W., G. Haddock, Y. Lee, Prajoto, Whitley, R., and V. Denny. 1974. *Historical Perspective on Risk for Large-Scale Technological Systems*. UCLA-ENG-7485. Los Angeles: UCLA School of Engineering and Applied Science.

Doro-on, A. M. 2009. "Risk assessment embedded with cumulative prospect theory for terrorist attacks on aquifer of karstic limestone and water supply system," PhD diss., University of Texas at San Antonio.

Kates, R. 1976. Risk Assessment of Environmental Hazard, SCOPE Report No. 8, International Council of Scientific Unions, Scientific Committee on Problems of the Environment, Paris, France.

Rowe, W. 1977. *An Anatomy of Risk*. New York: John Wiley & Sons, Inc.

San Antonio Water System. 2010. Aquifer Storage and Recovery. http://saws.org/our_water/WaterResources/projects/asr.shtml (accessed July 15, 2008).

Schreiber, A. M. 1982. Using Fault Trees and Event Trees. *Chemical Engineering*. v(89). Oct. 4, 1982, p. 115–120.

Shih, C. and Riojas, A. 1990. Risk and Its Acceptability for Groundwater Contamination by Hazardous Wastes. *Risk Assessment for Groundwater Pollution Control*. Editors: McTernan, W. and Kaplan, E., American Society of Civil Engineers. p. 126–157. ASCE.

Vesely, W., Golberg, F., Roberts, N., and Hassl, D. 1981. *Fault Tree Handbook*. NUREG-0492. Systems and Reliability Research Office of Nuclear Regulatory Research, U.S. Nuclear Regulatory Commission, Washington, D.C.

Chapter 9

Cumulative Prospect Theory and Risk Acceptability

9.1 Introduction

Risk acceptability is involved with the determination of what level of safety is required or what degree of risk can be permitted by society for specific risk situations. The problems of risk acceptability can be summarized in three questions: The first is "How safe is safe enough?" the second question is "Which risks are acceptable?" and the third question is "Acceptable to whom?" However, the first and second questions cannot be answered without answering the third question first; but often, the answer to the third question is only implicitly stated if at all. It is the objective of this book to develop a systematic approach to risk acceptability and provide answers to these questions.

Meanwhile, risk acceptability requires a clear definition and a systematic quantitative method to evaluate it. This chapter will provide risk acceptability analysis and risk assessment quantification embedded with *cumulative prospect theory*. The information required to estimate the risk includes the joint probability of a series of events leading to the consequence, the value of this consequence, and the functional relationship defining the risk.

9.1.1 Cumulative Prospect Theory of Kahneman and Tversky

Prospect theory was developed by Daniel Kahneman, who won a Nobel Memorial Prize in economic sciences in 2002, and Amos Tversky. This theory describes decisions between alternatives that involve risk, explicitly alternatives with uncertain outcomes, where the probabilities are known. This classical prospect theory explains the major violations of expected utility theory in choices between prospects with a small number of outcomes (Tversky and Kahneman 1986). According to Tversky and Kahneman (1992), the two significant key elements of this theory are as follows (see Sections 6.3.2 and 6.4)

- The value function of prospect theory is steeper for losses than for gains.
- There is a nonlinear transformation of the probability scale in prospect theory, which inflates small probabilities and deflates moderate and high probabilities.

That is why people are interested not only in the benefit they receive but also the benefit received by others. This hypothesis is consistent with psychological research into happiness, which finds that subjective measures of well-being are relatively stable over time, even in the face of large increases in the standard of living (Easterlin 1974; Frank 1997).

Meanwhile, *risk perception* is the perceived or subjective judgment that an individual or group of people make about the characteristics, condition, and severity of a risk. Decision makers do not always react with perfect rationality to prospects of loss and gain in the presence of risk because individual perception impacts decisions. Kahneman and Tversky (1979) tested this implication and found that subjects systematically preferred to accept risk when prospects were presented in terms of costs and risk avoidance than when the same prospects were presented in beneficial terms. Meanwhile, Kahneman and Tversky presented a new enhanced model of prospect theory in 1992 that gives rise to different evaluations of gains and losses, which are not distinguished in the standard cumulative model, and that provides a unified treatment of both risk and uncertainty. The critical adjustment and revision to classical prospect theory is that, as in rank-dependent expected utility theory, cumulative probabilities rather than the probabilities themselves are transformed. This brings us to the aforementioned inflating of extreme events (e.g., a coordinated series of terrorist attacks against United States infrastructure), which occur with small probability, rather than a deflating of all small probability events. The adjustment and improvement of prospect theory helps to prevent a violation of first-order stochastic dominance and makes an attainable generalization to arbitrary outcome distributions. The cumulative prospect theory is an expansion and variant of Kahneman and Tversky's prospect theory.

9.2 Public Perception of Risk

If every individual perceived the world around him or her in the same manner, there would be no difficulty in assessing the acceptability of a particular risk situation or event. In the real world, people often fail to perceive reality very clearly or in the same way. The risk problems of interest to authors are, in many different conditions, neither well established, nor documented. They are often surrounded by a large degree of uncertainty resulting from such diverse causes as limited knowledge and restricted measurement capabilities. Compounding these limitations is the complexity of the problem, not just based on the multiplicity of risk pathways but also because risk does not exist by itself. It is only one of many problems that must be considered as simply one factor in a morass of benefits and costs, which can be direct and indirect, that surround any public decision problem. Kahneman and Tversky (2000) concluded that the intuitive and cognitive abilities of the normal human being are clearly overwhelmed by this complexity, thereby forcing him or her to rely on simplified and standardized rules of thumb. These simplified information-straining and decision-making rules always create bias and erroneous judgments. For instance, one such heuristic as judging the probability of a risk based on the ease with which instances can be brought to mind can obviously lead to unjustified biases. This heuristic at least partially accounts for the media's capability to mislead or distort the public's perception of risk. Under these circumstances and conditions, it is not surprising that it is often difficult, if not impossible, to evaluate the public's acceptability of an assigned risk.

9.2.1 *Advanced Theory and Risk*

An anatomy of human perception and its influence on discretion or choice behavior based on experimental evidence is generalized in cumulative prospect theory. Under this enhanced theory, one can no more utilize an anticipated value (i.e., probability and consequence) to describe the preference ordering of options. Instead, one must also incorporate functions that account for the differences in perception due to the different ways in which problems are framed (i.e., the observer's conception of the problem, consequences, and contingencies). Therefore, instead of the common expected value of risk, one gets

$$\begin{aligned} V(f) &= V(f^+) + V(f^-) \\ V(f^+) &= 0 \, (\text{no gain for terrorists attack}) \end{aligned} \tag{9.1}$$

Therefore,

$$\begin{aligned} V(f) &= V(f^-) \\ V(f) &= \omega(p)\mathrm{v}(x) \end{aligned} \tag{9.2}$$

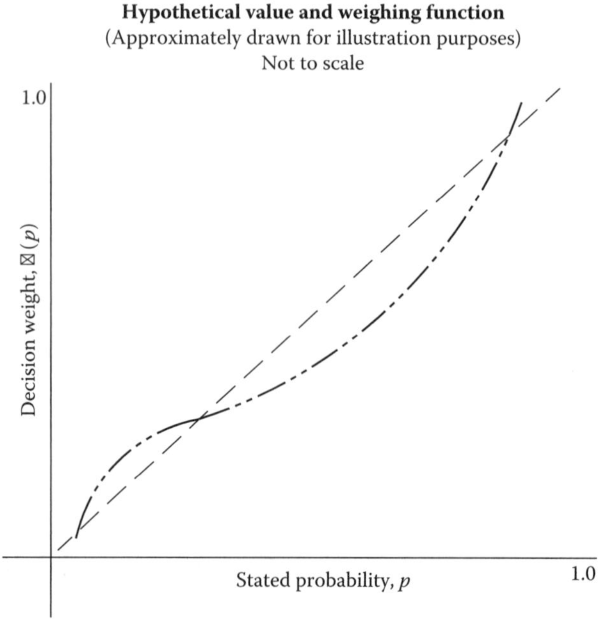

Figure 9.1 Hypothetical value and weighing Function. Hypothetical probability function. (Data from Doro-on, A. M., "Risk assessment embedded with cumulative prospect theory for terrorist attacks on aquifer of karstic limestone and water supply system," PhD diss., University of Texas at San Antonio, 2009; Shih, C. S., A. M. Doro-on, and G. A. Arroyo, *Risk Assessment of Terrorism Based on Prospect Theory for Groundwater Protection. Vol. 1 Environmental Science and Technology,* Houston: American Science Press, 2007; Kahneman, D., and A. Tversky, *Choices, Values and Frames,* New York: Russell Sage Foundation and Cambridge University Press, 2000.)

where

$\omega(p)$ = decision weight associated with the probability of occurrence
$v(x)$ = values associated with consequences
$V(f)$ = risk

Hypothetical value and decision weight functions derived from Kahneman and Tversky (2000) are depicted in Figure 9.1. If $\omega(p)$ and $v(x)$ were a straight line, an individual or a decision maker's choice would be exclusive of the problem's framing. However, due to characteristic nonlinearities, different frames can lead to different choices even though the expected values of the options remain the same (Shih and Riojas 1981; Shih, Doro-on, and Arroyo 2007; Doro-on 2009).

Besides the theoretical and experimental work done on cumulative prospect theory, a great deal has been done to determine the inferred or intuitive factors involved in the development of perception (Shih and Riojas 1981; Doro-on 2009).

Some of the most complete analyses, at least for the specific area of risk assessments, are presented in Sections 9.2.1.1 through 9.2.1.8.

9.2.1.1 Voluntary or Involuntary

Perception appears to be clearly affected by whether a risk is incurred by choice or not. For instance, one normally expects a worker at a hazardous waste facility such as an ammunition plant or nuclear power plant to be much more tolerant to risk than the surrounding people.

9.2.1.2 Discounting Time

An event currently happening tends to be valued higher than the same event occurring at some time in the past or in the future. This corresponds with the long-held financial concept that a dollar in the past is worth more than the same dollar today according to an inflationary perspective of the world. The length of time one is subjected to a risk also seems to affect the valuation process in the form of discounting risk (Nogami and Streufort 1973).

9.2.1.3 Identifiability of Taking a Statistical Risk

Whether a risk will be taken by or imposed on individuals or groups with which one identifies or in which one is just a "number in the crowd" influences one's perception of risk. A classic example of this can be seen in the expending of huge amounts of money to rescue trapped miners who have become identifiable while begrudging support to routine safety budgets. Known circumstances and conditions are more highly valued than hypothetical ones.

9.2.1.4 Controllability

People appear to accept higher risk when they comfortably feel that the situation is well controlled such as when they are driving an automobile.

9.2.1.5 Position in Hierarchy of Consequence

The desire to prevent an unwanted consequence depends heavily on the perceived undesirability, that is, position in a desirable–undesirable hierarchy, of the consequence as shown in Table 9.1. As a result, once would normally expect the threshold for acknowledging risk to be much lower for fatal or catastrophic situations than for ones involving risk to security.

9.2.1.6 Ordinary or Catastrophic

A large number of fatalities happening in a single accident have a greater impact than the equivalent number of fatalities spread randomly over a number of smaller

Table 9.1 Consequence Hierarchy

Lowest priority	*Self-actualization*
	Egocentric
	Belonging/love
	Security
	Exhaustible resources
	Survival factors
	Illness and disability
Highest priority	Death

Source: Data from Doro-on, A. M., "Risk assessment embedded with cumulative prospect theory for terrorist attacks on aquifer of karstic limestone and water supply system," PhD diss., University of Texas at San Antonio, 2009.

accidents within the same period. For instance, a greater risk tolerance is expressed by the public for automobile accidents (which are normally ordinary) versus commercial aviation accidents (which tend to be catastrophic).

9.2.1.7 Natural- or Man-Originated

Risks imposed by natural causes such as earthquakes tend to be much more easily tolerated versus man-originated risks (e.g. terrorist attacks) probably because man has always considered that natural disasters are attributed to acts of God.

9.2.1.8 Magnitude of Probability of Occurrence

The perceptions of a consequence are not continuously influenced by the degree of the probability of that consequence. This often results in very small probabilities being inflated and high level of risks are acceptable to an individual or group can be expected to vary. As a result, we see situations such as the nuclear power plant controversy on acceptability of a risk event.

If an individual desires to use subjective perceptions in one's assessments, then risk must be distinguished into categories that correspond with the variations between reality (or one's best discretion) and these perceptions.

9.3 Strategic Determination of Risk Acceptability

A number of possible strategies for addressing the questions "How safe is safe enough?" and "Which risks are acceptable?" have been proposed. Three basic approaches can be readily identified. The first is the formal analysis approach. The principal methods included in this category are benefit–cost analysis and decision analysis. This approach relies heavily on formal logic and optimization principles. Meanwhile, *cost-benefit analysis* involves the analysis of cost effectiveness of risk reduction, while *benefit-cost analysis* is involved when risk is a surrogate for social cost.

The next technique is the comparative analysis approach, which is composed of three distinct methods: (1) revealed preference, (2) expressed preference, and (3) natural standards. An absolute acceptable risk boundary is developed against which the estimated risk can be evaluated. The last crucial category is professional judgment (e.g., scientific and engineering judgment). This relies principally on the perceptive intelligence and experience of the professional individual or group. A detailed comparison of each approach utilizing the following five key characteristics is illustrated in Table 9.2: decision-making criteria, locus of wisdom, principal assumptions, possible decision attributes, and data requirements. Each of these techniques has strengths and weaknesses (Table 9.3). Benefit–cost analysis is very limited, since any element that cannot be transformed to economic terms is disregarded. The formal analysis methods, particularly the decision analysis, impart structure assessment. Both formal techniques, benefit–cost analysis and decision analysis, require large amounts of detailed and reliable data and information and failure to entrench public subjective perceptions of risk into the equation.

All the comparative analysis methodologies have the advantage of determining *absolute* risk boundaries. All three techniques are only intended to address risk and they are incapable of handling the overall decision problem. Furthermore, both revealed and expressed preference methods are dependent on the limitations of society and its citizens.

9.4 Quantitative Revealed Societal Preference Method

The quantitative revealed societal preference method examines existing databases relative to societal risk before using these data to calculate the relative impact of risk factors as risk referents for use in risk acceptability analysis. According to Rowe (1977), two aspects of risk valuation are addressed: (1) relative risk and (2) absolute risk. Relative risk provides an initial screening at the effect of risk factors or risk rates on risk valuation through the comparison of different risks, whereas absolute risk represents an effort to evaluate, analyze, and differentiate quantitatively the risk acceptance levels for all type of risks based on revealed societal preference. Meanwhile, accidental risks (e.g., nuclear power accidents) provide the most straightforward database, which can be used as a comparison to risks related to terrorist attacks, as there are no standard risk factors associated to terrorist attacks.

9.4.1 *Behavior and Risk Attitude*

Revealed societal preferences are used to generate a risk referent according to the typical notion that the societal behavior is acceptable no matter whether it is right

Table 9.2 Comparison of Techniques

Techniques	*Decision Making*				*Data Requirement*
Formal Analysis	*Criteria*	*Locus of Wisdom*	*Principal Assumption*	*Decision Attributes*	*Possible*
Benefit–cost analysis	Most favorable economic condition	Standardized intellectual procedure	Man should be a rational economic maximizer; decision is entirely objective	Anything that can be materialized to money	All significant events, conditions, and consequences related to money or wealth; precise probabilities and risk magnitudes for each
Decision analysis	Utility optimization	Formalized intellectual processes	Man is or should be a rational utility maximizer; decision should utilize the decision makers' value (professional) judgments	Any value	All significant events, conditions, and consequences related to money or wealth; precise probabilities and risk magnitudes for each
Revealed preference	Preservation of historical stability	Decisions were made historically by society	Decisions made historically by society were optimal; little or no change in the events	Risk only	Present risk; historical risk

Source: Data from Doro-on, A. M., "Risk assessment embedded with cumulative prospect theory for terrorist attacks on aquifer of karstic limestone and water supply system," PhD diss., University of Texas at San Antonio, 2009.

Table 9.3 Techniques' Strengths and Weaknesses

Techniques of Formal Analysis	*Strengths*	*Weaknesses*
Benefit–cost analysis	• Logical approach • Easy to perform analysis • Deal with all types of decision aspects	• Discounts attributes that cannot be easily transformed to economic terms and conditions • Requires ample detailed data and/or information • Difficulty to process subjective value judgments
Decision analysis	• Logical approach • Easy to perform analysis • Compliant or adaptable • Deal with all types of decision aspects • Incorporates decision maker's (professional) judgment • Handles uncertainty	• Requires ample detailed data and/or information • Cannot deal with public perceptions of risk
Revealed preference	• Establishes absolute boundaries • Incorporates historical experience and events	• Historical decisions often were not always optimal • Situations and events change rapidly with time • Disaggregated historical baseline hard to establish • Does not address the entire decision • Subject to inherent limitations of society and its citizens

Source: Data from Doro-on, A. M., "Risk assessment embedded with cumulative prospect theory for terrorist attacks on aquifer of karstic limestone and water supply system," PhD diss., University of Texas at San Antonio, 2009.

or wrong. Harry Otway (1975) pronounced the use of existing societal behavior of this nature the method of revealed societal preferences, which involves the psychological and psychometric study of behavior in identified groups or strata of society, and attempting to measure attitudes toward risk as opposed to risk behavior.

9.4.2 Establishing Risk Comparison Factors

Risk comparison factors can be determined for different types of consequences and risks. Risk data are given for fatalities, illnesses, property damage, life-shortening factors, and productive days lost, by some government agencies such as the U.S. Census Bureau (USCB) or U.S. Environmental Protection Agency (EPA), which can be analyzed and quantified. Nevertheless, there are not enough available risk data for terrorism; therefore, data of fatal automobile accidents and/or nuclear power accidents will be utilized for the risk analysis of terrorist attacks on groundwater and the water supply system in the United States. The magnitude of fear and consequences created by nuclear power accidents are comparable to terrorism. Additional data involves consequences of types that are less agreeable to objective standards, such as esthetic values and quality of life. The EPA actually listed four major life factors (USEPA Quality of Life Indicators 1973, 2009): (1) household and environmental economic condition (e.g., adequate income and job opportunities), (2) health (e.g., safety and environmental sustainability), (3) natural resources and amenities, and (4) vibrant community (e.g., attracts businesses and retirees). It is evident that humans accept different levels of risk for different types of risk (e.g., voluntary risk vs. involuntary risk) (Velimirovic 1975). Refer to Chapter 8, Section 8.3 for risk conversion factors.

9.4.3 Controllability of Risk

Controlling risk based on one's perception of controllability as an individual or group and the *degree of systematic control* provided by regulatory requirements, and technological and institutional processes can potentially increase the value of consequences and risk acceptability. Technological innovations to improve security from terrorist attacks, reduce water contamination and hazards, and prevent dam failure and accidental mishaps are commonplace. On the other hand, society is becoming increasingly aware and focused on requiring that sophisticated technology be used to protect the entire population. Therefore, terrorist attacks on aquifers and water supply systems can be mitigated when new sophisticated technology for security and surveillance are implemented. Reduction of risk is in itself considered to be a benefit. Three main classes of benefit are as follows: (1) materialistic (economic survival), (2) physical protection and security (e.g., protection against terrorist attacks), and (3) self-advancement (free from chaos and distress).

9.4.4 Perceived Degree of Control

The perceived *degree of control* (as opposed to the "real degree of control") to avoid a risk consequence by a valuing factor is a primary condition in defining consequence value. The degree of controllability, whether real or perceived, must be crucially considered.

9.4.5 System Control in Risk Reduction

A society concerned about exposure to risks from new or ongoing activities of humans or from natural causes can achieve the reduction of risk systematically. For example, flood control projects by the Federal Emergency Management Agency (FEMA) save many lives from naturally occurring flood conditions. Terrorist attacks against groundwater resources can be prevented if protection and security technology are installed on major recharge system areas or at the original water source.

9.4.5.1 Systemic Control of Risk

More formally, systemic control of risk as presented in Table 9.4 requires a standard procedure that must be implemented to control risks, which includes the following:

- A standard measure of controlling and reducing risk that is given the most emphasis in the design and operation of the technological system involved
- A regulatory requirement or policy of the overall system to assure maximum safety and security
- A system design that includes the following: quality control, redundancy for critical systems, training and educating of personnel involved, and ongoing screening of system performance to meet enforcement and auditing system goals in accordance with the regulatory or policy requirements

Table 9.4 Systemic Control of Risk

Systemic Control of Risk		
Positive	*Level*	*Negative*
1. Risk must be balanced with lesser value to ensure that the risk per unit of measure of technological system performance and operation is decreasing over time.	1. Risks increase over time no faster than the technological system's rate of development, either absolutely or relatively. Or risks maintain the same value over time.	1. When the systemic control concept is not considered and/or a technological system whose risk behavior is characterized by an increase in risk over time.

(Continued)

Table 9.4 Systemic Control of Risk (*Continued*)

Systemic Control of Risk		
Positive	*Level*	*Negative*
2. Technological systems that are designed and built with positive systemic control as a goal (e.g., weapons and defense systems)	2. Whose risk behavior is described by a equilibrium level of risk over time.	
Other Types of Control		
Control through Specific Design Features	*Control by Inspection and Regulation*	*Risk Management System*
Safety is achieved through special and specific design features of the technological system that provide safety and security (e.g., the use of an alternate reverse osmosis system to remove prescription drugs in the water supply is an example of specific design features to reduce risk).	Positive control is to be achieved, not through special and specific designs but through inspection and regulatory requirements for the technological systems (e.g., the recall of defective Toyota Yaris automobiles in the United States that occurred between 2009 and 2010 is an example of proposed positive control). Whether or not it is combined with specific design features, control by inspection and regulation is part of reducing risk in high-risk situations.	Combine control through special design features of the technological system involving inspection and regulation. It is a crucial measurement for achieving systemic control.

Source: Data from Doro-on, A. M., "Risk assessment embedded with cumulative prospect theory for terrorist attacks on aquifer of karstic limestone and water supply system," PhD diss., University of Texas at San Antonio, 2009; Rowe, W., *An Anatomy of Risk*, New York: John Wiley & Sons, 1977.

9.4.5.2 Control Factors

The four control factors required to give a degree of controllability value for every combination, are as follows: (1) control approach, (2) degree of control, (3) state of implementation, and (4) basis for control effectiveness.

9.4.6 Controllability of New Technological Systems

Controllability of new technological systems indicates the requirement for practice of systemic control of risk. Since no data are established to evaluate controllability, the effective calculated different levels of control are based entirely on judgment of value, in this condition, the author's engineering discretion.

The level of desirability of control is defined as

$$F3 = C_1 \times C_2 \times C_3 \times C_4 \tag{9.3}$$

where $F3$ = condition for a given risk with control (minimum of 0.01 and maximum of 1.0), C_1 = no control, C_2 = uncontrolled, and C_3 and C_4 are ignored (set at unity).

9.4.7 Cost–Benefit Analysis

A cost–benefit (loss–gain) analysis is consists of two processes, as follows: (a) first, overall comparability of gains and losses; and (b) second, a specific analysis to determine whether inequities have been improved. Richard Wilson (1975) presented a four-step process, as follows: (a) we must be sure that we understand the benefit and the risk and that the former outweighs the latter; (b) we must be sure we have chosen the method of achieving the benefit with the least risk; (c) we must be sure we are spending enough money to reduce the risk further; and (d) we go back and recheck our numbers with a new perspective from the preliminary calculations.

Rowe's (1977) technique is composed of four principal parts: (1) design an applicable risk classification scheme; (2) define an absolute risk reference for each category in the scheme; (3) using risk references as a basis, quantify the risk reference that performs as the acceptability boundary for particular conditions; and (4) examine and balance the estimated risk within an order of magnitude of the reference to be acceptable. As indicated in Table 9.5, these processes explicitly include the objective to subjective transformation factors. Risk assessments must be divided into different parts to understand the aspects that direct to subjective perception.

The fundamental classification scheme advocated by Rowe (1977) is perception. The basic classification scheme advocated by Rowe is shown in Table 9.6. In addition, the hierarchy of consequences as shown in Table 9.1 illustrates the value of a consequence is associated to life and health.

Once a classification scheme is applied, an absolute risk reference must be determined and defined for every category. These are approximated definitely from

Table 9.5 Transformation Factor Utilization in Risk Referents

Factors that are unequivocally integrated in the absolute risk reference determination • Voluntary or involuntary • Discounting of time • Identifiable statistical risk taker • Position in hierarchy of consequences • Ordinary or catastrophic • Natural or human-originated
Factors that are unequivocally integrated in the determination of risk referent • Controllability • Propensity for risk taking
Other factors • Magnitude of probability of occurrence

Source: Data from Doro-on, A. M., "Risk assessment embedded with cumulative prospect theory for terrorist attacks on aquifer of karstic limestone and water supply system," PhD diss., University of Texas at San Antonio, 2009; Shih, C., and A. Riojas, In *Risk Assessment for Groundwater Pollution Control*, Editors: McTernan, W., and Kaplan, E., American Society of Civil Engineers, 1990; Rowe, W., *An Anatomy of Risk,* New York: John Wiley & Sons, 1977.

historic societal risk data as *revealed preferences*. The risk references derived from the data provided by Rowe (1977) for immediate statistical accidents are shown in Table 9.7.

For the risk, $V(f)$, defined by Kahneman and Tversky's cumulative prospect theory, risk reference is essentially the value of consequence, $v(x)$ (Doro-on 2009). Using the revealed preference concepts, the $v(x)$ or risk reference is really the current incremental acceptable risk by U.S. society as shown in Figure 9.2 (Doro-on 2009).

9.4.8 Prerequisites for Risk Acceptance of Terrorist Attacks against Groundwater and the Water Supply System

Before undertaking the development of a methodology for risk acceptance, a number of questions must be asked. "Is there a need for risk acceptance?" "How and where shall it be used?" "What methods and techniques are currently available?" "What alternative approaches can be employed?"

Table 9.6 Classification of Acceptable Risk

Immediate statistical	
1. Natural	
a. Catastrophic	Involuntary
b. Ordinary	Involuntary
2. Man-originated	
a. Catastrophic	Voluntary and involuntary
b. Ordinary	Voluntary, regulated voluntary, and involuntary
3. Man-originated	
a. Catastrophic	Involuntary
b. Ordinary	Voluntary, regulated voluntary, and involuntary
Immediate identifiable (1)	
Delayed statistical (1)	
Delayed identifiable (1)	
(1) Same as immediate statistical	

Source: Data from Doro-on, A. M., "Risk assessment embedded with cumulative prospect theory for terrorist attacks on aquifer of karstic limestone and water supply system," PhD diss., University of Texas at San Antonio, 2009; Shih, C., and A. Riojas, In *Risk Assessment for Groundwater Pollution Control*, Editors: McTernan, W., and Kaplan, E., American Society of Civil Engineers, 1990; Rowe, W., *An Anatomy of Risk*, New York: John Wiley & Sons, 1977.

9.4 8.1 Need for a Methodology

Humans are naturally risk averse, but they are willing to take risks to achieve specific benefits and personal desires when the choice is under their direct control. When the risk is imposed by humans or nature as "acts of God" without immediate gain, however, risk averse action dictates. The subjects of news reports, a reflection of society's news preferences, make it evident that society is more concerned with controversial and undesirable consequences than with benefits. Disaster or terrorism reports and political controversy news overshadow news about achievements and health benefits.

Table 9.7 Summary of Risk References

Risk Classification			*Class of Consequences*	
	Fatality/ Year	*Health Effects/Year*	*Property Damage ($)/Year*	*Life Span Shortened/ Year*
Naturally occurring				
Catastrophic	9.5×10^{-7}	4.8×10^{-6}	0.02	2.8×10^{-2}
Ordinary	6.8×10^{-5}	3.8×10^{-4}	2.8	0.2
Man-originated catastrophic				
Voluntary	1.8×10^{-6}	1.8×10^{-6}	0.38	5.8×10^{-3}
Regulated Voluntary	2.8×10^{-5}	2.8×10^{-6}	0.38	5.8×10^{-2}
Involuntary	9.8×10^{-8}	4.8×10^{-7}	1.8	2.8×10^{-2}
Ordinary				
Involuntary	4.8×10^{-6}	2.8×10^{-5}	1	9.5×10^{-3}
Voluntary	5.8×10^{-4}	2.8×10^{-1}	200	1
Regulated Voluntary	9.5×10^{-5}	5.8×10^{-2}	40	0.1
Man-originated catastrophic				
Involuntary	1.8×10^{-7}	1×10^{-6}	3.8×10^{-2}	5.8×10^{-4}
Voluntary	3.8×10^{-6}	4×10^{-6}	0.75	5.8×10^{-3}
Ordinary				
Involuntary	9.8×10^{-6}			2.8×10^{-2}
Voluntary	9.8×10^{-4}			1.8
Regulated Voluntary	1.8×10^{-4}			0.18

Source: Data from Doro-on, A. M., "Risk assessment embedded with cumulative prospect theory for terrorist attacks on aquifer of karstic limestone and water supply system," PhD diss., University of Texas at San Antonio, 2009; Shih, C. S., A. M. Doro-on, and G. A. Arroyo, *Risk Assessment of Terrorism Based on Prospect Theory for Groundwater Protection. Vol. 1 Environmental Science and Technology*. Houston: American Science Press, 2007; Rowe, W., *An Anatomy of Risk*, New York: John Wiley & Sons, 1977.

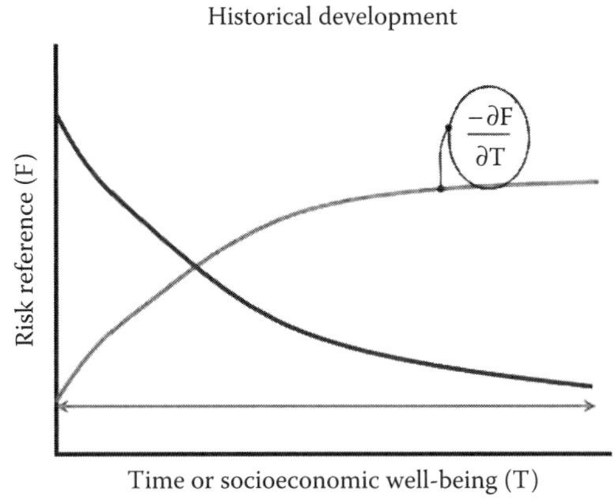

Figure 9.2 Risk reference versus socioeconomic well-being. (Data from Doro-on, A. M., "Risk assessment embedded with cumulative prospect theory for terrorist attacks on aquifer of karstic limestone and water supply system," PhD diss., University of Texas at San Antonio, 2009.)

The risk aversion of society, coupled with increasing awareness of new risks resulting from the side effects of new technology, has focused increased attention on technological risk. The side effects of new technology are probably irreversible, since the knowledge base for technology assessment and risk identification is available to everyone. Consideration of societal risk in all technological approaches in evaluating risk is estimated in two different theoretical models for regulatory approaches: (1) the rational model, and (2) the bureaucratic model.

9.5 Establishing the Risk Referent

Sections 9.5.1 through 9.5.6 illustrate the systematic procedures for developing the risk referent.

9.5.1 Multiple Risk Referents

Different types of risk can be analyzed through the absolute risk levels for involuntary risk and for regulated voluntary risk. Moreover, risks can be compared and balanced across equivalent indirect gains at a certain degree to create final risk values, for the activity is correlated with every equivalent type of risk as referent. When all quantified risks are less than their risk referent counterparts, the net calculated risks are acceptable. If any risks exceed the referents, then the net calculated risks are unacceptable, and therefore, risk reduction shall be employed to make them acceptable.

For the weighting factor of cumulative prospect theory, $\omega(p)$, the considerations of degree of voluntarism ($F1$), benefit–cost balance to society ($F2$), and controllability of risk ($F3$) will be included and quantified (Doro-on 2009). Furthermore, the risk as defined by cumulative prospect theory is essentially the risk referent, which is the incremental acceptable risk of U.S. society: risk referent = $F1 \times F2 \times F3 \times$ risk reference (Doro-on 2009).

9.5.2 Risk Proportionality Factor Derivation From Risk References

In utilizing risk references, it is expected that there is a proportion of total societal risk that is acceptable to society (societal value judgment) to gain indirect benefit, and this is called a *risk proportionality factor*. For example, a greatly beneficial plan to society such as termination of heart disease and breast cancer as a cause of death might shorten the total life span of those not affected by heart disease and breast cancer because the resultant lower death rate might increase the age of the population and the competition for limited resources.

As a value judgment, an extremely beneficial plan to society could be acceptable if the increase of net involuntary societal risks were less than 9% of the overall degree of involuntary risk. This value can be used as a top level for the risk proportionality factor for involuntary risk. In this case, the risk of terrorist attacks against aquifers including water supply systems will be compared to "accidents" to analyze risk and quantify risk acceptability.

If there are no other alternatives available, one expects to assume a greater proportion of risk, or it can be equivalent to all other risks. The author of this book has made a personal judgment of 0.09 for the risk proportionality factor for involuntary risk and a value of 1.0 for the regulated voluntary or voluntary risk.

There are two differences in voluntary risk: One group of risks involves the operator or controller of a technological system (e.g., a light rail transit, LRT). The second group involves the population that is voluntary risk, with appropriate alternatives available (e.g., LRT passengers). The first group of voluntary risks is the operator or driver in this case, and secondly, the society, passengers in this case. Although voluntary absolute risk levels are used for examining the second type of voluntary risk, a risk proportional factor of 0.09 is proposed. If the society is risk averse, it will reject large risks if alternatives are achievable.

9.5.3 Risk Proportionality Derating Factors

A second group of social value judgments is to identify the risk proportionality derating factors for smaller favorable indirect gain–loss balances. The five conditions of indirect gain–loss balance are presented in Table 9.8: (1) favorable balance, (2) marginal favorable balance, (3) indecisive balance, (4) marginal unfavorable balance, and (5) unacceptable balance. Note: A factor of 1.0 represents a doubling of existing risk for the new proposed scheme. A factor of 0.09 is 9% of the present risk. In this book, the derating functions shown in Table 9.9 have been selected as "straw men" values.

Table 9.8 Risk Proportionality (*F*1) and Derating Factors (*F*2)

	Involuntary	*Regulated*
Factor	*Risk*	*Voluntary*
Proportionality factor (*F*1)	0.09	1.0
Derating factor (*F*2)		
Balance		
Favorable	1.0	1.0
Marginal favorable	0.09	0.18
Indecisive	0.0081	0.09
Marginal unfavorable	0.00073	0.018
Unfavorable	0.000065	0.009

Source: Data from Doro-on, A. M., "Risk assessment embedded with cumulative prospect theory for terrorist attacks on aquifer of karstic limestone and water supply system," PhD diss., University of Texas at San Antonio, 2009; Rowe, W., *An Anatomy of Risk*, New York: John Wiley & Sons, 1977.

Table 9.9 Controllability Factors (*F*3)

Control Approach		*Degree of Control*		*State of Implementation*		*Control Effectiveness*	
Factor	*C1*	*Factor*	*C2*	*Factor*	*C3*	*Factor*	*C4*
Systematic control	1.0	Positive	1.0	Demonstrated	1.0	Absolute	1.0
Risk management system	0.80		0.80		0.80		0.80
Special design features	0.55		0.55	Proposed	0.55	Relative	0.55
Inspection and regulation	0.25	Level	0.25		0.30		0.30

(*Continued*)

Table 9.9 Controllability Factors (*F*3) (*Continued*)

Control Approach		*Degree of Control*		*State of Implementation*		*Control Effectiveness*	
Factor	*C1*	*Factor*	*C2*	*Factor*	*C3*	*Factor*	*C4*
		Negative	0.18		0.2		0.2
No control scheme	0.08	Uncontrolled	0.08	No action	0.08	None	0.08

Source: Data from Doro-on, A. M., "Risk assessment embedded with cumulative prospect theory for terrorist attacks on aquifer of karstic limestone and water supply system," PhD diss., University of Texas at San Antonio, 2009; Rowe, W., *An Anatomy of Risk*, New York: John Wiley & Sons, 1977.

9.5.4 Degree of Systemic Control

The degree of risk that society is tolerating in a current situation is not always acceptable to society; or the society may not be satisfied with the current level of risks. In this condition, the society will want to minimize the risk compared to the present level of risk. The author has used the values derived in Table 9.10 for the risk controllability factor *F*3, which is the product of the four factors listed in Table 9.10.

9.5.5 Conversion of a Risk Reference to a Risk Referent

The conversion of a risk reference to a risk referent requires three factors:

1. Establish the appropriate *risk proportionality factor*, that is, the fraction of existing societal risk or known as *risk reference*, that would be considered acceptable in a condition where there was a very favorable indirect benefit–cost balance, for both regulated voluntary (or voluntary) and involuntary risks (*F*1).
2. Establish a factor that is the *risk proportionality derating factor*, which can be applied in those conditions where the indirect benefit–cost balance is not as favorable, which transforms the *risk proportionality factor* in those identified conditions (*F*2).
3. Establish the modification factor related to the degree of risk controllability (*F*3).

Using the three aforementioned factors, calculate the risk referent, which is the incremental acceptable risk ("$V(f)$" in cumulative prospect theory) based on current socioeconomic well-being in the United States (or in another society, country, or nation).

$$\text{Risk referent} = \text{risk reference} \times F1 \times F2 \times F3 \tag{9.4}$$

The first two factors deal with the fundamental propensity of individuals and/or groups to take risks and integrate the additional decision aspect of indirect benefits/costs.

Table 9.10 Risk of Terrorism on Water Infrastructure

(General Overview)				
Illustration of Alternatives				
Alternative	*Degree of Voluntarism (F1)*	*Benefit–Cost Balance (F2)*	*Controllability (F3)*	*Risk Referent Values*
E1: Business as usual	0.09	0.000065	0.000041	2.4×10^{-10}
E2: Protection of major recharge zone	1.0	0.0081	0.41	3.2×10^{-10}
E3: Complete protection over recharge zone	1.0	1.0	0.41	4×10^{-8} (Acceptable)
General overview: Values for terrorist attacks against groundwater resources and water supply systems				
Risk reference = [catastrophic, man-originated: fatality/year] = 9.8×10^{-8}				
Proportionality by degree of voluntarism = 0.09				
Derating = cost–benefit balance = 6.5×10^{-5}				
Controllability = 4.1×10^{-5}				

This acknowledges the tendency for people to accept a higher level of risk if the benefit to them more than offsets the imposed risk or for people to be increasingly risk averse in the opposite case. All three of these factors are based on value judgments. The specific numbers in Table 9.8, risk proportionality and proportionality derating factor ($F1$ and $F2$), and Table 9.9, controllability factor ($F3$), are based on the straw men values originally posed by Rowe (1977) and modified in this book based on the author's scientific and engineering judgment.

The overall controllability factor is the result of multiplication of four subfactors ($F3 = C1 \times C2 \times C3 \times C4$). The four subfactors are as follows: (1) control approach (i.e., the type of risk control management used), (2) degree of control (i.e., effectiveness of risk control), (3) state of implementation, and (4) basis for control effectiveness.

Meanwhile, the relationship of cumulative prospect theory and risk referents is as follows (Doro-on 2009):

$$V(f) = \omega(p) \cdot \text{v}(x) \quad \text{(from Equation 9.2)}$$

$$\text{Risk referent} = \{F1 \times F2 \times F3\} \times \text{risk reference}$$

where

$V(f)$ = risk referent
$\omega(p) = \{F1 \times F2 \times F3\}$
$v(x)$ = risk reference

Therefore, incremental risk acceptability = $V(f)$ = risk referent:

$$V(f) = \{F1 \times F2 \times F3\} \times \text{risk reference} \tag{9.5}$$

An illustration of alternatives using the incremental risk acceptability in general overview is presented in Table 9.10.

9.5.6 Risk Estimation and Risk Acceptability for Water Infrastructure

Based upon the simple form of *event tree* analysis for terrorist attacks against the water supply system as shown in Figure 8.4, the risk of 4.33×10^{-5} is very high and the incremental risk acceptability of 2.35×10^{-17} is very low, which means that the existing groundwater protection policy and technological security system shall be improved and revised. The risk acceptability calculation based on Figure 8.4 is

= $[9.8 \times 10^{-8}] \times [9 \times 10^{-2}] \times [6.5 \times 10^{-5}] \times [4.1 \times 10^{-5}]$
= 2.35×10^{-17} (*not acceptable* incremental risk) $\ll 4.33 \times 10^{-5}$ (the estimated risk of potential terrorism, from Chapter 8)

The potential inflow of weapons from U.S. borders is illustrated in Figure 9.3. The detailed event tree analysis for water infrastructure terrorism using arsenic, cyanide, biological threats, prescription drugs, and endocrine disruptors are presented in Figures 9.4a through 9.8g. The designed probability scale based upon the author's refined engineering judgment for the risk estimation model applied in the event tree analysis is provided in Table 9.11. The risk rates for the detailed event tree analysis are provided in Tables 9.12 through 9.16, whereas Tables 9.17 through 9.19 present the calculated risk estimation and risk acceptability based upon the event tree analysis of Figures 9.4a through 9.8g. Currently, the risks related to terrorist attacks are not acceptable to society according to the quantified risks.

Among all the weapons presented in the event tree analysis in Figures 9.4a through 9.8g, the highest potential risk against water supply is posed by the combinations of prescription drugs, endocrine disruptors, and arsenic-/cyanide-based pesticides, since they are easy to acquire in the United States. Prescription drugs and endocrine disruptors are currently difficult to detect in the drinking water supply and they are difficult to treat using the municipal water treatment plant, as the analysis provided in Table 9.19 shows. Moreover, Table 9.20 shows a comparison of proposed alternatives regarding protective measures against terrorist attacks. If the

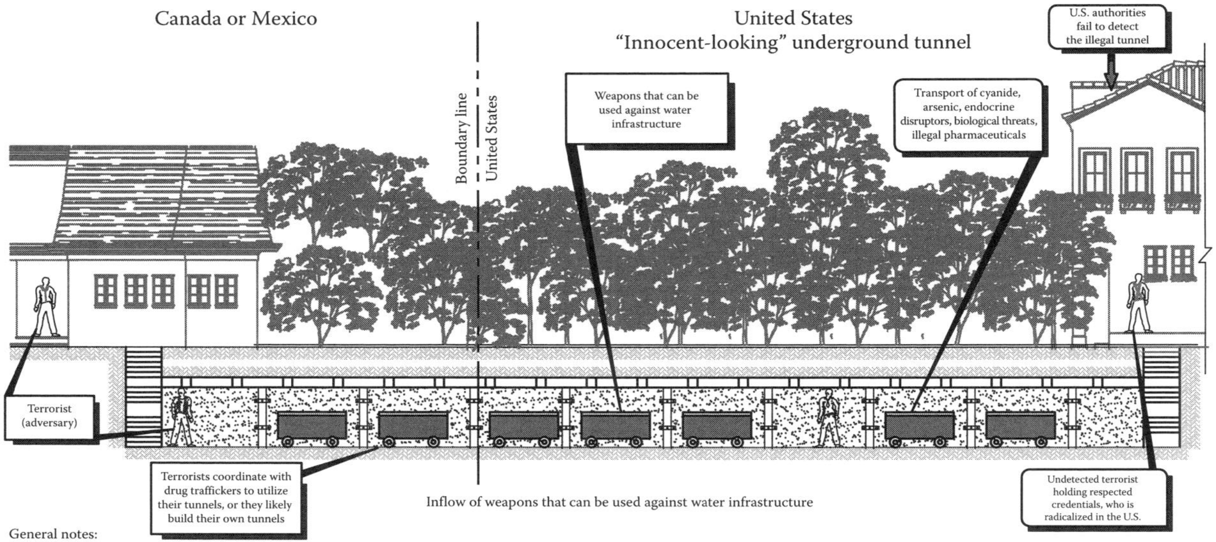

General notes:

1. Traffickers continue to construct tunnels in U.S. borders to transport drugs, illegal aliens, and other contraband, according to an internal briefing prepared by a U.S. Northern task force.
2. Overall, between 1990 and 2008, ninety-three (93) "cross-border" tunnels were discovered, thirty-five (35) of those were in California, fifty-seven (57) in Arizona, and one (1) in Washington state (FAS 2009).
3. Tunnels are typically built to smuggle illegal aliens or illegal drugs into the U.S.
4. Terrorists can intrude into the U.S. easily by using underground tunnels as presented here.
5. According to BBC News, dated July 22, 2005, the only U.S.-Canada tunnel discovered to date was located at Lynden, Washington and Aldergrove, British Columbia, just east of a legitimate port of entry. It served as a "conduit" for illegal drugs. The tunnel was equipped with ventilation and a cart.
6. Every year, about 500 million people cross these borders, and over two-thirds of them are non-U.S. citizens. Some countries have more "porous borders" than the U.S., but the U.S. borders are some of the most porous in terms of being well-traveled. There is one border agent for every five miles of border, and for each person who crosses legally, officials only have 12 seconds to determine their legitimate identity, whether they should be admitted, under what conditions, and for how long (O'connor 2008).

This document contains copyrighted material and confidential trade secret information belonging exclusively to CRC Press - Taylor & Francis. Any unauthorized use, disclosure, dissemination or duplication of any of the information contained herein may result in liability under applicable laws.	"SECRET" UNDERGROUND TUNNELS U.S. MEXICO/CANADA BORDERS DRAWN BY: ANNA M. DORO-ON, PH.D.	9.3 JANUARY 2011

The author and publisher take no responsibility for the actions of anyone as a result of this figure or presentation.

Figure 9.3 Inflow of weapons through tunnels underneath the United States borders.

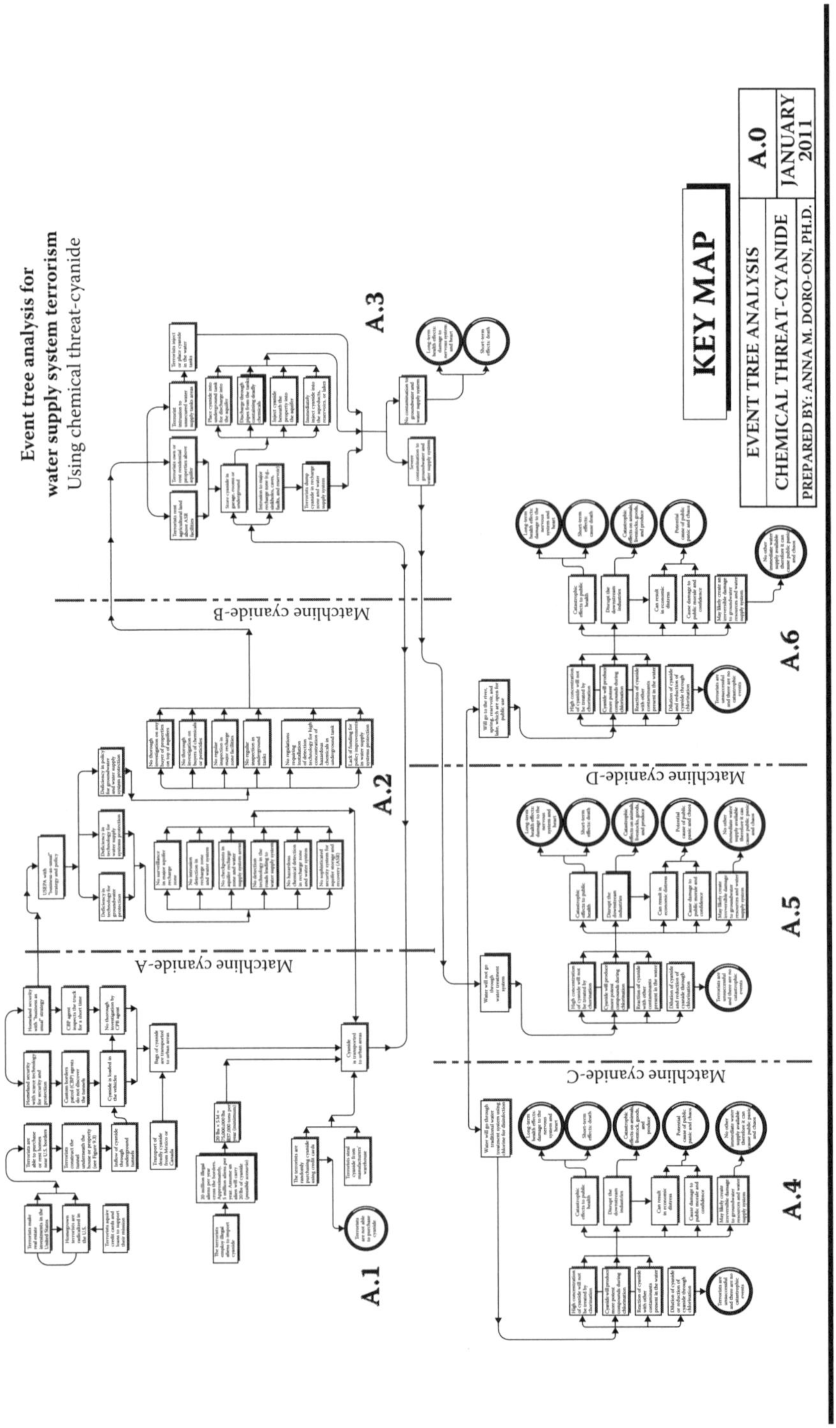

Figure 9.4a Plate A.0: Event tree analysis for water supply terrorism using cyanide.

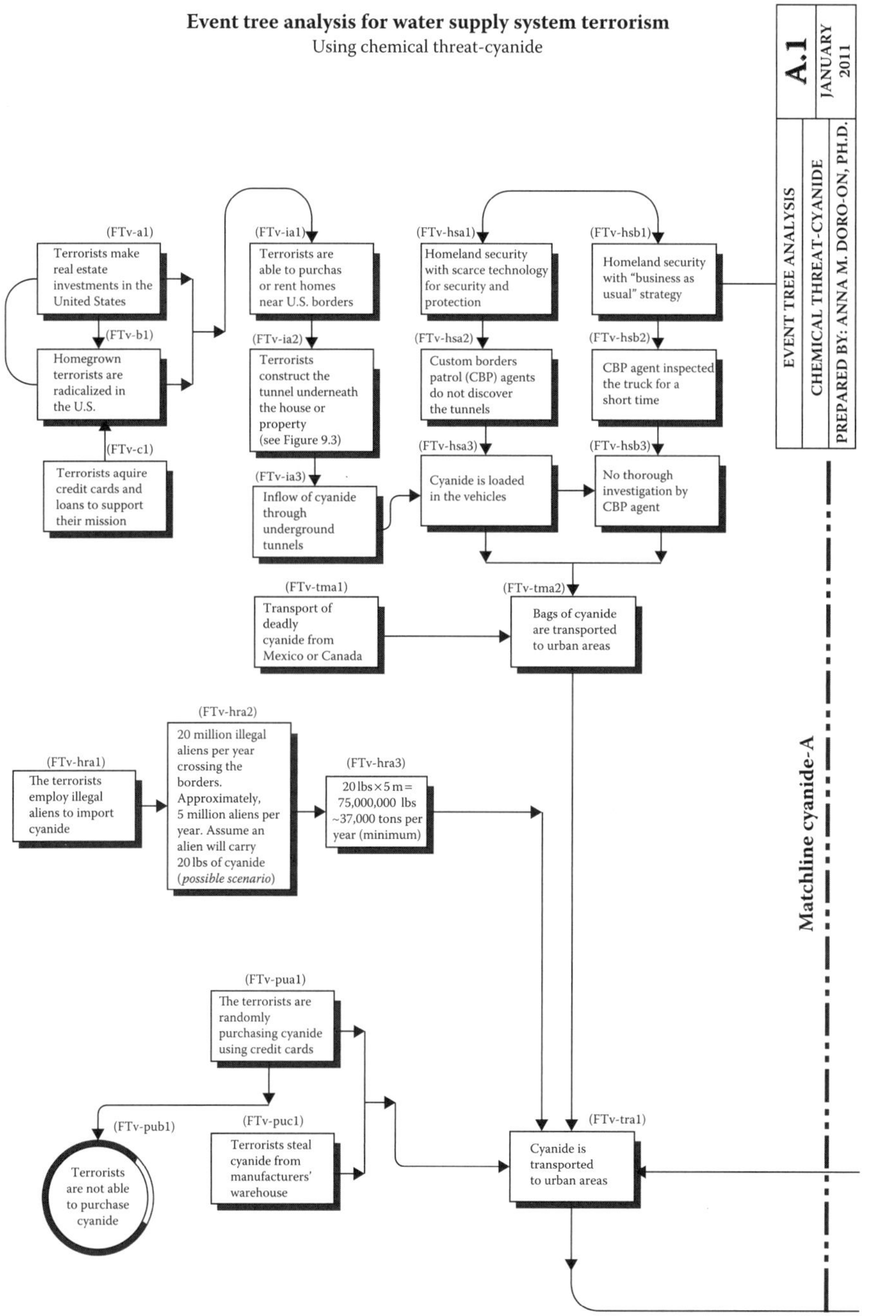
Event tree analysis for water supply system terrorism
Using chemical threat-cyanide
A.1
JANUARY 2011
EVENT TREE ANALYSIS
CHEMICAL THREAT-CYANIDE
PREPARED BY: ANNA M. DORO-ON, PH.D.
(FTv-a1)
Terrorists make real estate investments in the United States
(FTv-b1)
Homegrown terrorists are radicalized in the U.S.
(FTv-c1)
Terrorists aquire credit cards and loans to support their mission
(FTv-ia1)
Terrorists are able to purchas or rent homes near U.S. borders
(FTv-ia2)
Terrorists construct the tunnel underneath the house or property (see Figure 9.3)
(FTv-ia3)
Inflow of cyanide through underground tunnels
(FTv-hsa1)
Homeland security with scarce technology for security and protection
(FTv-hsa2)
Custom borders patrol (CBP) agents do not discover the tunnels
(FTv-hsa3)
Cyanide is loaded in the vehicles
(FTv-hsb1)
Homeland security with "business as usual" strategy
(FTv-hsb2)
CBP agent inspected the truck for a short time
(FTv-hsb3)
No thorough investigation by CBP agent
(FTv-tma1)
Transport of deadly cyanide from Mexico or Canada
(FTv-tma2)
Bags of cyanide are transported to urban areas
(FTv-hra1)
The terrorists employ illegal aliens to import cyanide
(FTv-hra2)
20 million illegal aliens per year crossing the borders. Approximately, 5 million aliens per year. Assume an alien will carry 20 lbs of cyanide (possible scenario)
(FTv-hra3)
20 lbs × 5 m = 75,000,000 lbs ~37,000 tons per year (minimum)
(FTv-pua1)
The terrorists are randomly purchasing cyanide using credit cards
(FTv-pub1)
Terrorists are not able to purchase cyanide
(FTv-puc1)
Terrorists steal cyanide from manufacturers' warehouse
(FTv-tra1)
Cyanide is transported to urban areas
Matchline cyanide-A

Figure 9.4b Plate A.1: Event tree analysis for water supply terrorism using cyanide.

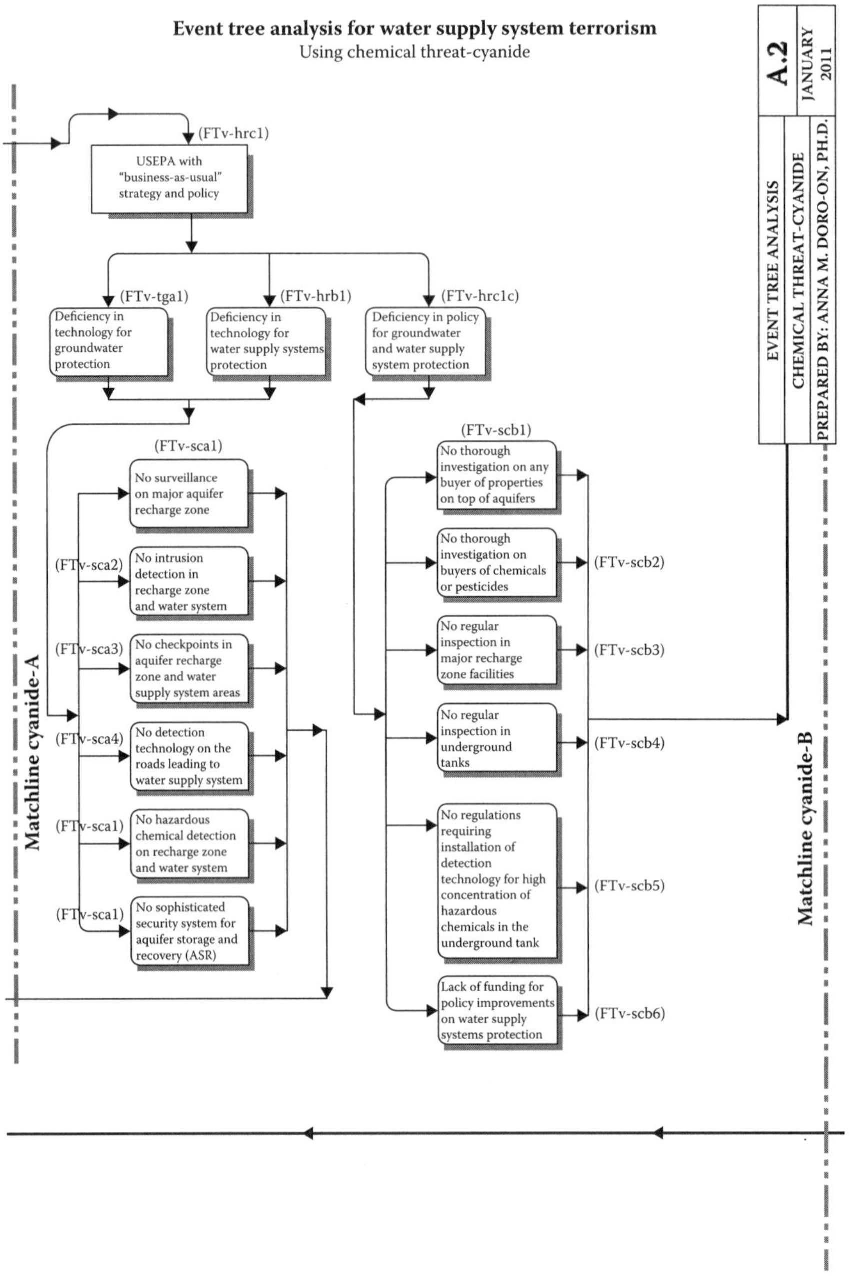

Figure 9.4c Plate A.2: Event tree analysis for water supply terrorism using cyanide.

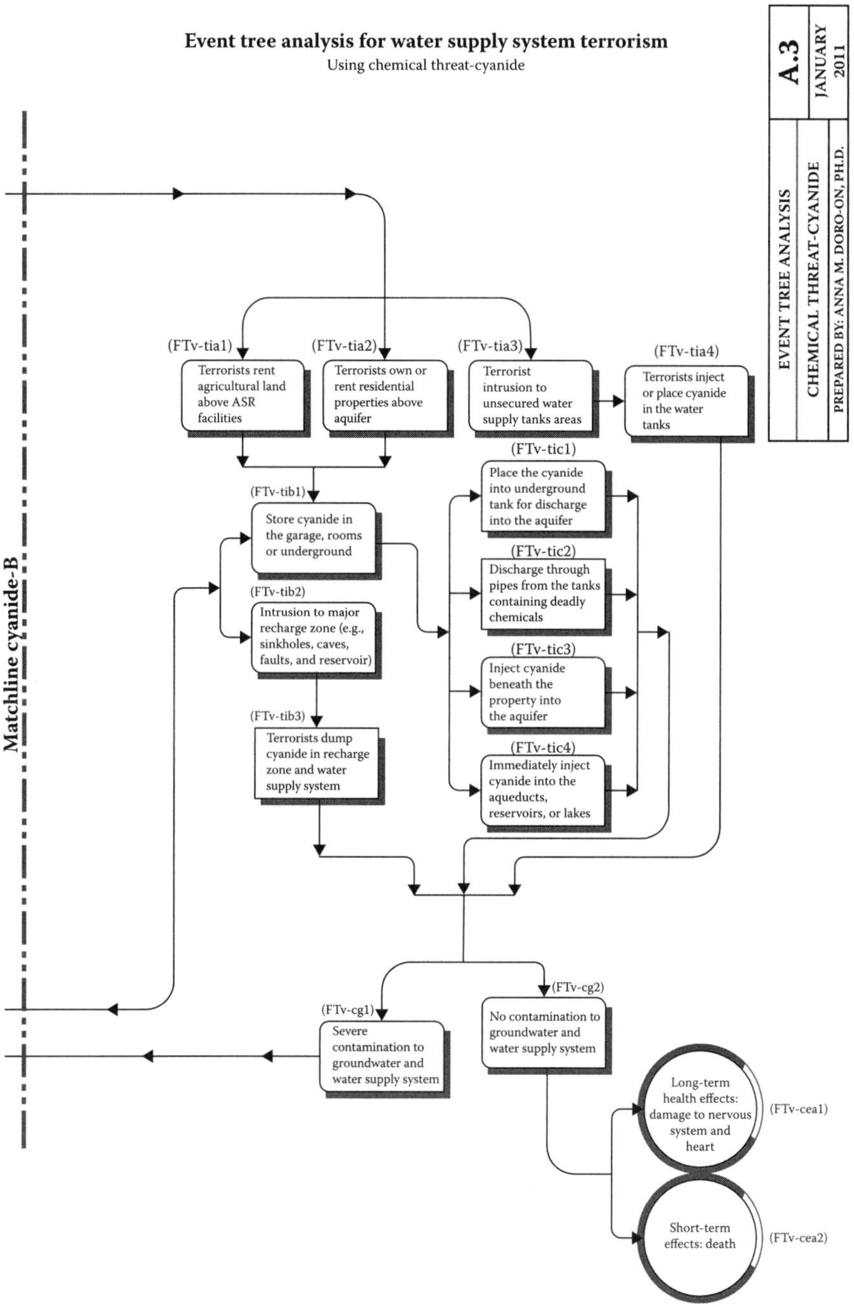

Figure 9.4d Plate A.3: Event tree analysis for water supply terrorism using cyanide.

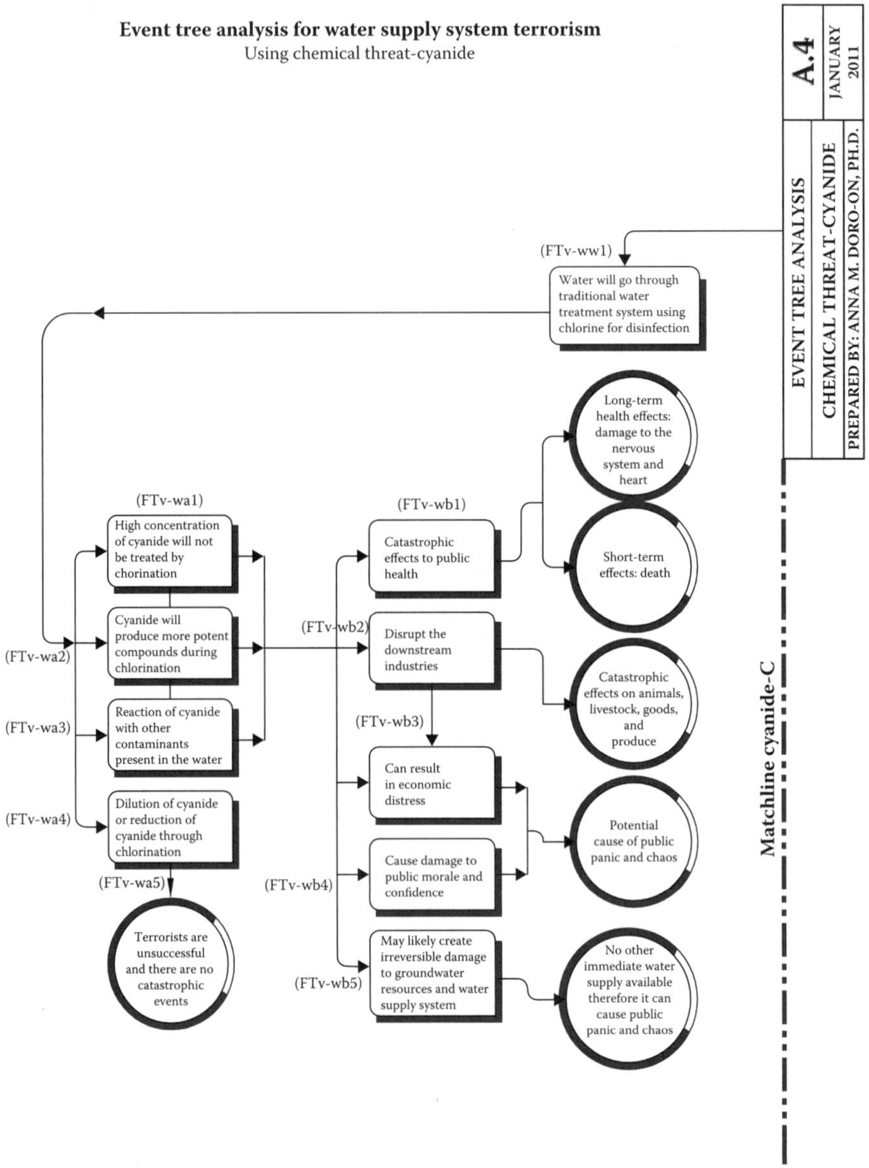

Figure 9.4e Plate A.4: Event tree analysis for water supply terrorism using cyanide.

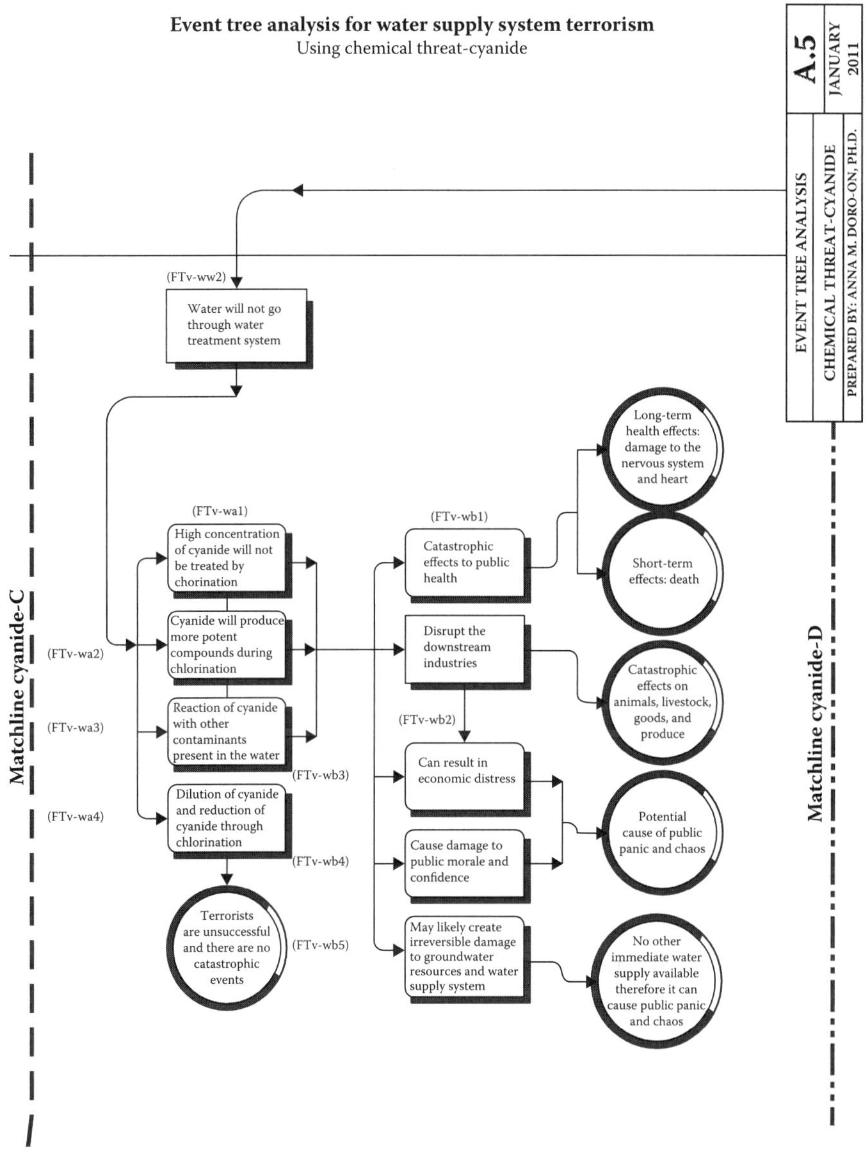

Figure 9.4f Plate A.5: Event tree analysis for water supply terrorism using cyanide.

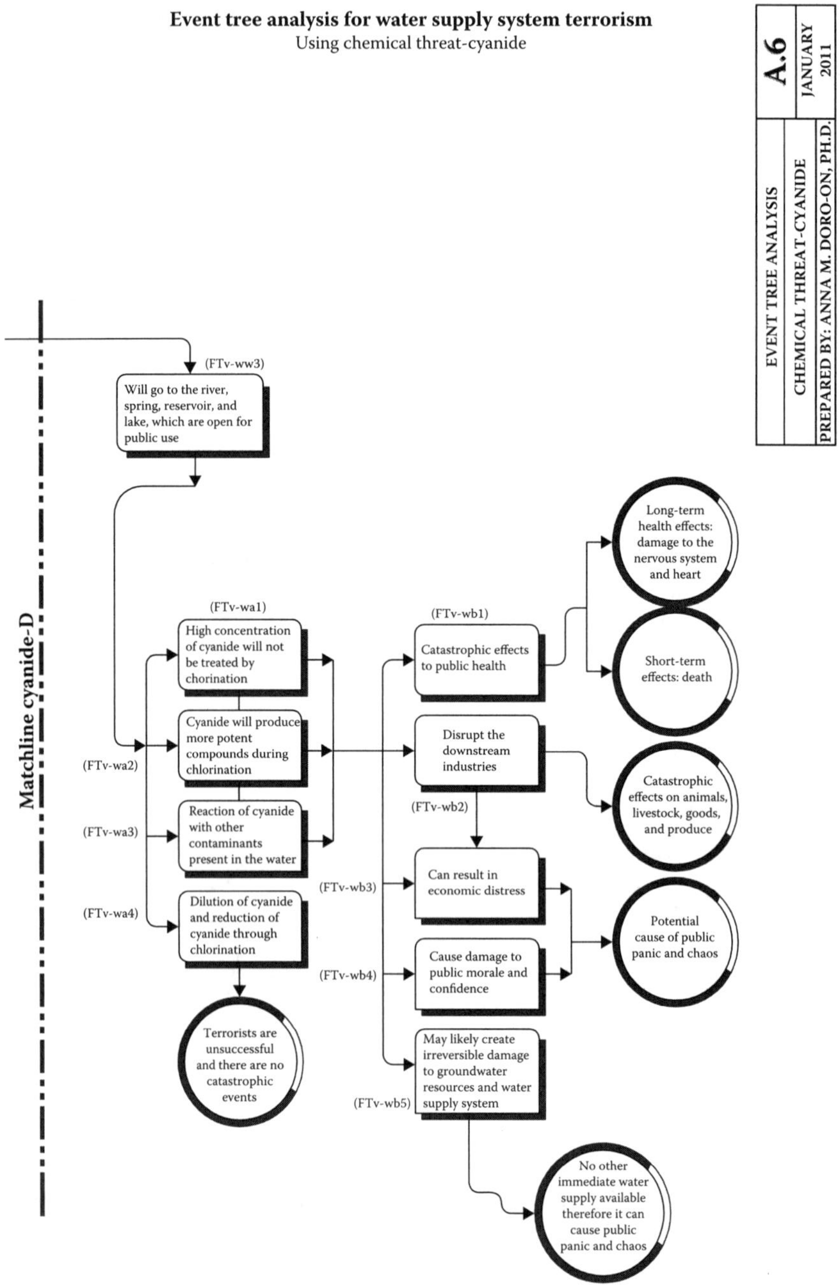

Figure 9.4g Plate A.6: Event tree analysis for water supply terrorism using cyanide.

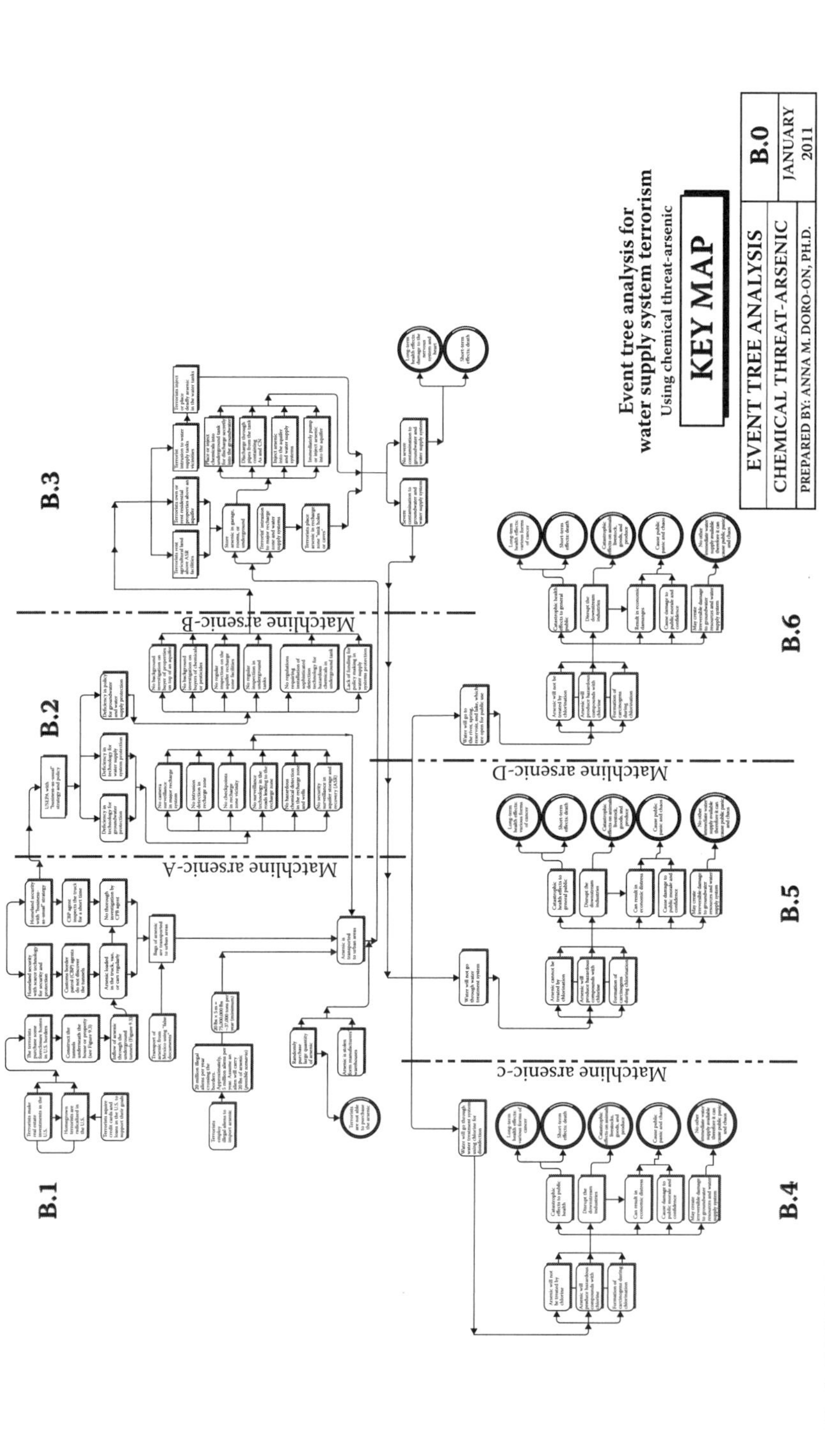

Figure 9.5a Plate B.0: Event tree analysis for water supply system terrorism using arsenic.

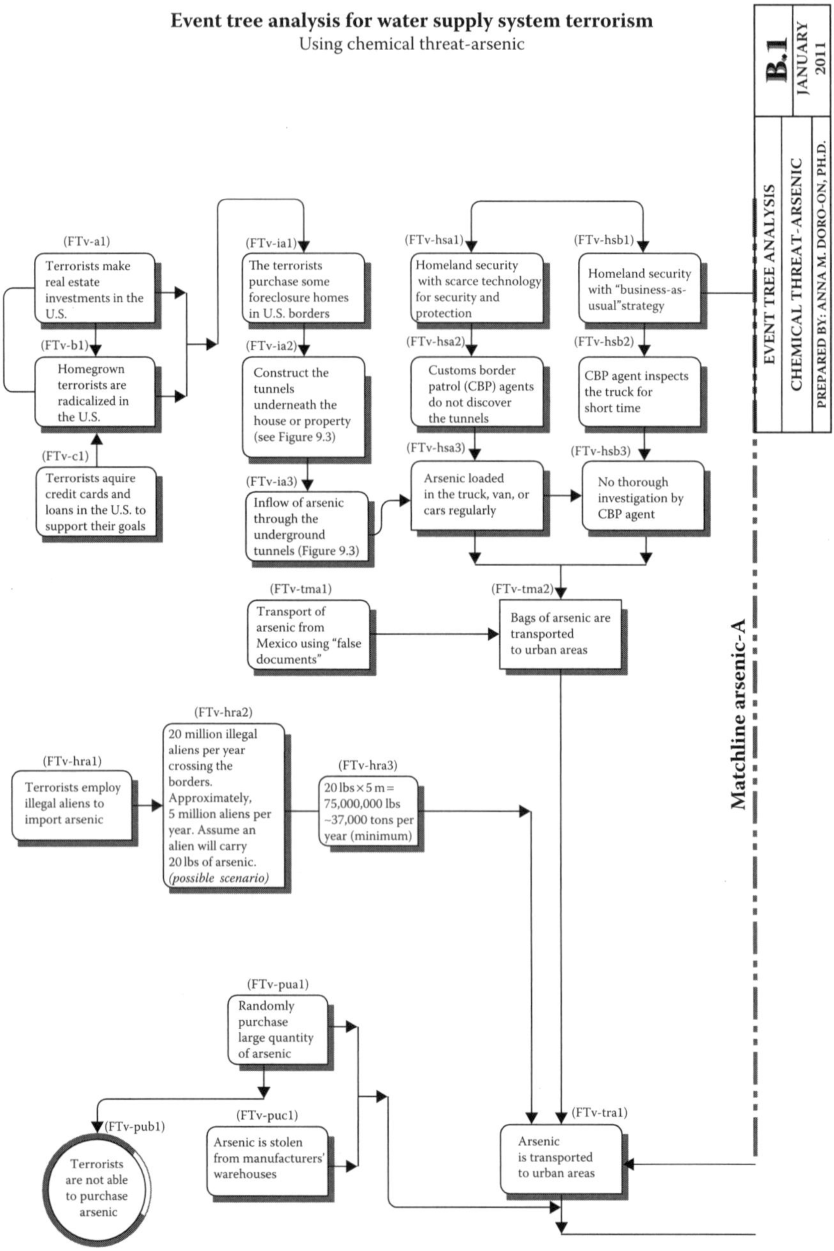

Event tree analysis for water supply system terrorism
Using chemical threat-arsenic
EVENT TREE ANALYSIS
CHEMICAL THREAT-ARSENIC
PREPARED BY: ANNA M. DORO-ON, PH.D.
B.1
JANUARY 2011
(FTv-a1)
Terrorists make real estate investments in the U.S.
(FTv-b1)
Homegrown terrorists are radicalized in the U.S.
(FTv-c1)
Terrorists aquire credit cards and loans in the U.S. to support their goals
(FTv-ia1)
The terrorists purchase some foreclosure homes in U.S. borders
(FTv-ia2)
Construct the tunnels underneath the house or property (see Figure 9.3)
(FTv-ia3)
Inflow of arsenic through the underground tunnels (Figure 9.3)
(FTv-hsa1)
Homeland security with scarce technology for security and protection
(FTv-hsa2)
Customs border patrol (CBP) agents do not discover the tunnels
(FTv-hsa3)
Arsenic loaded in the truck, van, or cars regularly
(FTv-hsb1)
Homeland security with "business-as-usual"strategy
(FTv-hsb2)
CBP agent inspects the truck for short time
(FTv-hsb3)
No thorough investigation by CBP agent
(FTv-tma1)
Transport of arsenic from Mexico using "false documents"
(FTv-tma2)
Bags of arsenic are transported to urban areas
(FTv-hra1)
Terrorists employ illegal aliens to import arsenic
(FTv-hra2)
20 million illegal aliens per year crossing the borders. Approximately, 5 million aliens per year. Assume an alien will carry 20 lbs of arsenic. (possible scenario)
(FTv-hra3)
20 lbs × 5 m = 75,000,000 lbs ~37,000 tons per year (minimum)
Matchline arsenic-A
(FTv-pua1)
Randomly purchase large quantity of arsenic
(FTv-pub1)
Terrorists are not able to purchase arsenic
(FTv-puc1)
Arsenic is stolen from manufacturers' warehouses
(FTv-tra1)
Arsenic is transported to urban areas

Figure 9.5b Plate B.1: Event tree analysis for water supply terrorism using arsenic.

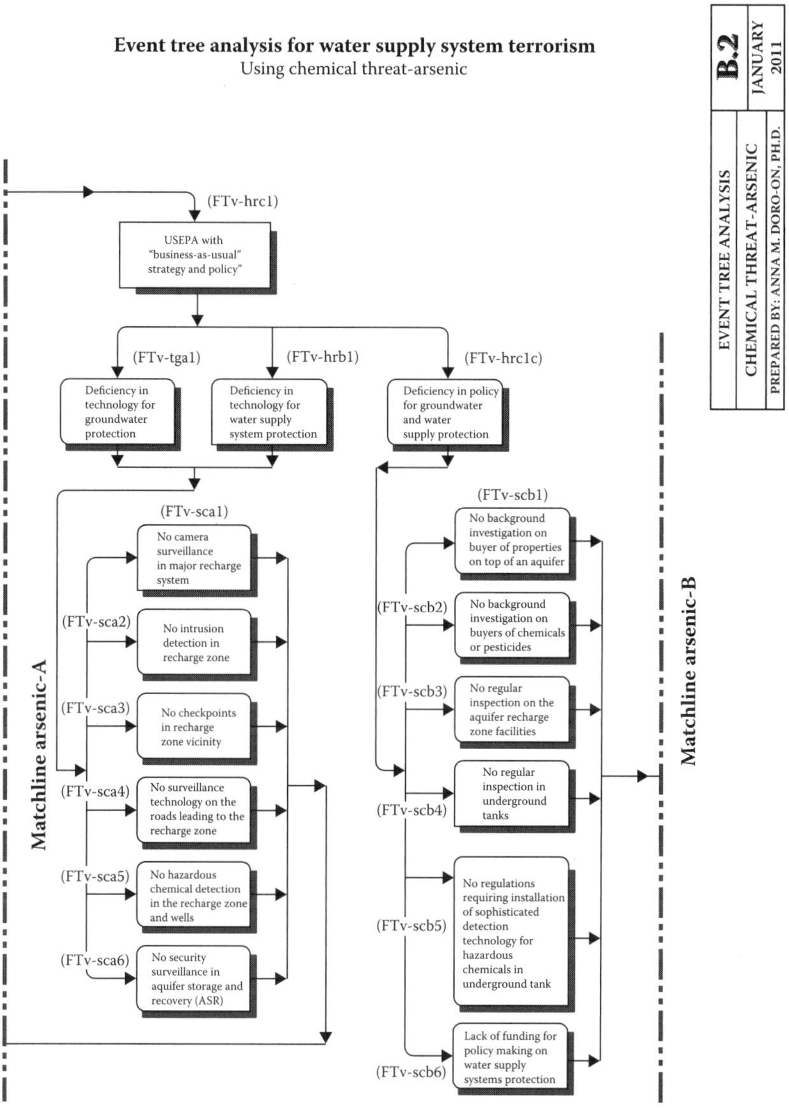

Figure 9.5c **Plate B.2: Event tree analysis for water supply terrorism using arsenic.**

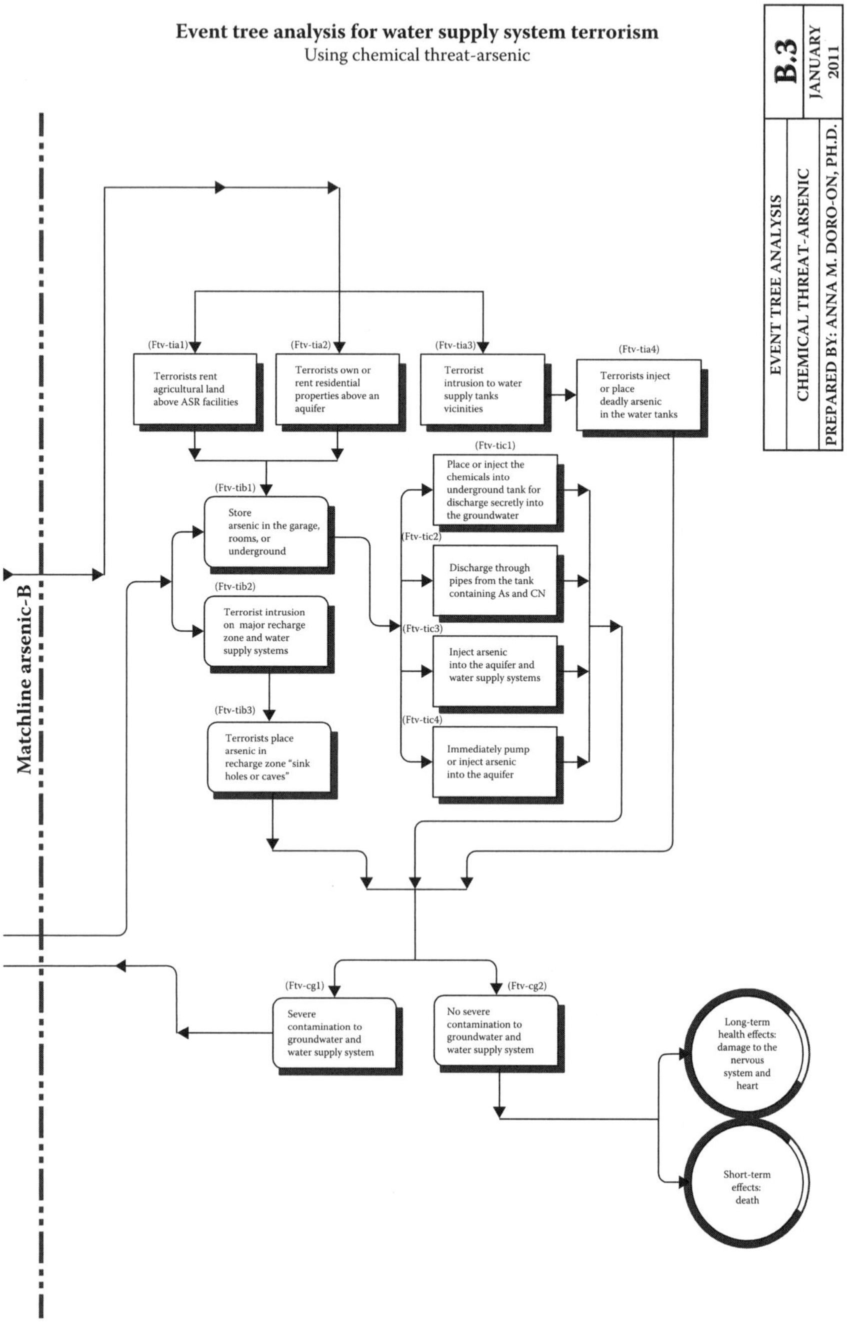

Figure 9.5d Plate B.3: Event tree analysis for water supply terrorism using arsenic.

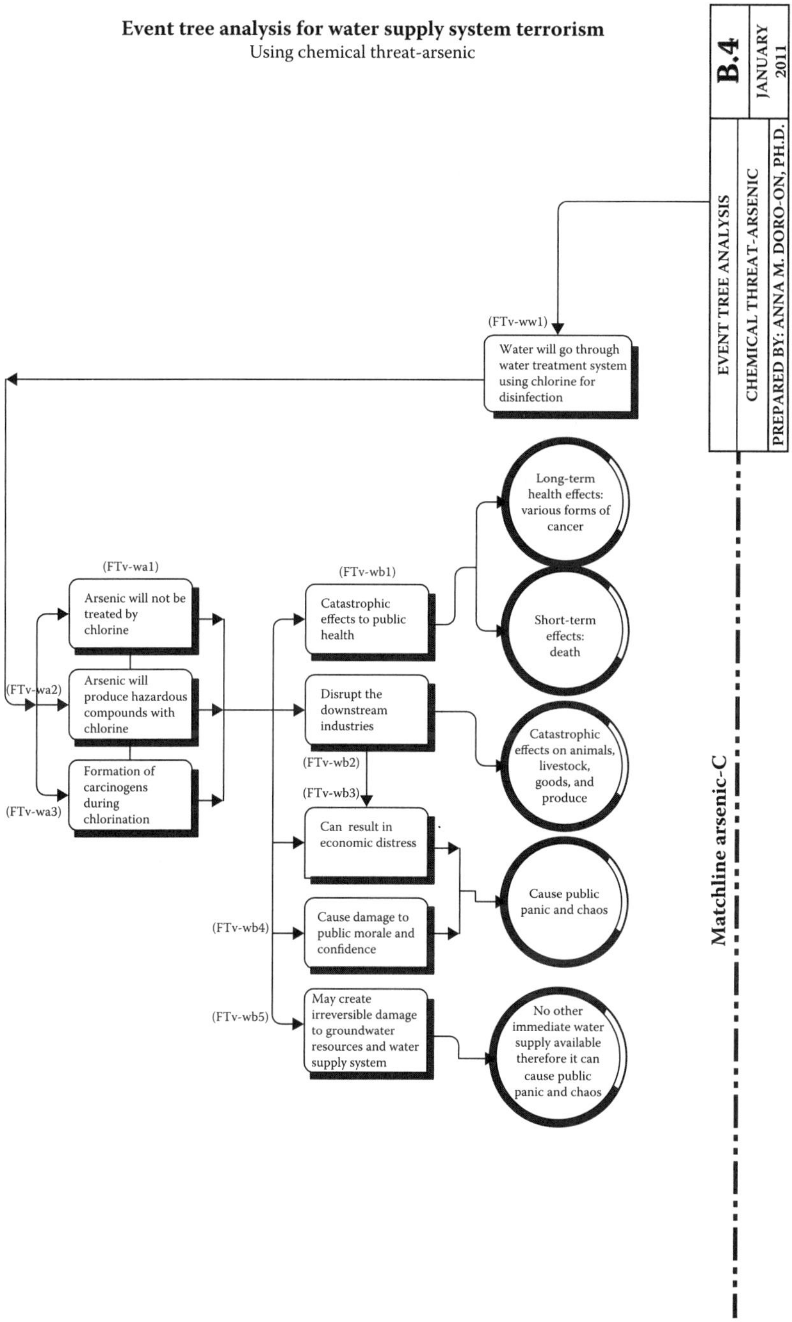

Figure 9.5e Plate B.4: Event tree analysis for water supply terrorism using arsenic.

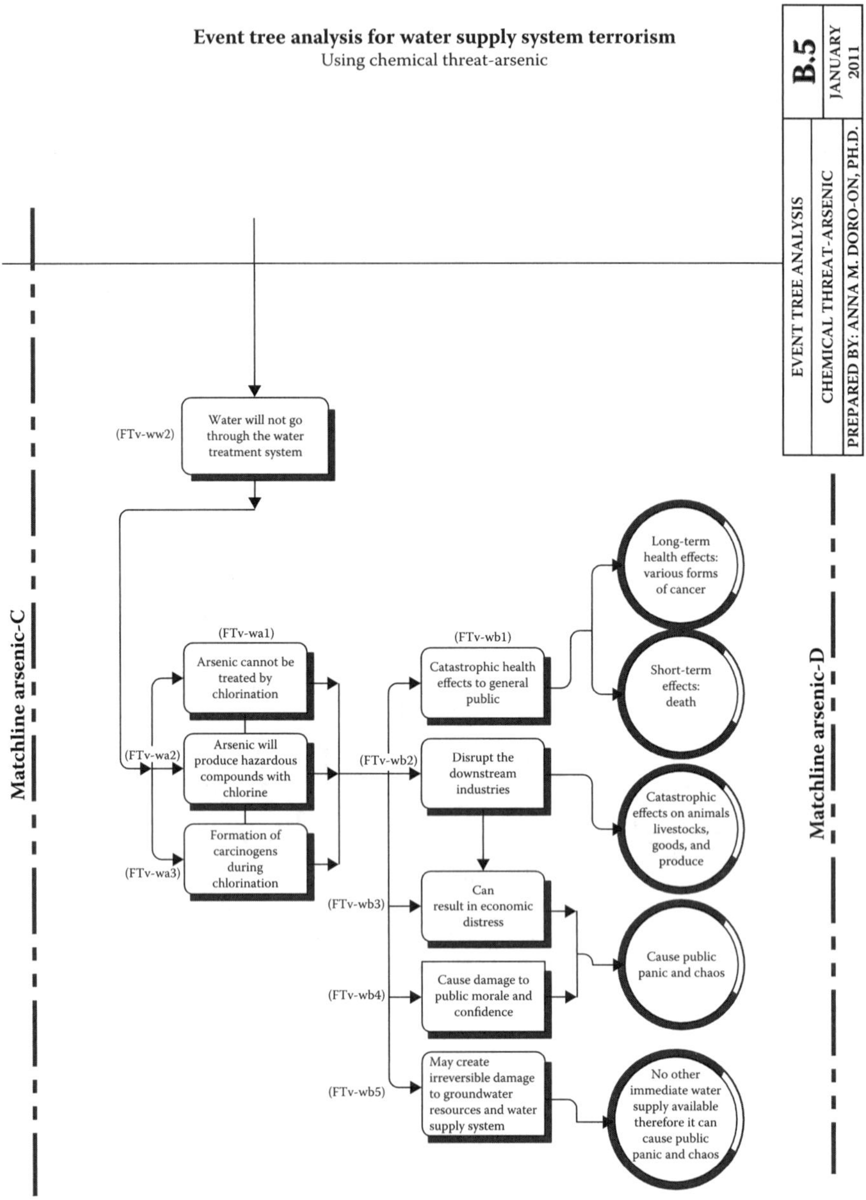

Figure 9.5f Plate B.5: Event tree analysis for water supply terrorism using arsenic.

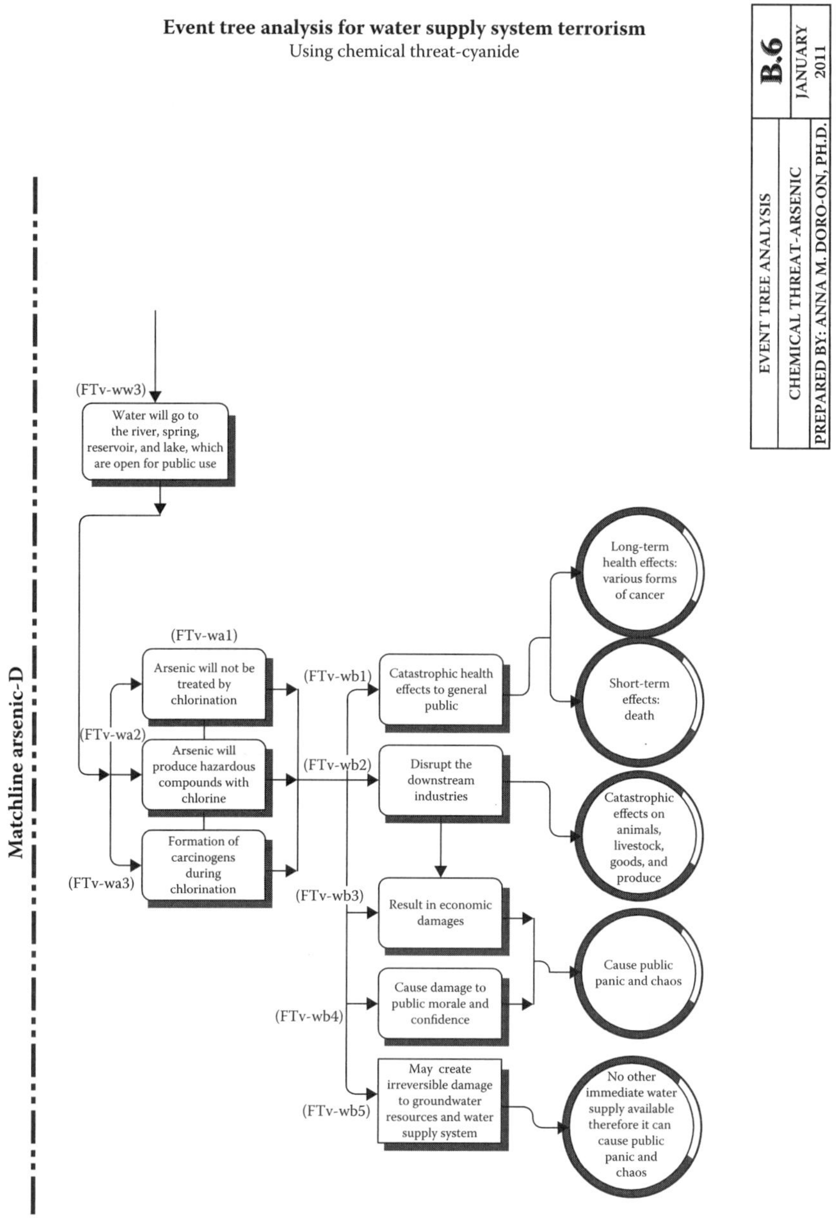

Figure 9.5g Plate B.6: Event tree analysis for water supply terrorism using arsenic.

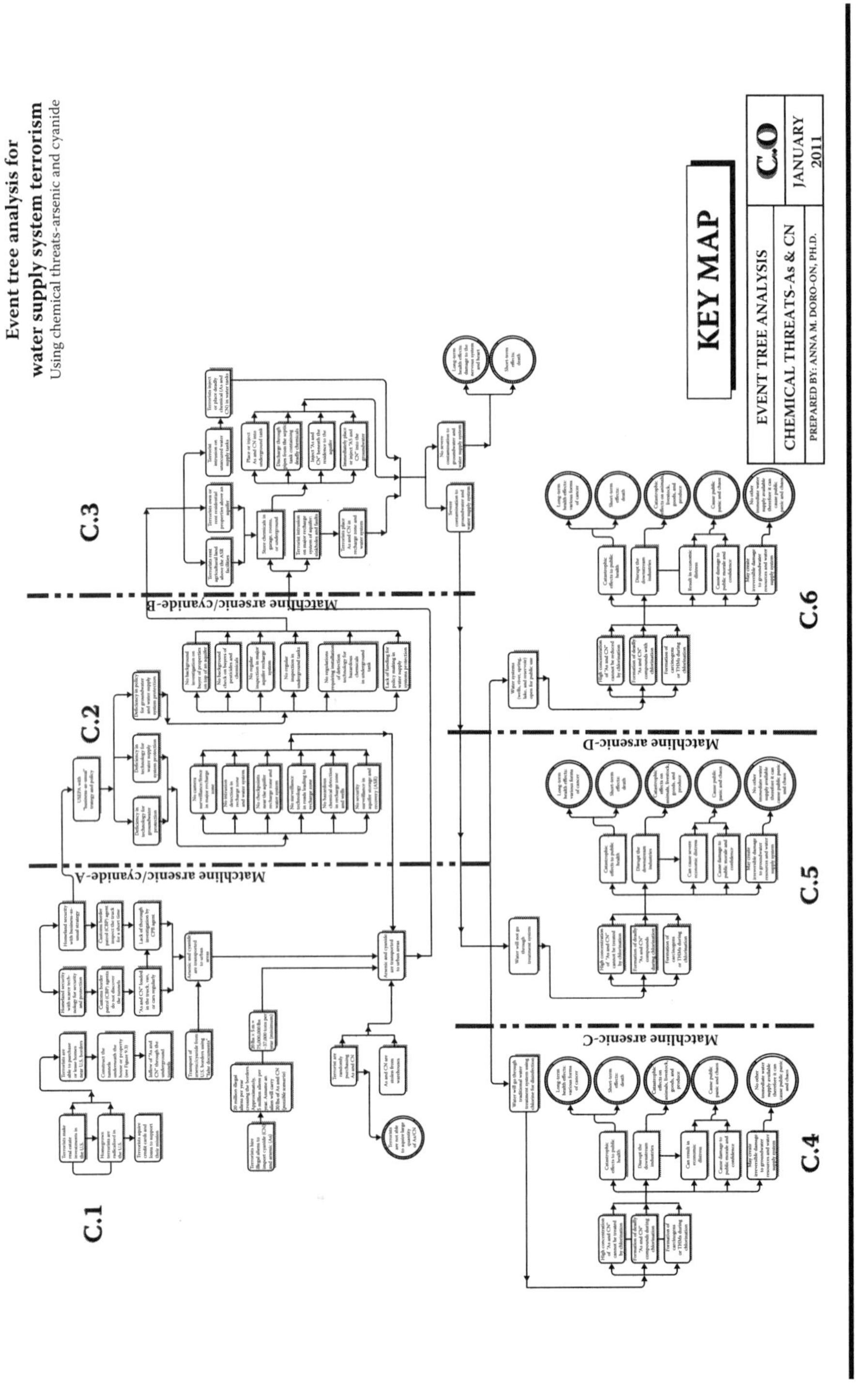

Figure 9.6a Plate C.0: Event tree analysis for water supply terrorism using arsenic and cyanide.

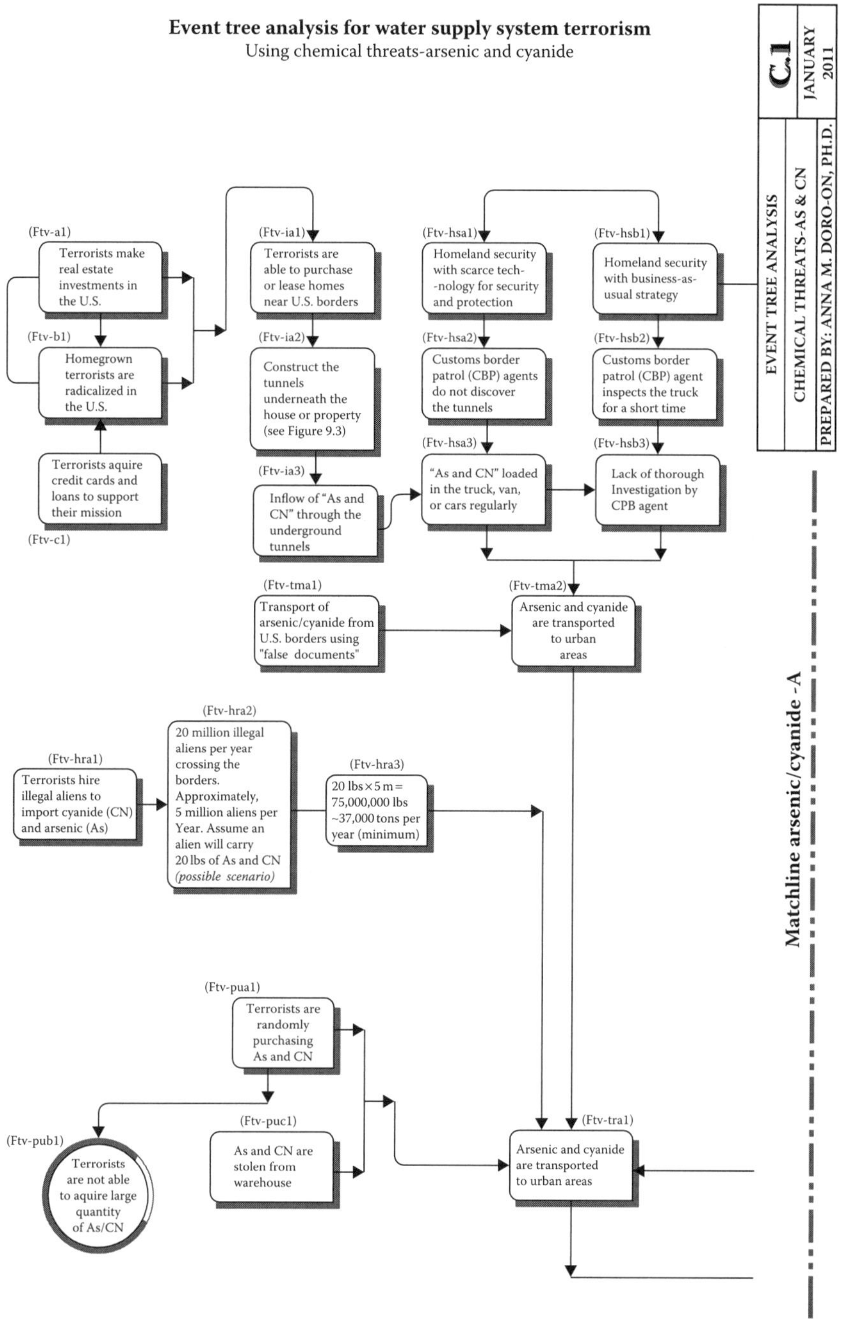
Event tree analysis for water supply system terrorism
Using chemical threats-arsenic and cyanide
(Ftv-a1)
Terrorists make real estate investments in the U.S.
(Ftv-b1)
Homegrown terrorists are radicalized in the U.S.
Terrorists aquire credit cards and loans to support their mission
(Ftv-c1)
(Ftv-ia1)
Terrorists are able to purchase or lease homes near U.S. borders
(Ftv-ia2)
Construct the tunnels underneath the house or property (see Figure 9.3)
(Ftv-ia3)
Inflow of "As and CN" through the underground tunnels
(Ftv-hsa1)
Homeland security with scarce tech- -nology for security and protection
(Ftv-hsa2)
Customs border patrol (CBP) agents do not discover the tunnels
(Ftv-hsa3)
"As and CN" loaded in the truck, van, or cars regularly
(Ftv-hsb1)
Homeland security with business-as-usual strategy
(Ftv-hsb2)
Customs border patrol (CBP) agent inspects the truck for a short time
(Ftv-hsb3)
Lack of thorough Investigation by CPB agent
(Ftv-tma1)
Transport of arsenic/cyanide from U.S. borders using "false documents"
(Ftv-tma2)
Arsenic and cyanide are transported to urban areas
(Ftv-hra1)
Terrorists hire illegal aliens to import cyanide (CN) and arsenic (As)
(Ftv-hra2)
20 million illegal aliens per year crossing the borders. Approximately, 5 million aliens per Year. Assume an alien will carry 20 lbs of As and CN (possible scenario)
(Ftv-hra3)
20 lbs × 5 m = 75,000,000 lbs ~37,000 tons per year (minimum)
(Ftv-pua1)
Terrorists are randomly purchasing As and CN
(Ftv-pub1)
Terrorists are not able to aquire large quantity of As/CN
(Ftv-puc1)
As and CN are stolen from warehouse
(Ftv-tra1)
Arsenic and cyanide are transported to urban areas
EVENT TREE ANALYSIS
CHEMICAL THREATS-AS & CN
PREPARED BY: ANNA M. DORO-ON, PH.D.
C.1
JANUARY 2011
Matchline arsenic/cyanide -A

Figure 9.6b Plate C.1: Event tree analysis for water supply terrorism using arsenic and cyanide.

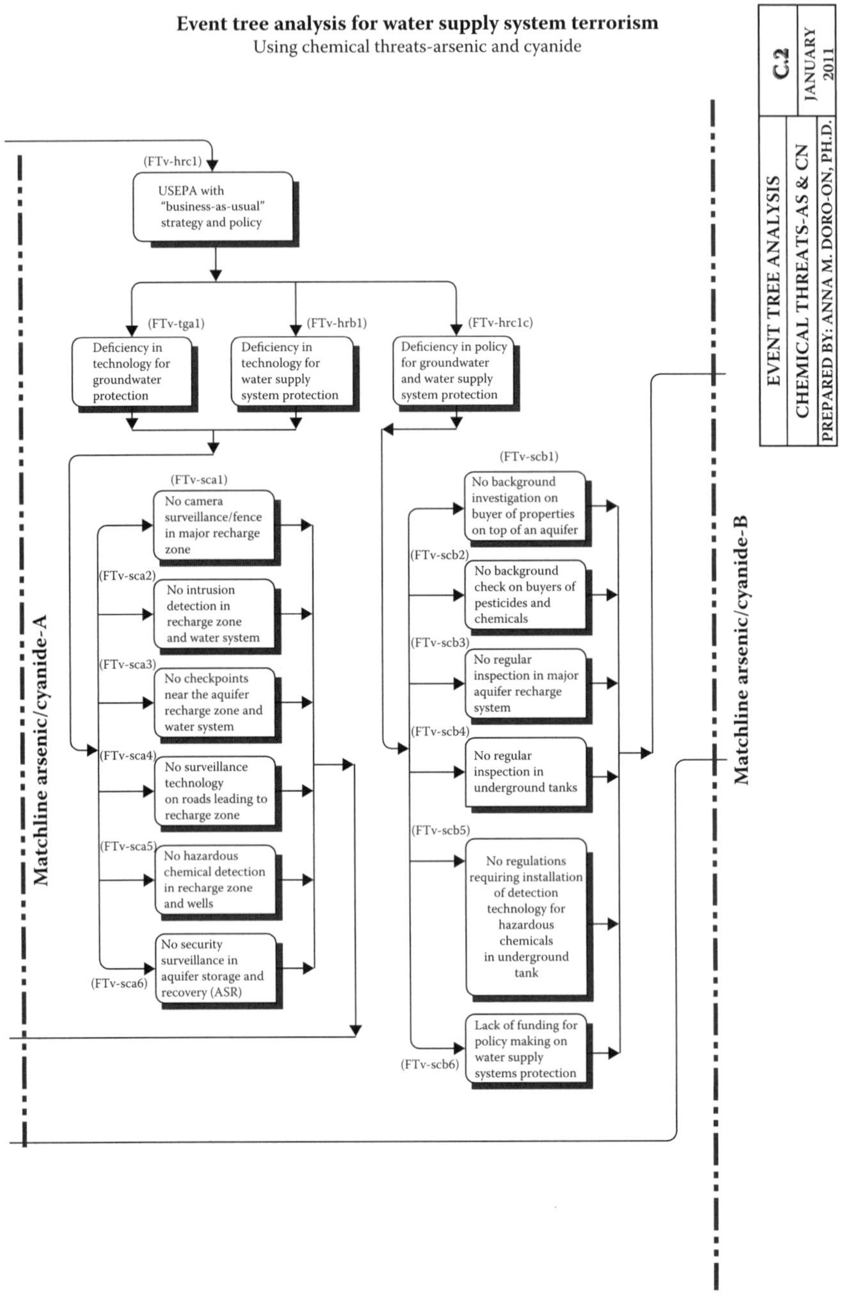

Figure 9.6c Plate C.2: Event tree analysis for water supply terrorism using arsenic and cyanide.

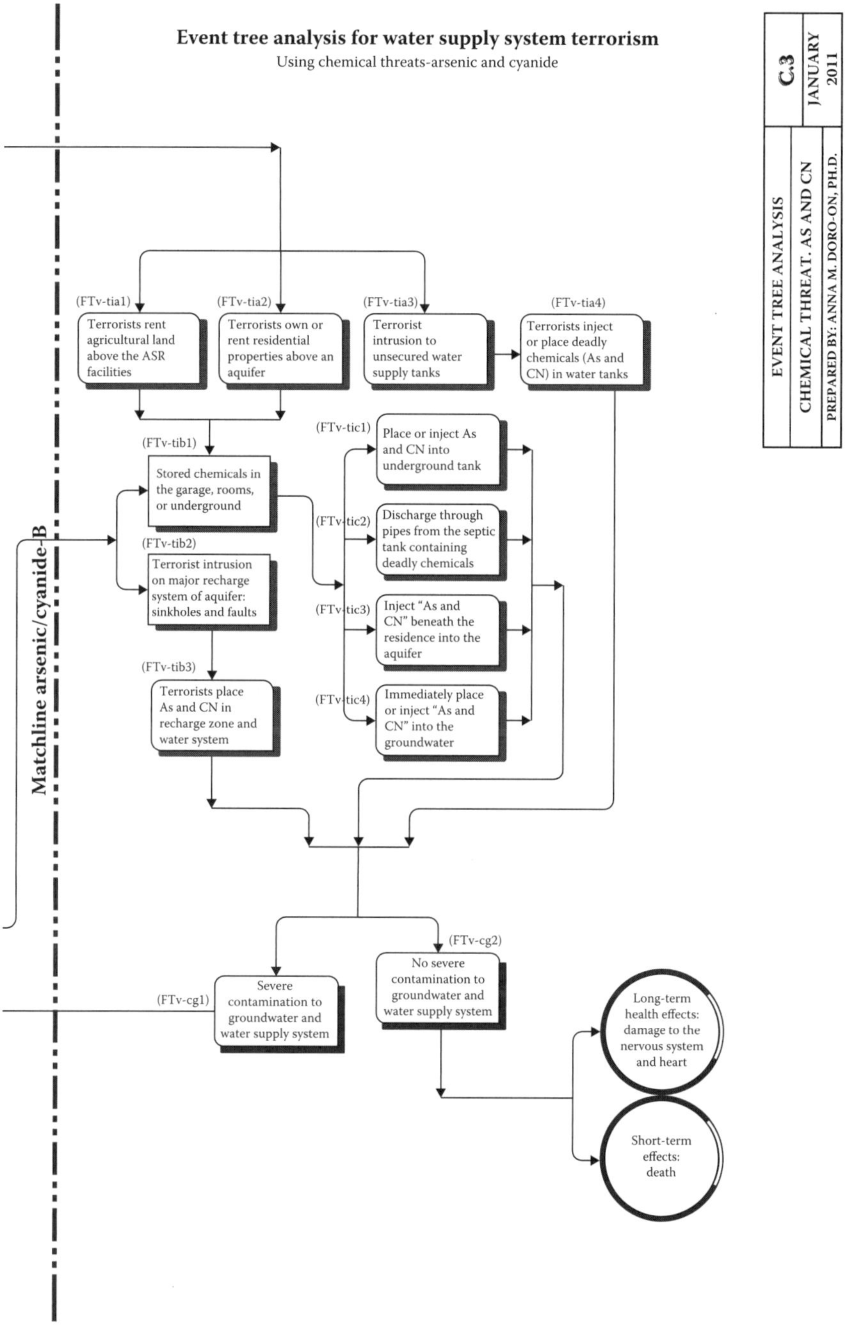

Figure 9.6d Plate C.3: Event tree analysis for water supply terrorism using arsenic and cyanide.

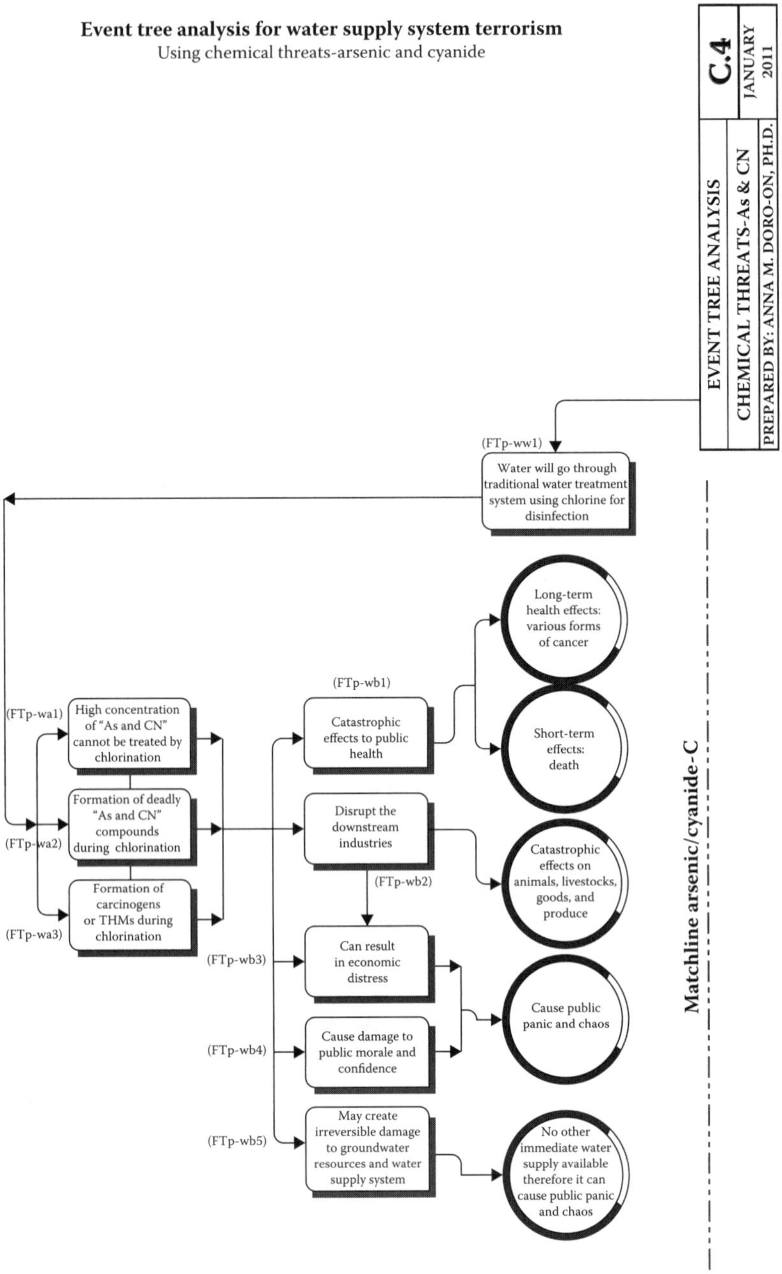

Figure 9.6e Plate C.4: Event tree analysis for water supply terrorism using arsenic and cyanide.

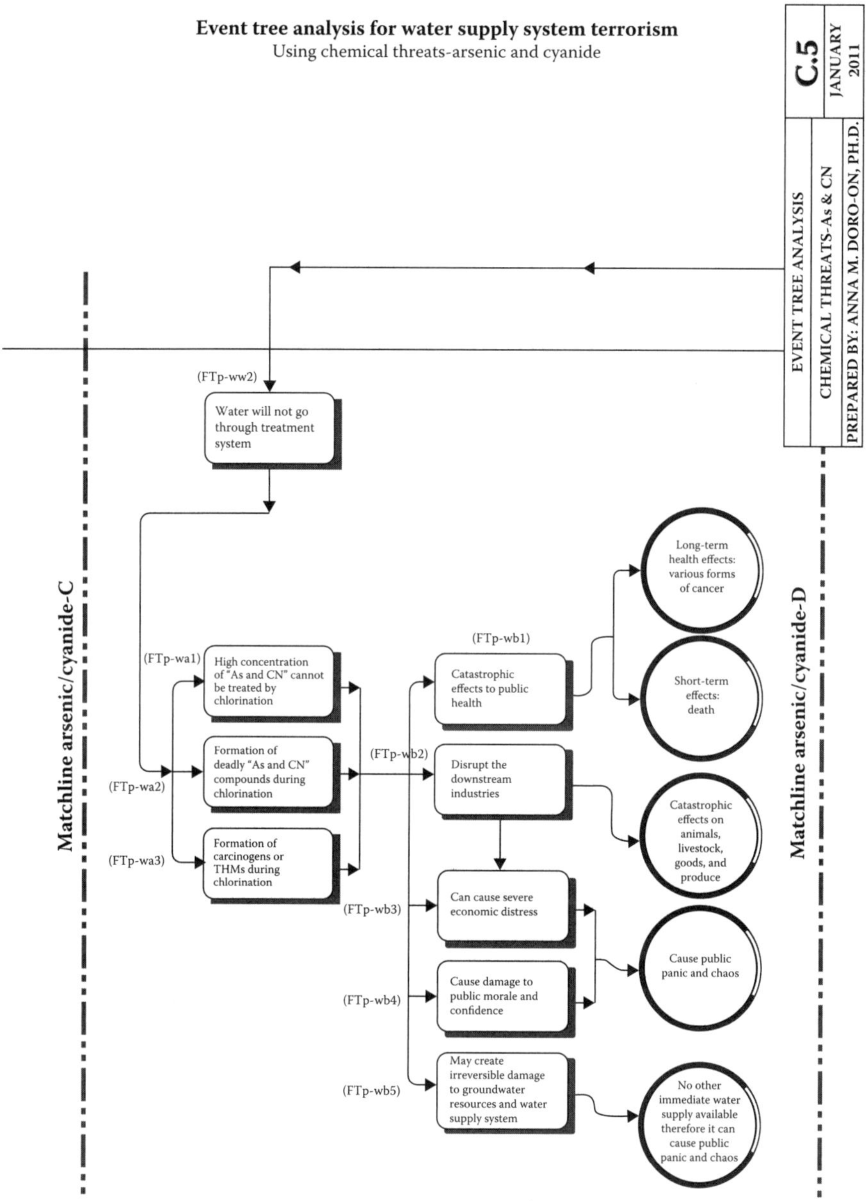

Figure 9.6f Plate C.5: Event tree analysis for water supply terrorism using arsenic and cyanide.

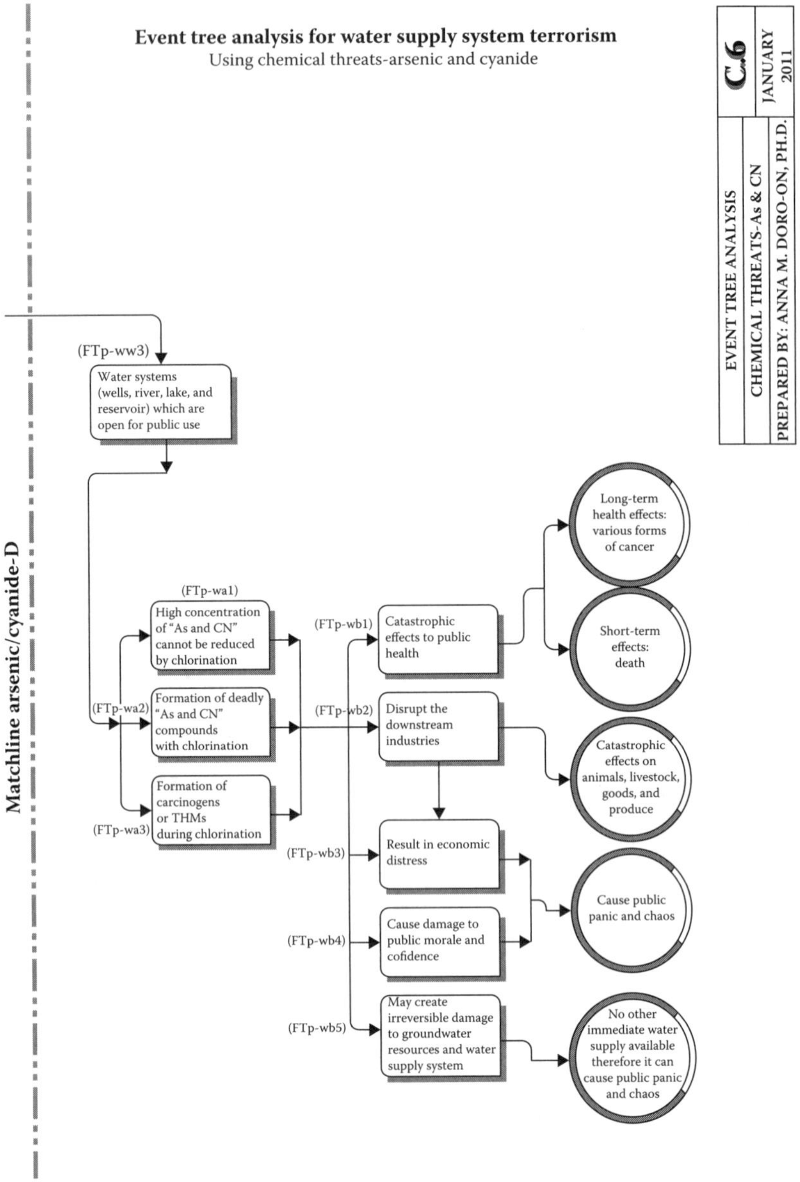

Figure 9.6g Plate C.6: Event tree analysis for water supply terrorism using arsenic and cyanide.

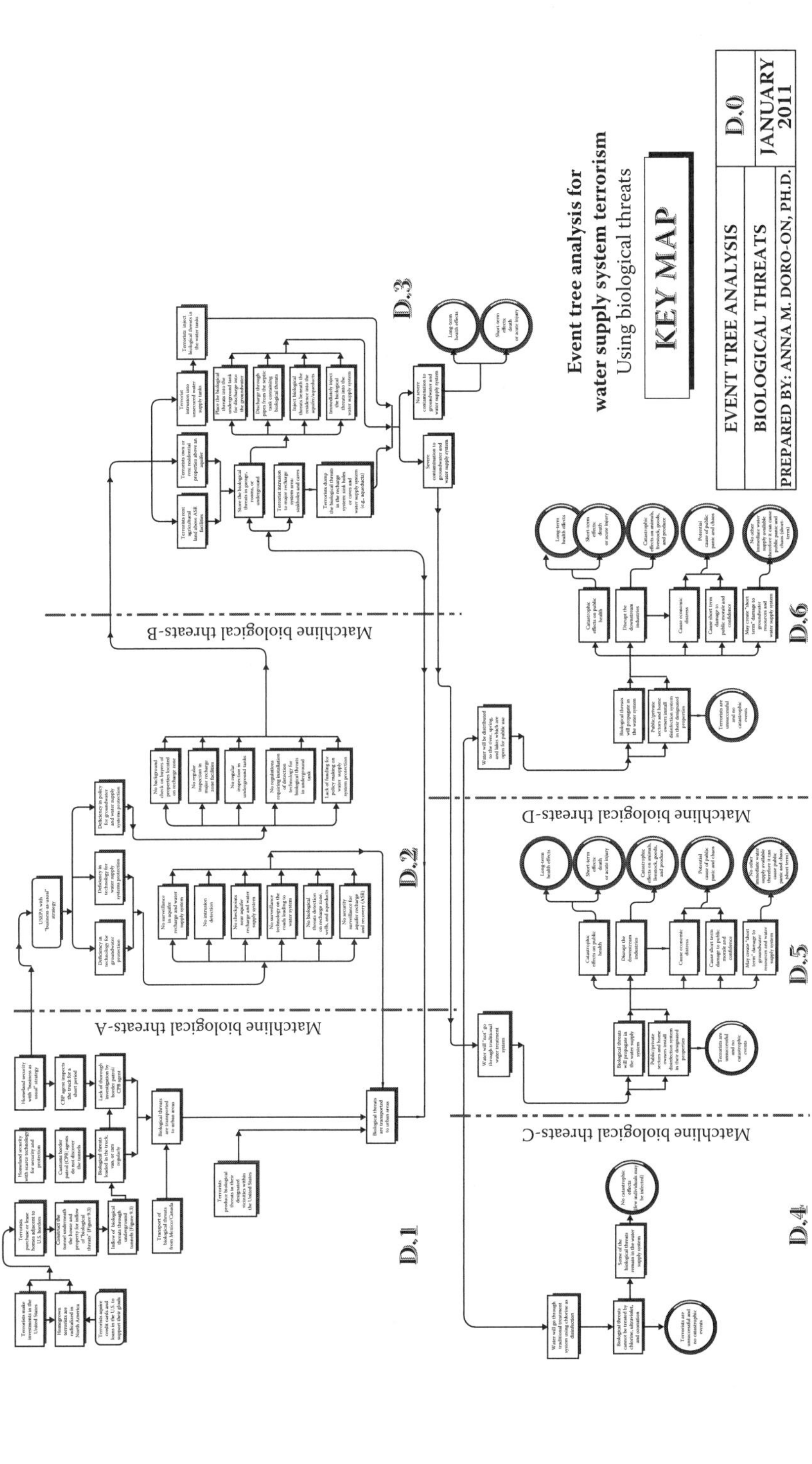

Figure 9.7a Plate D.0: Event tree analysis for water supply terrorism using biological threats.

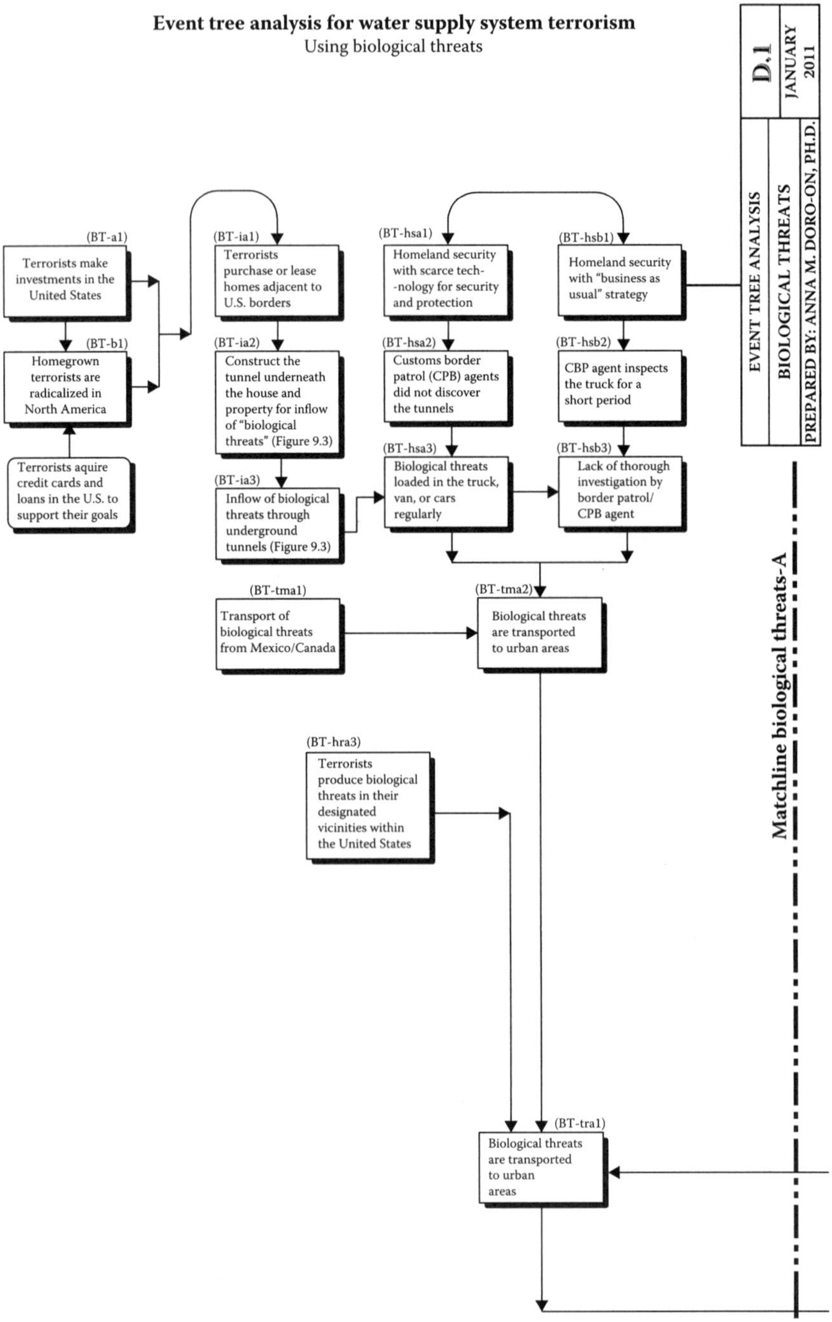

Event tree analysis for water supply system terrorism
Using biological threats
(BT-a1)
Terrorists make investments in the United States
(BT-b1)
Homegrown terrorists are radicalized in North America
Terrorists aquire credit cards and loans in the U.S. to support their goals
(BT-ia1)
Terrorists purchase or lease homes adjacent to U.S. borders
(BT-ia2)
Construct the tunnel underneath the house and property for inflow of "biological threats" (Figure 9.3)
(BT-ia3)
Inflow of biological threats through underground tunnels (Figure 9.3)
(BT-hsa1)
Homeland security with scarce tech--nology for security and protection
(BT-hsa2)
Customs border patrol (CPB) agents did not discover the tunnels
(BT-hsa3)
Biological threats loaded in the truck, van, or cars regularly
(BT-hsb1)
Homeland security with "business as usual" strategy
(BT-hsb2)
CBP agent inspects the truck for a short period
(BT-hsb3)
Lack of thorough investigation by border patrol/ CPB agent
(BT-tma1)
Transport of biological threats from Mexico/Canada
(BT-tma2)
Biological threats are transported to urban areas
(BT-hra3)
Terrorists produce biological threats in their designated vicinities within the United States
(BT-tra1)
Biological threats are transported to urban areas
EVENT TREE ANALYSIS
BIOLOGICAL THREATS
PREPARED BY: ANNA M. DORO-ON, PH.D.
D.1
JANUARY 2011
Matchline biological threats-A

Figure 9.7b Plate D.1: Event tree analysis for water supply terrorism using biological threats.

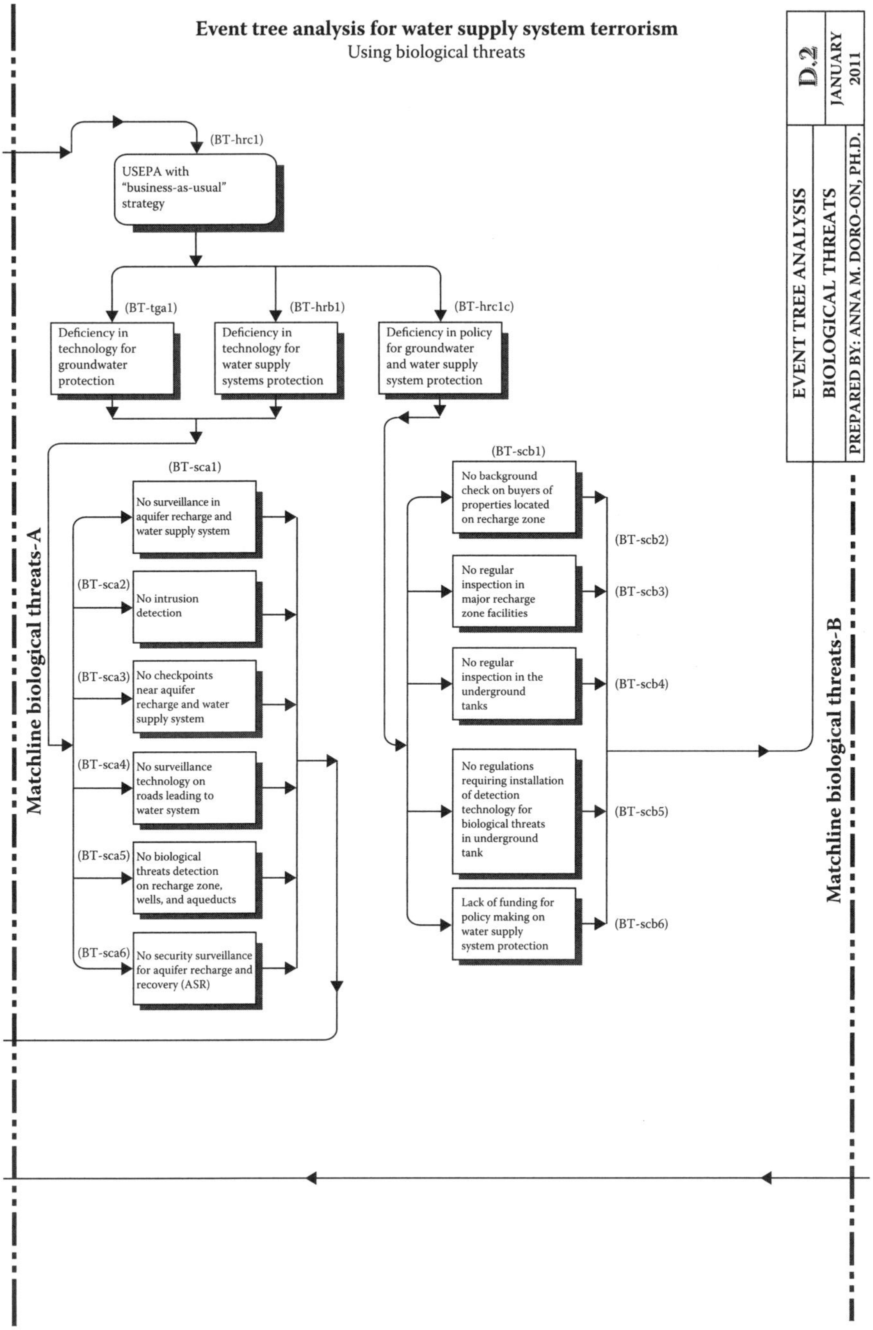

Figure 9.7c Plate D.2: Event tree analysis for water supply terrorism using biological threats.

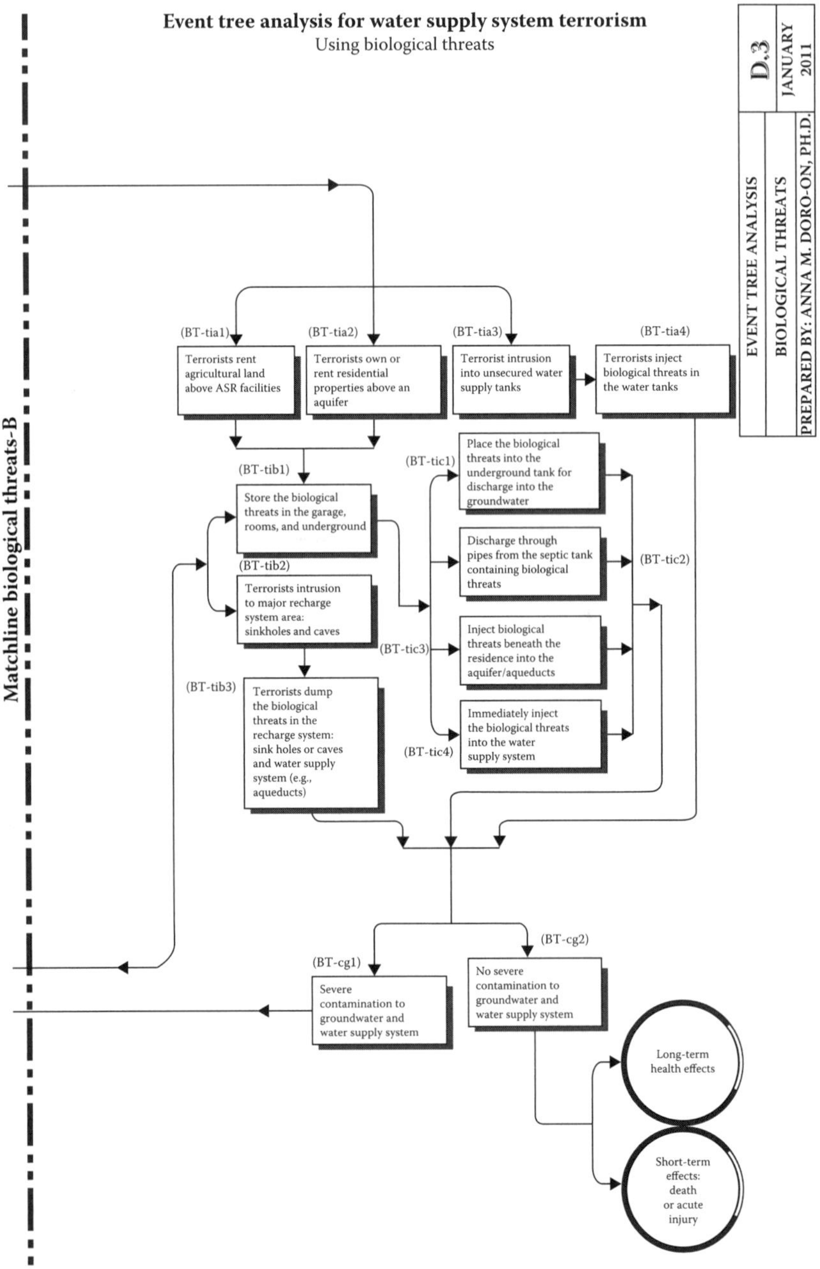

Figure 9.7d Plate D.3: Event tree analysis for water supply terrorism using biological threats.

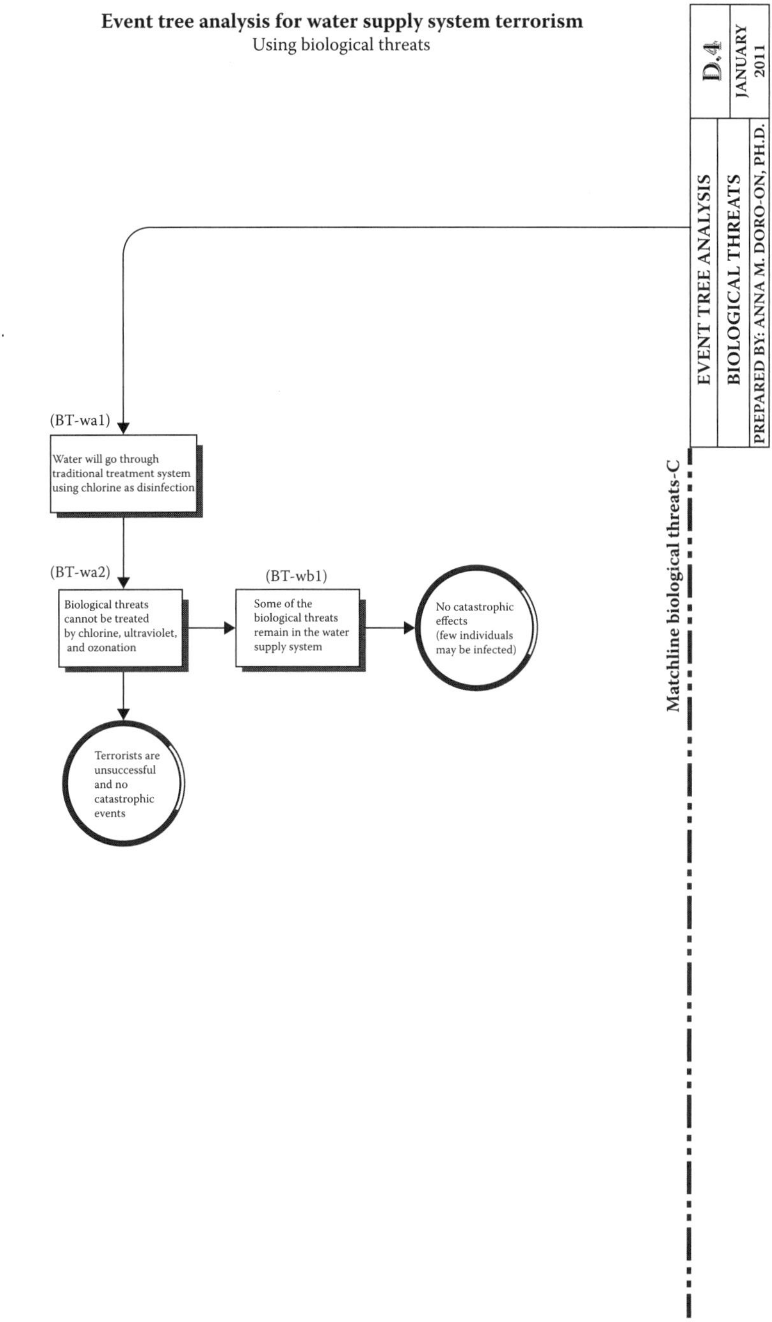
Event tree analysis for water supply system terrorism
Using biological threats
D.4
JANUARY 2011
EVENT TREE ANALYSIS
BIOLOGICAL THREATS
PREPARED BY: ANNA M. DORO-ON, PH.D.
(BT-wa1)
Water will go through traditional treatment system using chlorine as disinfection
(BT-wa2)
Biological threats cannot be treated by chlorine, ultraviolet, and ozonation
(BT-wb1)
Some of the biological threats remain in the water supply system
No catastrophic effects (few individuals may be infected)
Terrorists are unsuccessful and no catastrophic events
Matchline biological threats-C

Figure 9.7e Plate D.4: Event tree analysis for water supply terrorism using biological threats.

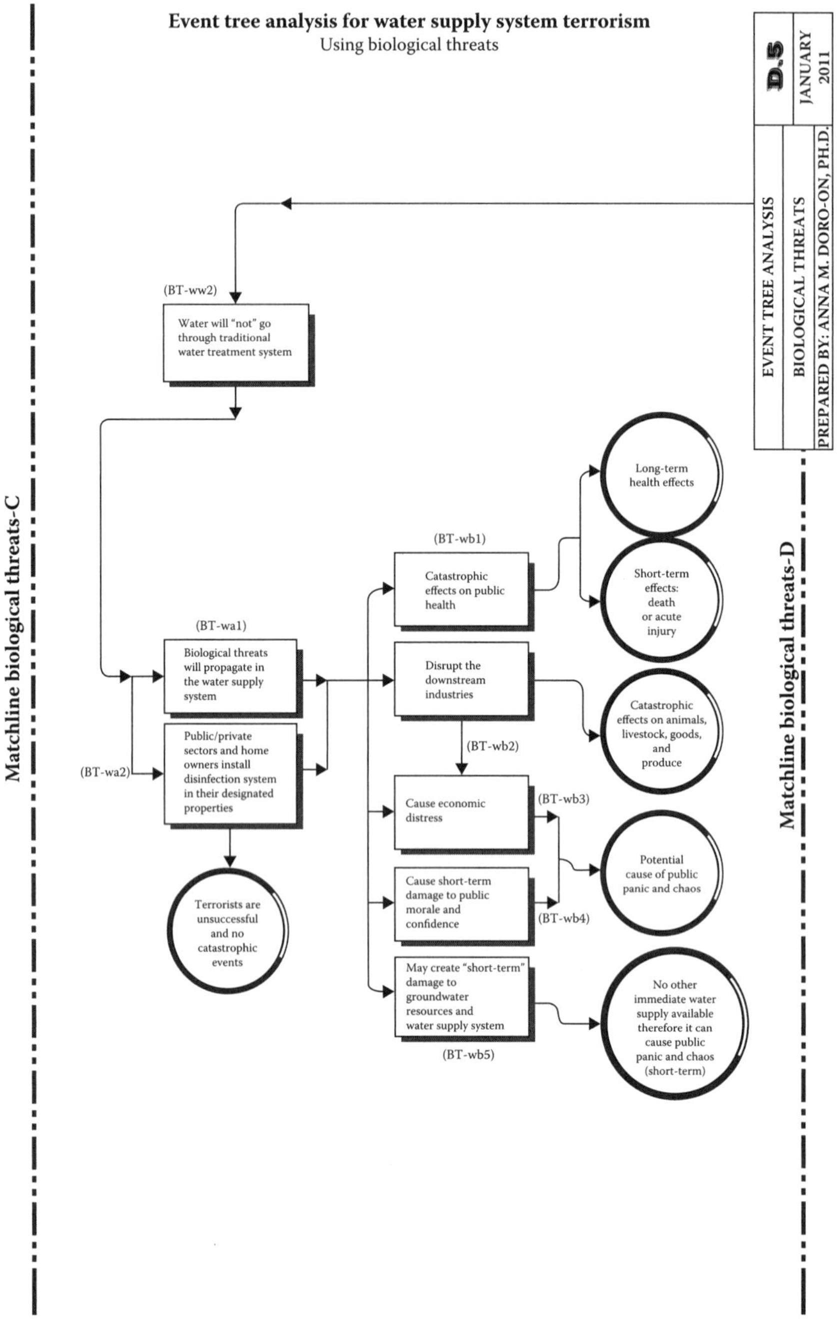

Figure 9.7f Plate D.5: Event tree analysis for water supply terrorism using biological threats.

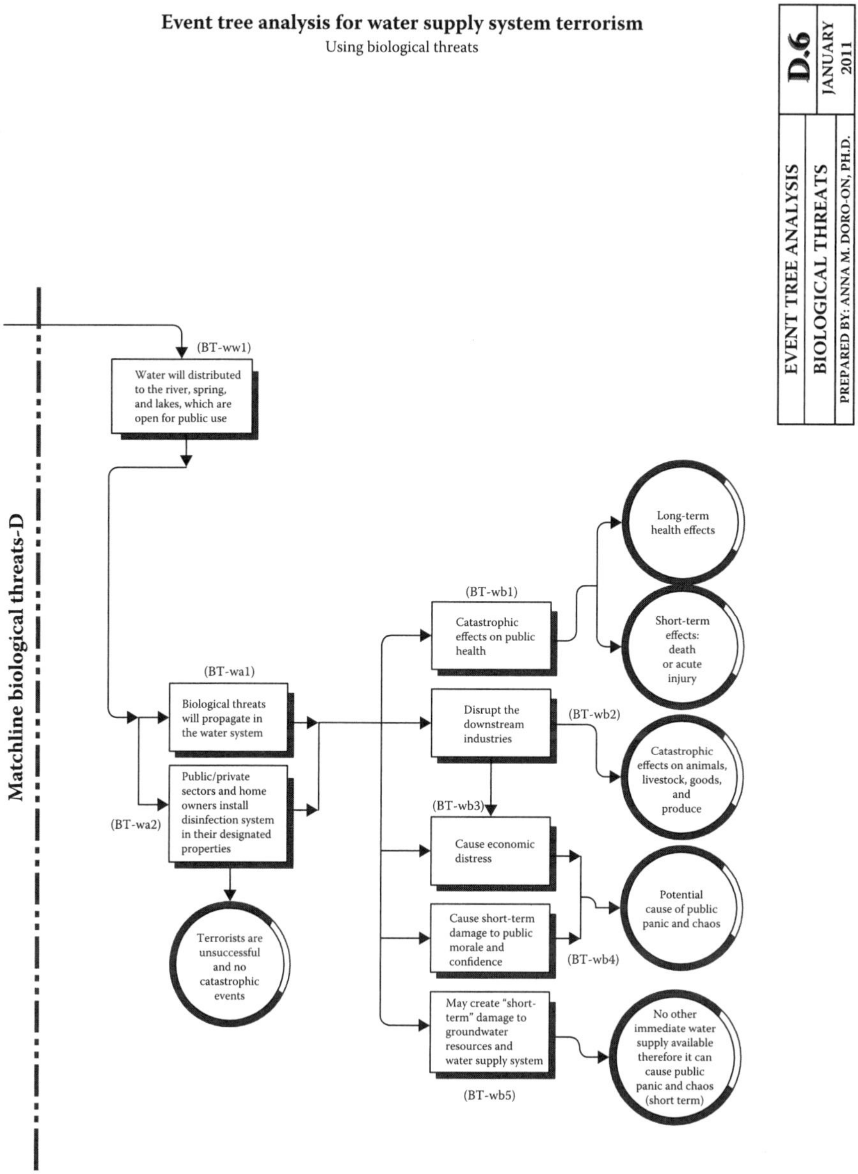

Figure 9.7g Plate D.6: Event tree analysis for water supply terrorism using biological threats.

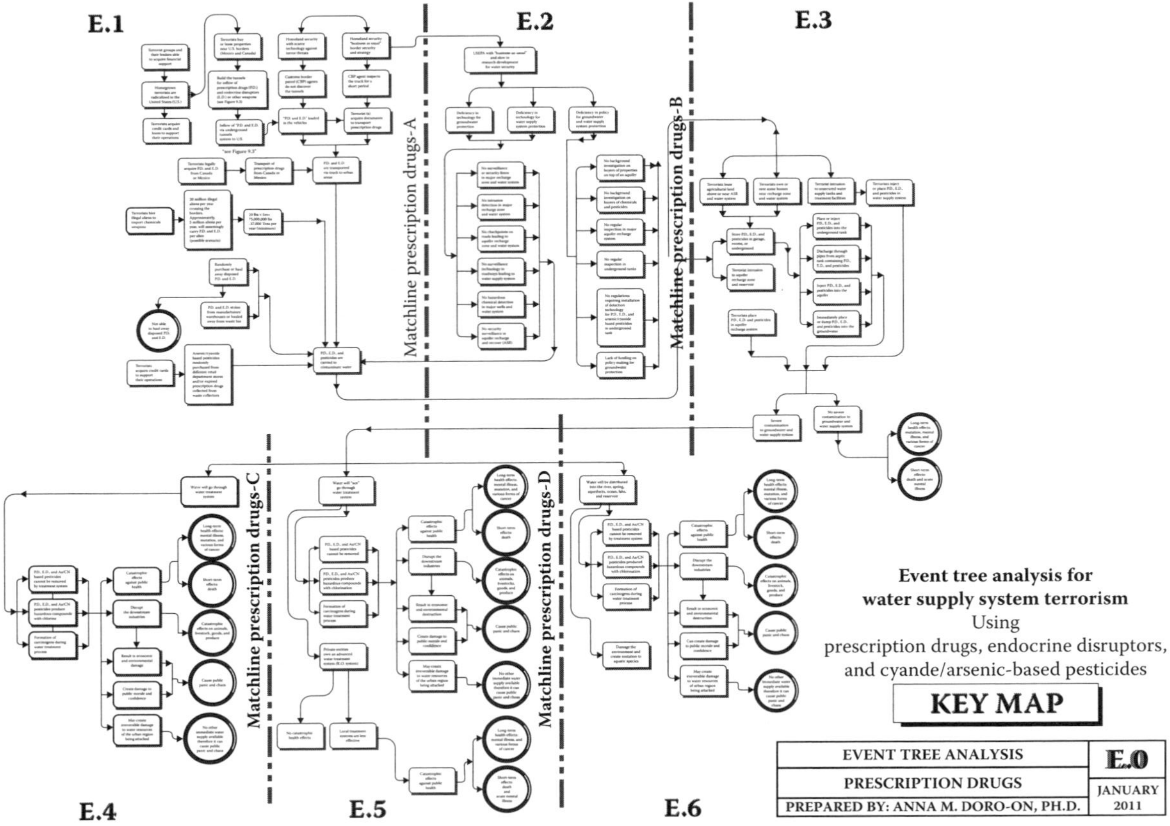

Figure 9.8a Plate E.0: Event tree analysis for water supply terrorism using prescription drugs, endocrine disruptors, and cyanide-/arsenic-based pesticides.

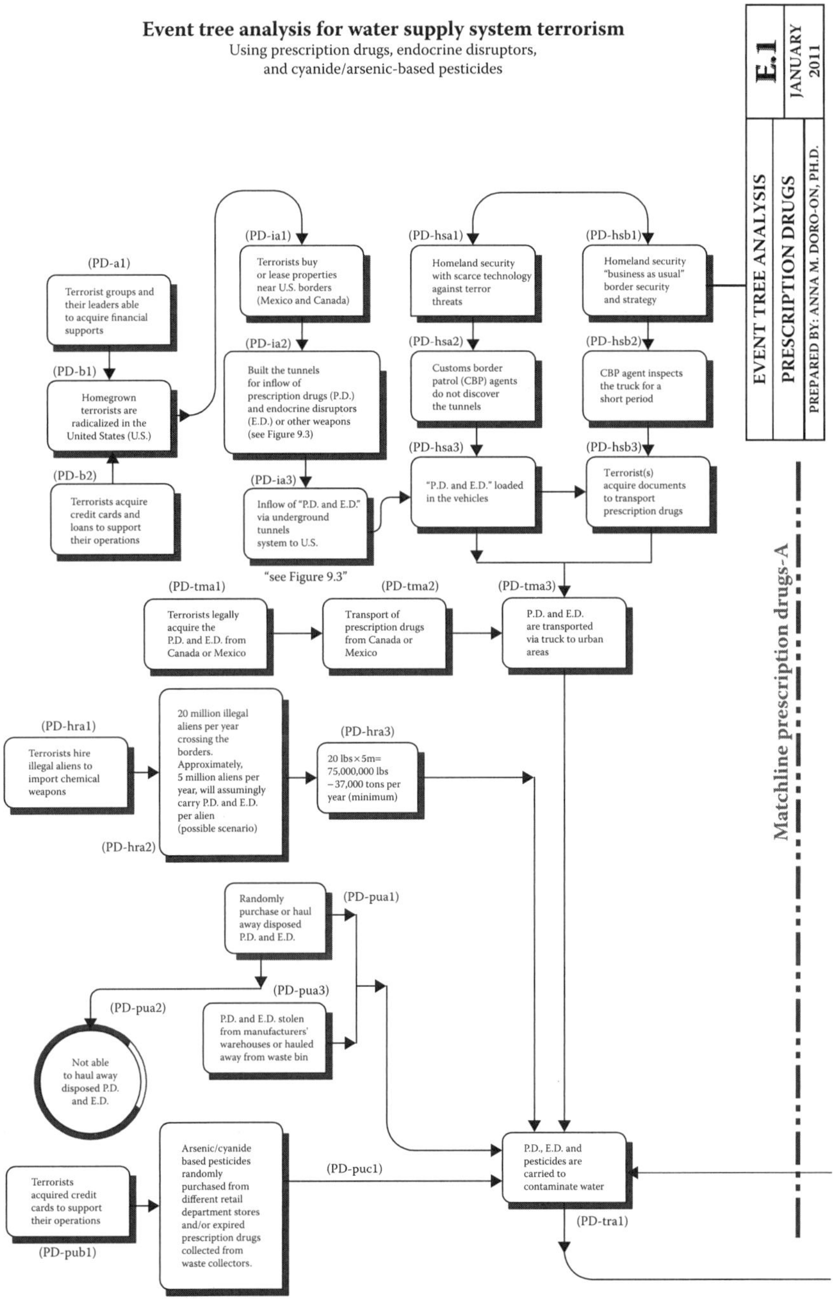

Figure 9.8b Plate E.1: Event tree analysis for water supply terrorism using prescription drugs, endocrine disruptors, and cyanide-/arsenic-based pesticides.

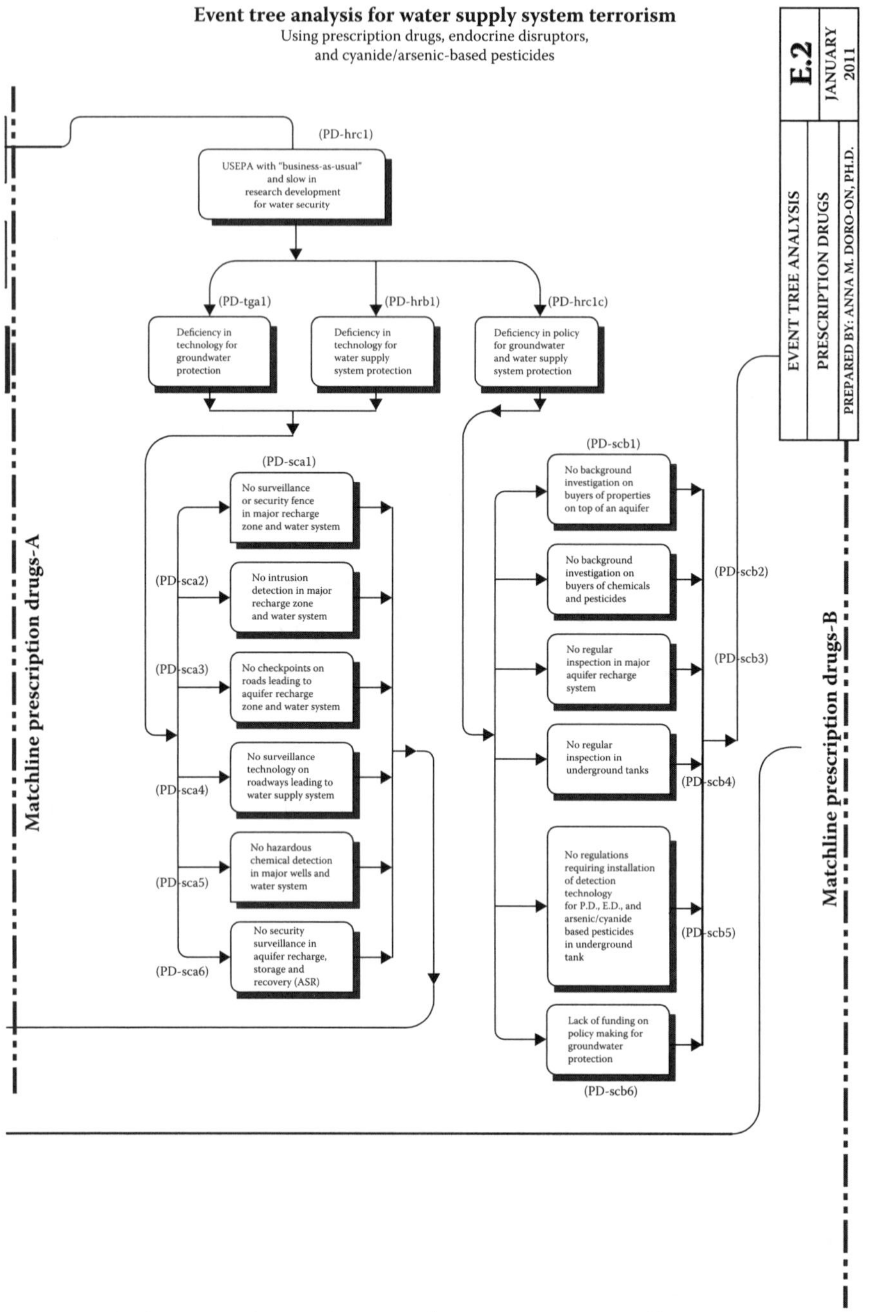

Figure 9.8c Plate E.2: Event tree analysis for water supply terrorism using prescription drugs, endocrine disruptors, and cyanide-/arsenic-based pesticides.

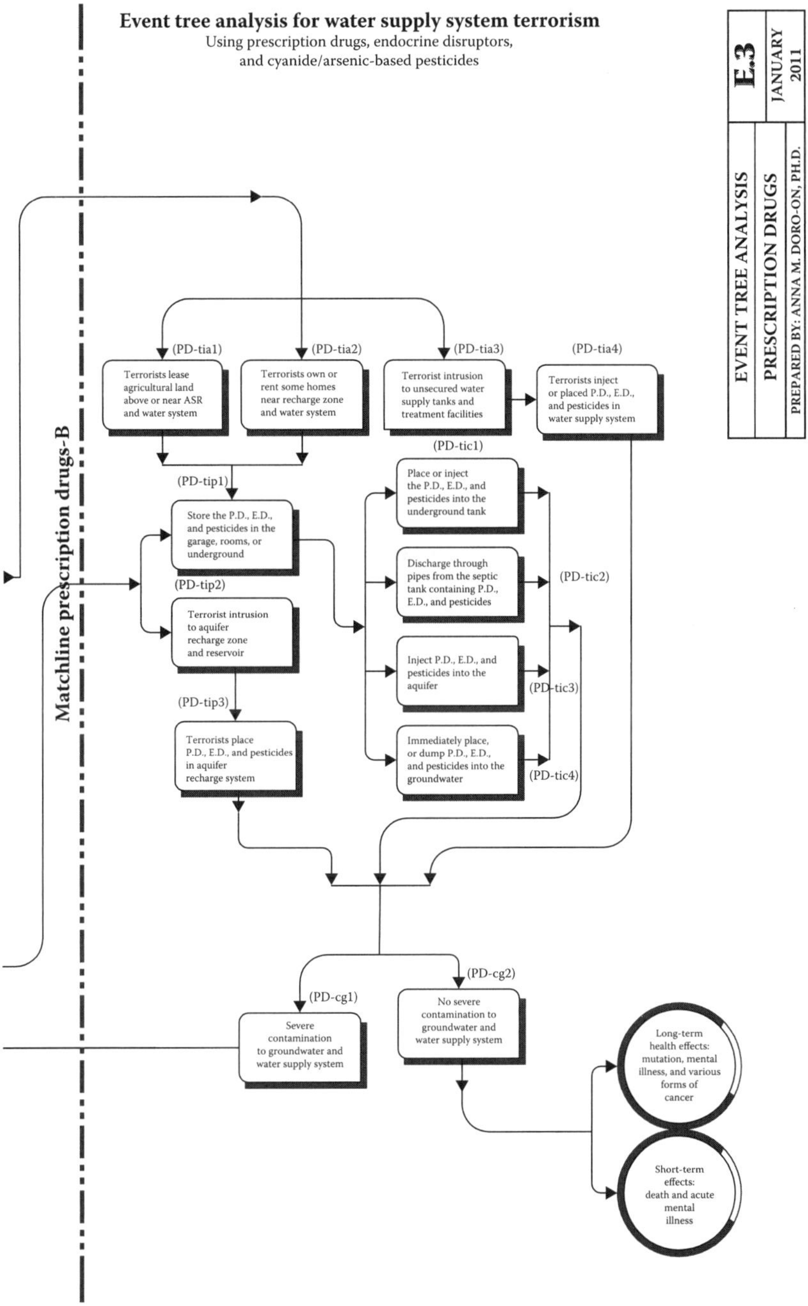

Figure 9.8d Plate E.3: Event tree analysis for water supply terrorism using prescription drugs, endocrine disruptors, and cyanide-/arsenic-based pesticides.

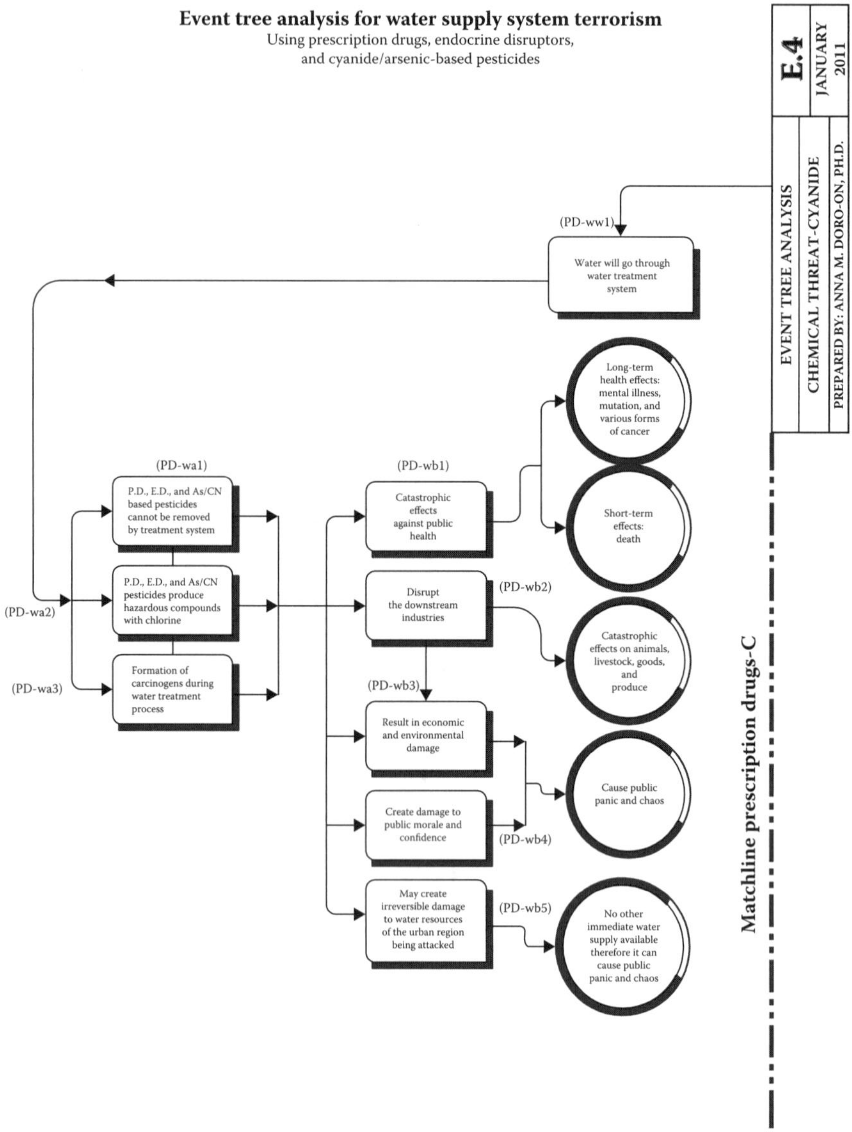

Figure 9.8e Plate E.4: Event tree analysis for water supply terrorism using prescription drugs, endocrine disruptors, and cyanide-/arsenic-based pesticides.

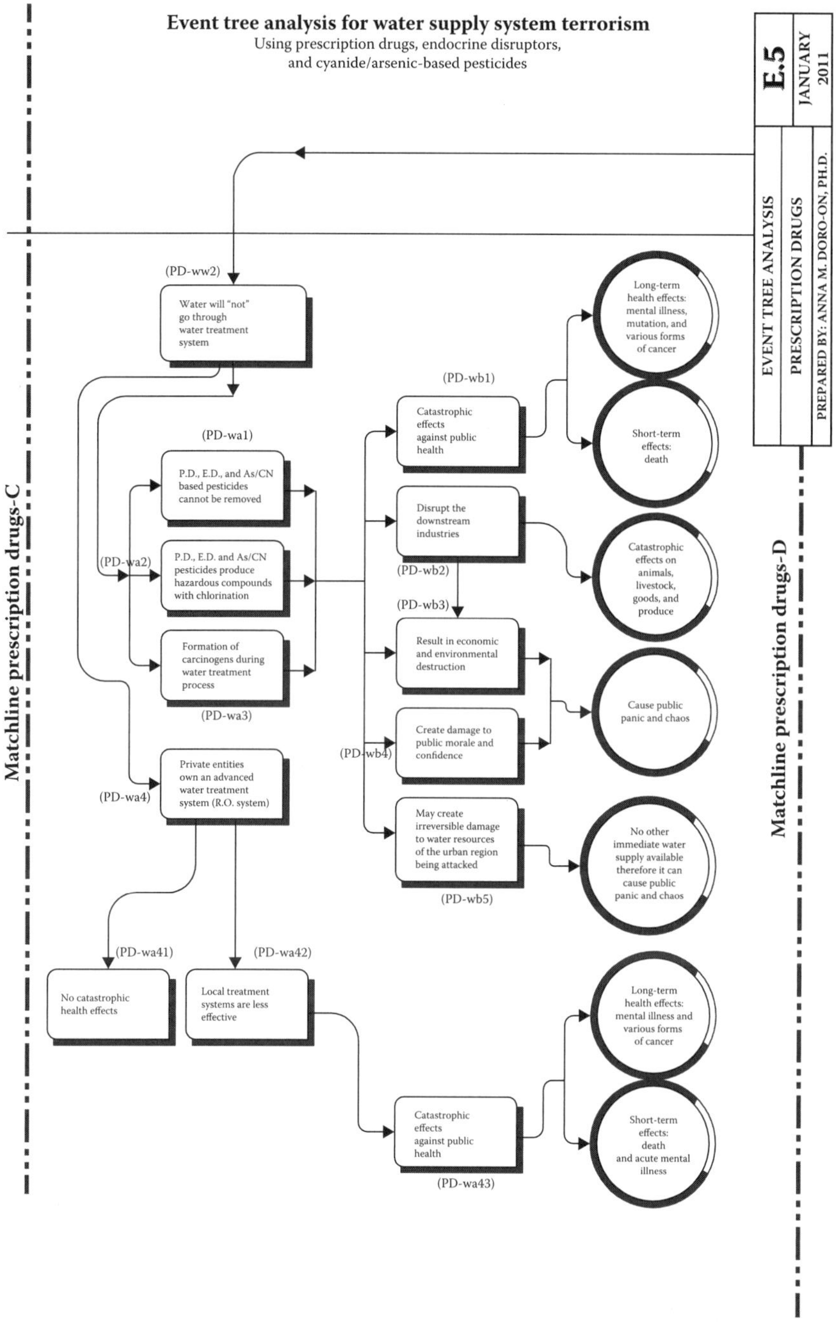

Figure 9.8f Plate E.5: Event tree analysis for water supply terrorism using prescription drugs, endocrine disruptors, and cyanide-/arsenic-based pesticides.

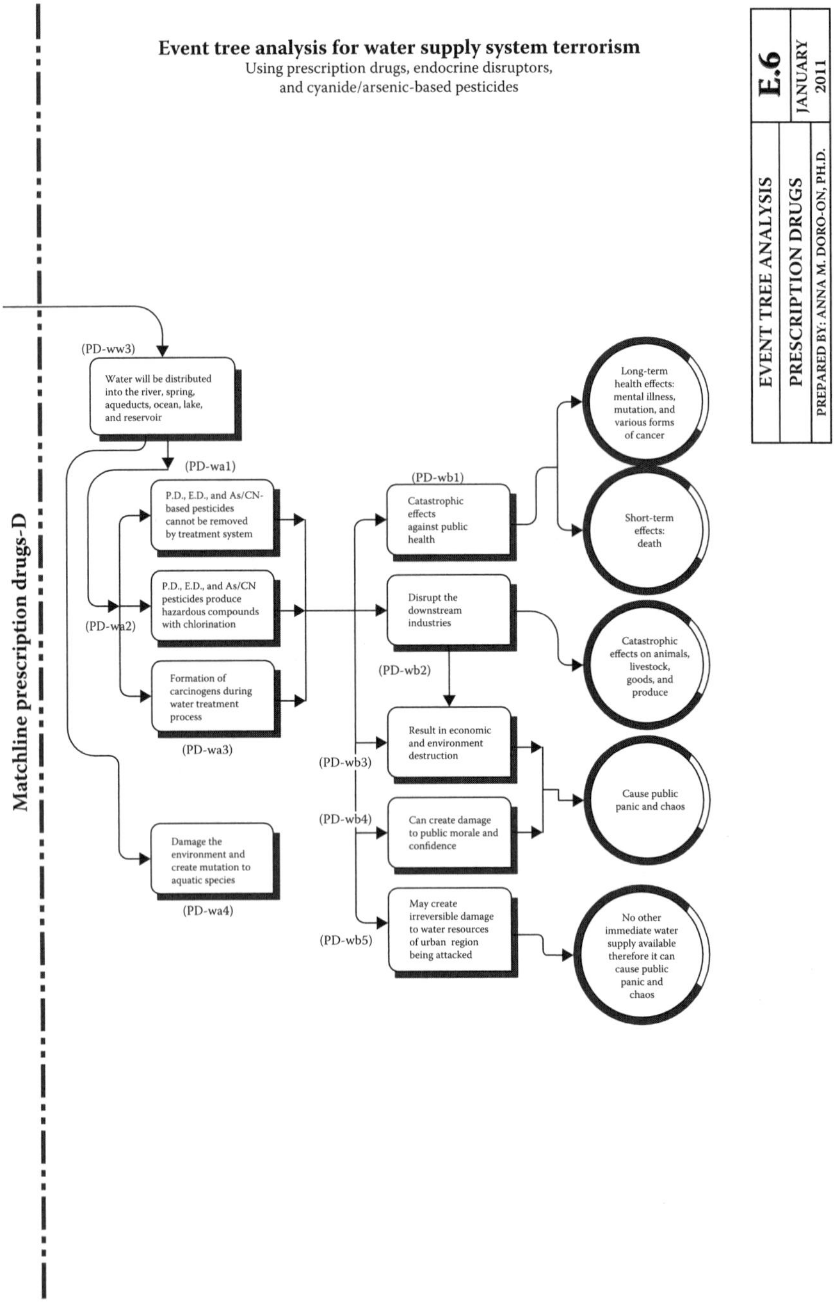

Figure 9.8g Plate E.6: Event tree analysis for water supply terrorism using prescription drugs, endocrine disruptors, and cyanide-/arsenic-based pesticides.

Table 9.11 Probability Scale for Risk Estimation Model

Probability Scale	
Description	*Probabilities in Decimal Description*
Very high (indicates that there are no effective policy or protective measures currently in place to deter, detect, delay, and respond to the threat)	0.90–1.00
High (there are some policy and protective measures to deter, detect, delay, defeat or respond to the asset but not a complete or effective application of these security strategies)	0.80–0.89
	0.71–0.79
Medium high (indicates that although there are some effective policy and protective measures, there is not a complete and effective application of these security strategies)	0.61–0.70
Medium low (indicates that although there are some effective policy and protective measures, there is not a complete and effective application of these security strategies)	0.40–0.60
Low (indicates that there are effective protective measures in place; however, at least one weakness exists such that an adversary would be able to defeat the countermeasure)	0.20–0.398
	0.10–0.198
Less likely—very low probability (indicates no credible evidence of capability)	0.05–0.099
	0.025–0.049
	<0.024

Table 9.12 Risk Rates Using Engineering Judgment of the Event Tree Analysis for Water Supply System Terrorism Using Cyanide

Event Tree List of Events		
Symbol	*Description*	*Risk Rate*
Ftv-a1	Terrorists make real estate investments in the United States	0.800
Ftv-b1	Homegrown terrorists—Americans who are radicalized in North America	0.800
Ftv-ia1	Terrorists purchase foreclosure homes and lands adjacent to U.S. borders	0.800
Ftv-ia2	Terrorists construct/build tunnels underneath houses for inflow of cyanide and illegal aliens (see Figure 9.3)	0.800
Ftv-ia3	Successful inflow of cyanide through the tunnel/pipeline system on the cross-borders	0.800
Ftv-hsa1	Homeland security with scarce technology for security and protection	0.800
Ftv-hsa2	U.S. Customs and Border Protection (CBP) agents do not discover the tunnels on U.S. borders	0.800
Ftv-hsa3	Cyanide is loaded in trucks/vehicles regularly	0.800
Ftv-hsb1	Homeland Security with business-as-usual strategy for security	0.800
Ftv-hsb2	Border patrols/CBP agents stop the vehicle with cyanide for short and temporary inspection	0.200
Ftv-hsb3	No thorough investigation by border patrols/CBP agents	0.200
FTv-tma1	Transport of deadly cyanide from Mexico with false documents	0.040
FTv-tma2	Cyanide is transported to urban areas	0.800
FTv-hra1	Illegal aliens hired by terrorists to import cyanide into the United States	0.800
FTv-hra2	20 million illegal aliens crossing U.S. borders per year; assume approximately 5 million aliens per year will carry some cyanide	0.200
FTv-hra3	20 lb × 5 million = 75,000,000 pounds or 37,000 tons per year (minimum of weapons being transported into the United States)	0.023

Table 9.12 Risk Rates Using Engineering Judgment of the Event Tree Analysis for Water Supply System Terrorism Using Cyanide (*Continued*)

Event Tree List of Events		
Symbol	*Description*	*Risk Rate*
FTv-pua1	Terrorists randomly purchase large quantity of cyanide	—
FTv-pub1	Terrorists are not able to purchase large quantity of cyanide	0.001
FTv-puc1	Stealing of cyanide from manufacturers' warehouses, laboratory facilities, or plants	0.001
FTv-tra1	Cyanide is transported to urban areas	0.800
FTv-hrc1	Presidential Directive 7: U.S. EPA with business-as-usual strategy and policy	0.800
FTv-tga1	Deficiency in technology for groundwater protection from cyanide	0.800
FTv-hrb1	Deficiency in technology for other water supply system protection from cyanide	0.800
FTv-hrc1c	Deficiency in policy on groundwater and water supply system protection	0.800
FTv-sca1	No surveillance on and no fence around major aquifer recharge zones and water supply systems	0.200
FTv-sca2	No intrusion detection on the sensitive aquifer recharge zone	0.200
FTv-sca3	No checkpoints within the major aquifer recharge zone and water supply system facilities (e.g., aqueducts, lakes, reservoirs)	0.200
FTv-sca4	No surveillance and detection technology on roads leading to the recharge zone or water supply system	0.200
FTv-sca5	No cyanide detection on wells and tanks	0.200
FTv-sca6	No surveillance or security for the land above aquifer storage and recovery facility	0.200

(*Continued*)

Table 9.12 Risk Rates Using Engineering Judgment of the Event Tree Analysis for Water Supply System Terrorism Using Cyanide (*Continued*)

Event Tree List of Events		
Symbol	*Description*	*Risk Rate*
FTv-scb1	No background investigation on buyers of properties located above the aquifer recharge zone or the underground reservoir	0.800
FTv-scb2	No background investigation on the chemical buyers or haul-away companies (these are the companies that collect the pharmaceutical waste)	0.800
FTv-scb3	No regular inspection on major aquifer recharge zones that are open for public use	0.230
FTv-scb4	No regular inspection on underground tanks	0.230
FTv-scb5	No regulations requiring installation of detection technology for high concentration of cyanide in tanks	0.800
FTv-scb6	Lack of funding for policy making on groundwater and water supply protection	0.500
FTv-tia1	Terrorists rent some agricultural land above aquifer storage and recovery (ASR) vicinity	0.800
FTv-tia2	Terrorists own or rent residential properties located above the aquifer	0.800
FTv-tia3	Terrorist intrusion into unsecured and secured water supply system facilities	0.300
FTv-tia4	Terrorists inject deadly chemical into the water tank	0.300
FTv-tib1	Terrorists store the cyanide in the garage, rooms, or underground tanks	0.001
FTv-tib2	Terrorist intrusion to major recharge zone areas such as sinkholes, faults, caves, wells	0.800
FTv-tib3	Terrorist dump the cyanide in the recharge zone, for example, sinkholes, faults, caves, wells	0.800
FTv-tic1	Terrorists inject or pump the cyanide into the underground tank to be discharged into the groundwater indirectly	0.800

Table 9.12 Risk Rates Using Engineering Judgment of the Event Tree Analysis for Water Supply System Terrorism Using Cyanide (*Continued*)

Event Tree List of Events		
Symbol	*Description*	*Risk Rate*
FTv-tic2	Discharge through pipes from the underground tank containing cyanide	0.800
FTv-tic3	Terrorists inject the cyanide beneath the residence to the aquifer	0.950
FTv-tic4	Immediately inject the cyanide beneath the ground	0.950
FTv-cg1	Severe contamination to groundwater and water supply system	0.800
FTv-cg2	No severe contamination to groundwater and water supply system	0.0001

Table 9.13 Risk Rates Using Engineering Judgment of the Event Tree Analysis for Water Supply System Terrorism Using Arsenic

Event Tree List of Events		
Symbol	*Description*	*Risk Rate*
Ftv-a1	Terrorists make real estate investments in the United States	0.800
Ftv-b1	Homegrown terrorists—Americans who are radicalized in North America	0.800
Ftv-ia1	Terrorists purchase foreclosure homes and land adjacent to U.S. borders	0.800
Ftv-ia2	Terrorists construct/build tunnels underneath the house for inflow of arsenic and illegal aliens (see Figure 9.3)	0.800
Ftv-ia3	Successful inflow of arsenic through the tunnel/pipeline system on the cross-borders	0.800
Ftv-hsa1	Homeland Security with scarce technology for security and protection	0.800
Ftv-hsa2	U.S. CBP agents do not discover the tunnels on the U.S. borders	0.800

(*Continued*)

Table 9.13 Risk Rates Using Engineering Judgment of the Event Tree Analysis for Water Supply System Terrorism Using Arsenic (*Continued*)

Event Tree List of Events		
Symbol	*Description*	*Risk Rate*
Ftv-hsa3	Arsenic is loaded in trucks/vehicles regularly	0.800
Ftv-hsb1	Homeland Security with business-as-usual strategy for security	0.800
Ftv-hsb2	Border patrols/CBP agents stop the vehicle with arsenic for short and temporary inspection	0.200
Ftv-hsb3	No thorough investigation by border patrols/CBP agents	0.200
FTv-tma1	Transport of deadly arsenic from Mexico with false documents	0.040
FTv-tma2	Arsenic is transported to urban areas	0.800
FTv-hra1	Illegal aliens hired by terrorists to import arsenic into the United States	0.800
FTv-hra2	20 million illegal aliens per year crossing U.S. borders; assume approximately 5 million aliens per year will carry illegal drugs	0.200
FTv-hra3	20 lb × 5 million = 75,000,000 pounds or 37,000 tons per year (minimum of weapons being transported into the United States)	0.023
FTv-pua1	Terrorists randomly purchase large quantities of arsenic	0.023
FTv-pub1	Terrorists are not able to purchase large quantities of arsenic	0.001
FTv-puc1	Stealing of arsenic from manufacturers' warehouses, laboratory facilities, or plants	0.001
FTv-tra1	Arsenic is transported to urban areas	0.800
FTv-hrc1	Presidential Directive 7: U.S. EPA with business-as-usual strategy and policy	0.800
FTv-tga1	Deficiency in technology for other water supply system protection from arsenic	0.800

Table 9.13 Risk Rates Using Engineering Judgment of the Event Tree Analysis for Water Supply System Terrorism Using Arsenic (*Continued*)

Event Tree List of Events		
Symbol	*Description*	*Risk Rate*
FTv-hrb1	Deficiency in policy on groundwater and water supply system protection	0.800
FTv-hrc1c	No surveillance on and no fence around major aquifer recharge zones and water supply systems	0.800
FTv-sca1	No intrusion detection on the sensitive aquifer recharge zone	0.200
FTv-sca2	No checkpoints within the major aquifer recharge zone and water supply system facilities (e.g., aqueducts, lakes, reservoirs)	0.200
FTv-sca3	No surveillance and detection technology on roads leading to the recharge zone or water supply system	0.200
FTv-sca4	No arsenic detection in wells and tanks	0.200
FTv-sca5	No surveillance or security for the land above ASR facility	0.200
FTv-sca6	No background investigation on buyers of properties located above the aquifer recharge zone or the underground reservoir	0.200
FTv-scb1	No background investigation on the chemical buyers or haul-away companies (these are the companies that collect the pharmaceutical waste)	0.800
FTv-scb2	No regular inspection on major aquifer recharge zones that are open for public use	0.800
FTv-scb3	No regular inspection on underground tanks	0.230
FTv-scb4	No regulations requiring installation of detection technology for high concentration of arsenic in tanks	0.230
FTv-scb5	Lack of funding for policy making on groundwater and water supply protection	0.800
	Terrorists rent some agricultural land above the ASR vicinity	—

(*Continued*)

Table 9.13 Risk Rates Using Engineering Judgment of the Event Tree Analysis for Water Supply System Terrorism Using Arsenic (*Continued*)

Event Tree List of Events		
Symbol	*Description*	*Risk Rate*
FTv-scb6	Terrorists own or rent residential properties located above the aquifer	0.500
FTv-tia1	Terrorist intrusion to unsecured and secured water supply system facilities	0.800
FTv-tia2	Terrorists inject arsenic into the water tank	0.800
FTv-tia3	Terrorists store the arsenic in the garage, rooms, or underground tanks	0.300
FTv-tia4	Terrorist intrusion to major recharge zone areas such as sinkholes, faults, caves, wells	0.300
FTv-tib1	Terrorists dump the arsenic in the recharge zone, for example, sinkholes, faults, caves, wells	0.001
FTv-tib2	Terrorists inject or pump the arsenic into the underground tank to be discharged into the groundwater indirectly	0.800
FTv-tib3	Discharge through pipes from the underground tank containing arsenic	0.800
FTv-tic1	Terrorists inject the arsenic beneath the residence to the aquifer	0.800
FTv-tic2	Immediately inject the cyanide beneath the ground	0.800
FTv-tic3	Severe contamination to groundwater and water supply system	0.950
FTv-tic4	No severe contamination to groundwater and water supply system	0.950
FTv-cg1	Deficiency in technology for groundwater protection from arsenic	0.800
FTv-cg2	Deficiency in technology for other water supply system protection from arsenic	0.0001

Table 9.14 Risk Rates Using Engineering Judgment of the Event Tree Analysis for Water Supply System Terrorism Using Arsenic and Cyanide

Event Tree List of Events		
Symbol	*Description*	*Risk Rate*
Ftv-a1	Terrorists make real estate investments in the United States	0.800
Ftv-b1	Homegrown terrorists—Americans who are radicalized in North America	0.800
Ftv-ia1	Terrorists purchase foreclosure homes and land adjacent to U.S. borders	0.800
Ftv-ia2	Terrorists construct/build tunnels underneath houses for inflow of arsenic, cyanide, and illegal aliens (see Figure 9.3)	0.800
Ftv-ia3	Successful inflow of arsenic and cyanide through the tunnel/pipeline system on cross-borders	0.800
Ftv-hsa1	Homeland Security with scarce technology for security and protection	0.800
Ftv-hsa2	U.S. CBP agents do not discover the tunnels on U.S. borders	—
Ftv-hsa3	Arsenic is loaded in trucks/vehicles regularly	0.800
Ftv-hsb1	Homeland Security with business-as-usual strategy for security	0.800
Ftv-hsb2	Border patrols/CBP agents stop the vehicle with arsenic and cyanide for short and temporary inspection	0.800
Ftv-hsb3	No thorough investigation by border patrols/CBP agents	0.200
FTv-tma1	Transport of deadly arsenic and cyanide from Mexico with false documents	0.200
FTv-tma2	Arsenic and cyanide are transported to urban areas	0.040
FTv-hra1	Illegal aliens hired by terrorists to import arsenic and cyanide into the United States	0.800
FTv-hra2	20 million illegal aliens crossing the U.S. borders per year; assume approximately 5 million aliens per year will carry illegal drugs	0.800

(*Continued*)

Table 9.14 Risk Rates Using Engineering Judgment of the Event Tree Analysis for Water Supply System Terrorism Using Arsenic and Cyanide (*Continued*)

Event Tree List of Events		
Symbol	*Description*	*Risk Rate*
FTv-hra3	20 lb × 5 million = 75,000,000 pounds or 37,000 tons per year (minimum of weapons being transported into the United States)	0.200
FTv-pua1	Terrorists randomly purchase large quantities of arsenic and cyanide	0.023
FTv-pub1	Terrorists are not able to purchase large quantities of arsenic and cyanide	0.023
FTv-puc1	Stealing of arsenic and cyanide from manufacturers' warehouses, laboratory facilities, or plants	0.001
FTv-tra1	Arsenic and cyanide are transported to urban areas	0.001
FTv-hrc1	Presidential Directive 7: U.S. EPA with business-as-usual strategy and policy	0.800
FTv-tga1	Deficiency in technology for groundwater protection from arsenic and cyanide contamination	0.800
FTv-hrb1	Deficiency in technology for water supply system protection from arsenic and cyanide contamination	0.800
FTv-hrc1c	Deficiency in policy for groundwater and water supply system protection from arsenic and cyanide contamination	0.800
FTv-sca1	No surveillance, no security, and no fence for major recharge zones and water supply systems (e.g., aqueducts and reservoirs)	0.200
FTv-sca2	No intrusion detection on the sensitive aquifer recharge zone and water supply system	0.200
FTv-sca3	No checkpoints within the major recharge facilities	0.200
FTv-sca4	No surveillance and no technology detection on roads leading to recharge zone and water supply system	0.200
FTv-sca5	No arsenic and cyanide detection on recharge wells and tanks	0.200

Table 9.14 Risk Rates Using Engineering Judgment of the Event Tree Analysis for Water Supply System Terrorism Using Arsenic and Cyanide (*Continued*)

Event Tree List of Events		
Symbol	*Description*	*Risk Rate*
FTv-sca6	No security and no surveillance on artificial aquifer recharge/ASR facility	0.200
FTv-scb1	No background investigation prior to purchasing properties above the aquifer	0.800
FTv-scb2	No background investigation prior to purchasing large amount of chemicals	0.800
FTv-scb3	No regular inspection on underground tanks (e.g., septic tank, water tank)	0.230
FTv-scb4	No regular inspection on underground (septic or water) tanks	0.230
FTv-scb5	No regulations requiring the installation of technology detecting pure/high concentration of chemicals	0.800
FTv-scb6	Lack of funding for policy making on groundwater protection	0.500
FTv-tia1	Terrorists rent agricultural land above artificial aquifer recharge	0.800
FTv-tia2	Terrorists own or rent residential properties (including land) above aquifer	0.800
FTv-tia3	Terrorist intrusion on unsecured water supply	0.300
FTv-tia4	Terrorists inject or dump deadly chemicals into the water tank	0.300
FTv-tib1	Hide/store the chemicals in the garage, rooms, or underground	0.001
FTv-tib2	Terrorist intrusion to major recharge zone areas like sinkholes, faults, caves, wells	0.800
FTv-tib3	Terrorists dump the arsenic and cyanide in the recharge zone, for example, sinkholes, faults, caves, wells	0.800
FTv-tic1	Inject chemicals into the underground septic tank	0.800
FTv-tic2	Discharge through pipes from the septic tank containing deadly chemicals	0.800

(*Continued*)

Table 9.14 Risk Rates Using Engineering Judgment of the Event Tree Analysis for Water Supply System Terrorism Using Arsenic and Cyanide (*Continued*)

Event Tree List of Events		
Symbol	*Description*	*Risk Rate*
FTv-tic3	Inject the chemicals beneath the residence to the aquifer	0.950
FTv-tic4	Immediately pump or inject chemicals beneath the ground	0.950
FTv-cg1	Severe contamination to groundwater and water supply system	0.800
FTv-cg2	No severe contamination to groundwater and water supply system	0.800

Table 9.15 Risk Rates Using Engineering Judgment of the Event Tree Analysis for Water Supply System Terrorism Using Biological Threats

Event Tree List of Events		
Symbol	*Description*	*Risk Rate*
BT-a1	Terrorists purchase properties such as foreclosure homes for investments in the United States	1.0
BT-b1	Homegrown terrorists—Americans who are radicalized in North America	1.0
BT-ia1	Terrorists buy/purchase homes and land adjacent to U.S. borders	0.80
BT-ia2	Construct tunnels or install pipes underneath the houses for inflow of biological threats and illegal aliens	0.75
BT-ia3	Inflow of biological threats through tunnels/pipeline system to the United States	0.60
BT-hsa1	Homeland Security with scarce technology for security and protection	0.80
BT-hsa2	U.S. CBP agents do not discover tunnels/pipes on U.S. borders	0.20
BT-hsa3	Biological threats are loaded in trucks/vehicles	1.00

Table 9.15 Risk Rates Using Engineering Judgment of the Event Tree Analysis for Water Supply System Terrorism Using Biological Threats (*Continued*)

Event Tree List of Events		
Symbol	*Description*	*Risk Rate*
BT-hsb1	Homeland Security with business-as-usual security strategy	0.80
BT-hsb2	Border patrols/CBP agents stop the vehicle for short and temporary inspection	0.20
BT-hsb3	Lack of thorough investigation by border patrols/CBP agents	0.50
BT-tma1	Transport of biological threats from Mexico with false documents	0.60
BT-tma2	Biological threats are transported to urban areas	0.80
FTv-hra3	Terrorists produce biological threats in their designated vicinities within the United States	0.90
BT-tra1	Biological threats are transported to urban areas	0.90
BT-hrc1	Presidential Directive 7: U.S. EPA typical or business-as-usual strategy and policy	0.90
BT-ww1	Water systems are open for public use	1.00
BT-ww2	Water will not go through the water treatment system	0.70
BT-tga1	Deficiency in technology for groundwater protection	0.90
BT-hrb1	Deficiency in technology for protecting other water supply systems	0.90
BT-hrc1c	Deficiency in policy on groundwater and water supply system	0.90
BT-sca1	No camera surveillance/fence on major recharge zone of an aquifer	0.95
BT-sca2	No intrusion detection on sensitive and major recharge zone	0.95
BT-sca3	No checkpoints within the major recharge facilities	0.95
BT-sca4	No surveillance and no detection technology on roads leading to recharge zone	0.95

(*Continued*)

Table 9.15 Risk Rates Using Engineering Judgment of the Event Tree Analysis for Water Supply System Terrorism Using Biological Threats (*Continued*)

Event Tree List of Events		
Symbol	*Description*	*Risk Rate*
BT-sca5	No biological threats detection on recharge zone wells	0.95
BT-sca6	No security/no surveillance for artificial aquifer recharge	0.95
BT-scb1	No background investigation on buyers of properties located on the ground above an aquifer	0.95
BT-scb2	No background investigation on buyers of chemicals	0.95
BT-scb3	No regular inspection on underground tanks (e.g., septic tank, water tank)	0.95
BT-scb4	No regular inspection on underground (septic or water) tanks	0.95
BT-scb5	No regulations requiring the installation of technology detecting pure/high concentration of chemicals in the tank or vessel	0.95
BT-scb6	Lack of funding for policy making on groundwater protection	0.90
BT-tia1	Terrorists rent the agricultural land above aquifer recharge and recovery facility	0.90
BT-tia2	Terrorists purchase or rent the residential properties above the aquifer	0.90
BT-tia3	Terrorist intrusion to unsecured water supply	0.95
BT-tia4	Terrorists inject or dump biological threats in the water tank	0.95
BT-tib1	Store the biological threats in the garage, rooms, or underground	0.95
BT-tib2	Terrorist intrusion to major recharge zone areas like sinkholes, faults, caves, wells	0.95
BT-tib3	Terrorists dump biological threats in the recharge zone, for example, sinkholes, faults, wells	0.95
BT-wa1	Water will go through treatment system with disinfection using chlorine oxidation	0.75
BT-wa11	Biological threats will propagate in the water	0.50

Table 9.15 Risk Rates Using Engineering Judgment of the Event Tree Analysis for Water Supply System Terrorism Using Biological Threats (*Continued*)

Event Tree List of Events		
Symbol	*Description*	*Risk Rate*
BT-wa21	Public/private sectors install disinfection system (chlorination) in their facilities	0.50
BT-wa2	Chlorination in the traditional water treatment system will remove biological threats	0.05
BT-wb1	Some of the biological threats remain in the water supply	0.05
BT-tic1	Inject the biological threats into the underground tank to indirectly discharge biological threats into the aquifer	0.95
BT-tic2	Discharge through pipes from the underground tank containing biological threats	0.95
BT-tic3	Inject the chemicals beneath the residence to the aquifer	0.95
BT-tic4	Immediately inject the biological threats beneath the ground	0.95
BT-cg1	Severe contamination to groundwater and water supply system	0.35
BT-cg2	No severe contamination to groundwater and water supply system	0.65

Table 9.16 Risk Rates Using Engineering Judgment of the Event Tree Analysis for Water Supply System Terrorism Using Prescription Drugs, Endocrine Disruptors, Cyanide-/Arsenic-Based Pesticides

Event Tree List of Events		
Symbol	*Description*	*Risk Rate*
PD-a1	Terrorists purchase some of the foreclosure homes for investments in the United States	1.0
PD-b1	Homegrown terrorists—Americans who are radicalized in North America	1.0
PD-b2	Terrorists acquire credit cards and loans in the United States to support their operations or missions	1.0

(*Continued*)

Table 9.16 Risk Rates Using Engineering Judgment of the Event Tree Analysis for Water Supply System Terrorism Using Prescription Drugs, Endocrine Disruptors, Cyanide-/Arsenic-Based Pesticides (*Continued*)

Event Tree List of Events		
Symbol	*Description*	*Risk Rate*
PD-ia1	Terrorists purchase or lease homes and land adjacent to U.S. borders	0.95
PD-ia2	Terrorists construct tunnels or install pipes underneath houses for inflow of weapons against water infrastructure such as expired prescription drugs, endocrine disruptors, arsenic, and cyanide	0.95
PD-ia3	Inflow of weapons through tunnel/pipeline system to the United States	0.95
PD-hsa1	Homeland Security with scarce technology for security and protection against terrorism	0.90
PD-hsa2	CBP agents do not discover the tunnels/pipes on U.S. borders	0.80
PD-hsa3	Chemical weapons are loaded in trucks/vehicles regularly	0.95
PD-hsb1	Homeland Security with business-as-usual security strategy	0.95
PD-hsb2	Border patrols/CBP agents stop the vehicle for short and temporary inspection	0.90
PD-hsb3	No thorough investigation by border patrols/CBP agents	0.60
PD-tma1	Transport of (expired) prescription drugs and endocrine disruptors from Mexico/Canada with false documents on U.S. borders	0.60
PD-tma2	(Expired) prescription drugs, endocrine disruptors, arsenic, and cyanide are transported to urban areas	0.75
PD-hra1	Illegal aliens are hired by terrorists to import weapons into the United States	0.50
PD-hra2	20 million illegal aliens per year crossing the U.S. borders; assume approximately 5 million aliens per year will carry illegal drugs	0.50
PD-hra3	20 lb × 5 million = 75,000,000 pounds or 37,000 tons per year (minimum of weapons being transported into the United States)	0.50

Table 9.16 Risk Rates Using Engineering Judgment of the Event Tree Analysis for Water Supply System Terrorism Using Prescription Drugs, Endocrine Disruptors, Cyanide-/Arsenic-Based Pesticides (*Continued*)

Event Tree List of Events		
Symbol	*Description*	*Risk Rate*
PD-pua1	Randomly buy or haul away disposed expired prescription drugs and endocrine disruptors	0.50
PD-pua2	Terrorists are not able to buy or haul away the expired prescription drugs and endocrine disruptors	0.50
PD-pua3	Terrorists steal prescription drugs and endocrine disruptors from manufacturers' or vendors' warehouses	0.50
PD-pub1	Terrorists acquire credit cards from department stores	0.95
PD-puc1	Terrorists randomly purchase arsenic-/cyanide-based pesticides from different local department stores; also, they collect expired prescription drugs from waste collectors	0.95
PD-tra1	Prescription drugs, endocrine disruptors, and pesticides are transported to contaminate the water supply or water resources	0.95
PD-hrc1	Presidential Directive 7: U.S. EPA with business-as-usual strategy and policy	0.90
PD-tga1	Deficiency in technology for groundwater protection from prescription drugs and endocrine disruptors	0.95
PD-hrb1	Deficiency in technology for water supply system protection from prescription drugs and endocrine disruptors	0.95
PD-hrc1	Deficiency in policy for protecting groundwater and water supply system from prescription drugs and endocrine disruptors	0.95
PD-sca1	No camera surveillance and no fence on major aquifer recharge zone and water supply system	0.95
PD-sca2	No intrusion detection on sensitive and major aquifer recharge zone	0.95
PD-sca3	No checkpoints within the major aquifer recharge zone and water supply system	0.95

(Continued)

Table 9.16 Risk Rates Using Engineering Judgment of the Event Tree Analysis for Water Supply System Terrorism Using Prescription Drugs, Endocrine Disruptors, Cyanide-/Arsenic-Based Pesticides (*Continued*)

Event Tree List of Events		
Symbol	*Description*	*Risk Rate*
PD-sca4	No surveillance and no detection technology on roads leading to recharge zone and water supply system	0.95
PD-sca5	No detection technology on recharge zone wells and water supply tanks	0.95
PD-sca6	No security/no surveillance located within the artificial aquifer recharge areas	0.95
PD-scb1	No background investigation on buyers of properties located on the ground above an aquifer	0.95
PD-scb2	No background investigation on buyers of chemicals	0.95
PD-scb3	No regular inspection on underground tanks (e.g., septic tank, water tank)	0.95
PD-scb4	No regular inspection on underground (septic or water) tanks	0.95
PD-scb5	No regulations on installing detection technology for prescription drugs and endocrine disruptors including arsenic-/cyanide-based pesticides	0.95
PD-scb6	Lack of funding for policy making on groundwater and water supply protection	0.95
PD-tia1	Terrorists rent agricultural land above the artificial aquifer recharge/aquifer recharge and recovery area	0.95
PD-tia2	Terrorists own or rent residential properties (including land) above aquifer	0.95
PD-tia3	Terrorist intrusion to unsecured and secured water supply facilities	0.95
PD-tia4	Terrorists inject or dump expired prescription drugs and endocrine disruptors into the water tank	0.95
PD-tib1	Hide/store the chemicals in the garage, rooms, or underground	0.95
PD-tib2	Terrorist intrusion on major recharge zone areas like sinkholes, faults, caves, wells	0.95

Table 9.16 Risk Rates Using Engineering Judgment of the Event Tree Analysis for Water Supply System Terrorism Using Prescription Drugs, Endocrine Disruptors, Cyanide-/Arsenic-Based Pesticides (*Continued*)

Event Tree List of Events		
Symbol	*Description*	*Risk Rate*
PD-tib3	Terrorists dump chemicals in the recharge zone, for example, sinkholes, faults, caves, wells	0.95
PD-tic1	Pump or inject chemicals into the underground septic tank	0.95
PD-tic2	Discharge through pipes from the septic tank containing deadly chemicals	0.95
PD-tic3	Inject the chemicals beneath the residence to the aquifer	0.95
PD-ww1	Water will go through the water treatment system	1.0
PD-wa1	Prescription drugs, endocrine disruptors, cyanide, and arsenic cannot be removed by the traditional water treatment system	0.95
PD-wa2	Prescription drugs, endocrine disruptors, cyanide, and arsenic produce hazardous compounds with chlorine in the water treatment process	1.0
PD-wa3	Formation of carcinogens during traditional treatment process	1.0
PD-wb1	Consequence	—
PD-wb2	Consequence	—
PD-wb3	Consequence	—
PD-wb4	Consequence	—
PD-wb5	Consequence	—
PD-wa41	Consequence	—
PD-wa42	Consequence	—
PD-tic4	Immediately pump or inject chemical beneath the ground	0.95
PD-cg1	Severe contamination to groundwater and water supply system	0.80
PD-cg2	No severe contamination to groundwater and water supply system	0.20

Table 9.17 Risk Acceptability Analysis for Terrorist Attacks against Water Supply Systems Using Chemical Threats

Description	*Risk Reference*	*Proportionality by Degree of Voluntarism*	*Derating = Cost–Benefit Balance*	*Controllability*	*Risk Acceptability $V(f) = \omega(p) \cdot v(x)$ (Fatality/Year)*
Installation of underground tanks to store and discharge chemical threats into the aquifer and water supply system	9.8×10^{-8} Man-originated Catastrophic Involuntary	9×10^{-2}	7.3×10^{-4}	2.5×10^{-4}	1.4×10^{-16} (Unacceptable)
Injection/discharge of chemical threats from Mexico/Canada borders into the U.S. groundwater and/or water supply systems	9.8×10^{-8} Man-originated Catastrophic Involuntary	9×10^{-2}	6.5×10^{-5}	1.3×10^{-3}	1.78×10^{-15} (Unacceptable)
Injection/discharge of chemical threats in the recharge zone of an aquifer (e.g., sinkholes, faults, and cracks)	9.8×10^{-8} Man-originated Catastrophic Involuntary	9×10^{-2}	6.5×10^{-5}	2.5×10^{-4}	1.4×10^{-16} (Unacceptable)
Immediately dump truckloads of cyanide and arsenic directly to unsecured aquifer recharge facilities	9.8×10^{-8} Man-originated Catastrophic Involuntary	9×10^{-2}	6.5×10^{-5}	2.5×10^{-4}	1.4×10^{-16} (Unacceptable)

Terrorist intrusion to water supply systems (e.g., aqueducts, reservoirs, lakes)	9.8×10^{-8} Man-originated Catastrophic Involuntary	9×10^{-2}	6.5×10^{-5}	9.1×10^{-2}	5.24×10^{-14} (Unacceptable)
Injection of chemical threats into the ASR facility (also known as the underground reservoir)	9.8×10^{-8} Man-originated Catastrophic Involuntary	9×10^{-2}	6.5×10^{-5}	2.9×10^{-4}	1.67×10^{-15} (Unacceptable)
Contamination of groundwater and water supply systems using various hazardous chemicals	9.8×10^{-8} Man-originated Catastrophic Involuntary	9×10^{-2}	6.5×10^{-5}	2.5×10^{-4}	1.4×10^{-16} (Unacceptable)

Table 9.18 Risk Estimation for Terrorist Attacks against Water Supply Systems Using Arsenic and Cyanide

Description	*Arsenic*	*Cyanide*	*Arsenic and Cyanide*	*Risk Acceptability* $V(f) = \omega(p) \cdot v(x)$ *(Fatality/Year)*
Terrorist intrusion into water supply tank facilities for contamination	1.36×10^{-7}	1.31×10^{-8}	2.0×10^{-7}	1.78×10^{-15} (Unacceptable)
Installation of underground tanks to store and discharge chemical threats into the aquifer and water supply system	1.37×10^{-5}	5.2×10^{-5}	1.3×10^{-5}	1.4×10^{-16} (Unacceptable)
Injection/discharge of chemical threats from Mexico/Canada borders into the U.S. groundwater and/or water supply systems	2.7×10^{-5}	2.6×10^{-5}	1.3×10^{-5}	1.78×10^{-15} (Unacceptable)
Injection/discharge of chemical threats in the recharge zone of an aquifer (e.g., sinkholes, faults, and cracks)	1.7×10^{-5}	2.21×10^{-5}	8.6×10^{-5}	1.4×10^{-16} (Unacceptable)

Immediately dump truckloads of cyanide and arsenic directly to an unsecured aquifer recharge facilities	1.7×10^{-5}	2.21×10^{-5}	8.6×10^{-5}	1.4×10^{-16} (Unacceptable)
Injection of chemical threats into the ASR facility (also known as the artificial underground reservoir)	4.2×10^{-6}	5.2×10^{-5}	6.9×10^{-5}	1.78×10^{-15} (Unacceptable)
Terrorist intrusion to the water supply systems (e.g., aqueducts, reservoirs, lakes)	6.9×10^{-8}	6.7×10^{-8}	1×10^{-7}	5.24×10^{-14} (Unacceptable)

Notes:

1. Refer to Figures 9.4a through 9.4g (risk rate values in Table 9.12).
2. Refer to Figures 9.5a through 9.5g (risk rate values in Table 9.13).
3. Refer to Figures 9.6a through 9.6g (risk rate values in Table 9.14).
4. Risk rate for breaking into security fence on U.S. borders = 0.01.
5. Risk rate for intruding on security fence of water supply system, for example, water storage tanks, reservoirs, etc. = 0.01.

Table 9.19 Risk Estimation for Terrorist Attacks against Water Supply System Using Biological Threats, Prescription Drugs, Endocrine Disruptors, and Cyanide-/Arsenic-Based Pesticides

Description	*Biological Threats*	*Combination of Prescription Drugs, Endocrine Disruptors, and Cyanide-/Arsenic-Based Pesticides*	*Risk Acceptability* $V(f) = \omega(p) \cdot v(x)$
Terrorist intrusion into the water supply tank facilities for contamination	2.1×10^{-5}	5.0×10^{-1}	1.78×10^{-15} (Unacceptable)
Installation of underground tanks to store and discharge chemical threats into the aquifer and water supply system	9.6×10^{-4}	7.7×10^{-3}	1.4×10^{-16} (Unacceptable)
Injection/discharge of biological or chemical threats from Mexico/Canada borders into the U.S. groundwater and/or water supply systems	9×10^{-6}	9.0×10^{-1}	1.78×10^{-15} (Unacceptable)
Injection/discharge of biological or chemical threats in the recharge zone of an aquifer (e.g., sinkholes, faults, and cracks)	2.2×10^{-6}	5.6×10^{-2}	1.78×10^{-15} (Unacceptable)

Immediately dump truckloads of biological or chemical threats directly to an unsecured aquifer recharge facilities	2.5×10^{-6}	7.3×10^{-2}	1.4×10^{-16} (Unacceptable)
Injection of biological or chemical threats into the ASR facility (also known as the artificial underground reservoir)	2.5×10^{-6}	4.5×10^{-1}	1.78×10^{-15} (Unacceptable)
Terrorist intrusion to the water supply systems (e.g., aqueducts, reservoirs, lakes)	2.5×10^{-6}	4.8×10^{-1}	1.4×10^{-16} (Unacceptable)

Notes:

1. Refer to Figures 9.4a through 9.4g (risk rate values in Table 9.12).
2. Refer to Figures 9.5a through 9.5g (risk rate values in Table 9.13).
3. Refer to Figures 9.6a through 9.6g (risk rate values in Table 9.14).
4. Risk rate for breaking into security fence on U.S. borders = 0.01.
5. Risk rate for intruding on security fence on water supply system, for example, water storage tanks, reservoirs, etc. = 0.01.

Table 9.20 Comparison of Alternatives

Alternative (Based upon Proposed Protective Measures)	*Risk*	*Risk Acceptability*
Business as usual	4.33×10^{-8}	2.35×10^{-17}
Groundwater and water supply system source points should be fenced, be well lighted, and have a perimeter that is monitored by surveillance cameras and motion detectors with chemical threat detectors on wells.	9.5×10^{-16}	6.5×10^{-12}
(a) Monitoring system (e.g., monitoring wells) with chemical threat detector/controls notifying authorities and governing agencies shall be installed on aquifer recharge system areas and surface waters located along U.S. borders. (b) Shut-off systems shall be installed or constructed to prevent discharge and flow or transport of contaminants to surface water systems, storage systems, and water pipelines when severe contamination is detected. (c) Mandatory inspection by governing agencies of all major recharge zone and water supply system vicinities/facilities. (d) Implementation of an advanced water treatment technology for emergency treatment system. (e) Secure the U.S. borders and detect illegal underground tunnels. (f) Thorough investigation on individuals or groups of people purchasing properties near recharge zone. (g) Thorough investigation on individuals purchasing large quantity of chemicals including pesticides. (h) Individual households shall install chemical threat detection systems or install small advanced treatment systems such as reverse osmosis technology.	9.5×10^{-18}	4×10^{-8}

Source: Data from Doro-on, A. M., "Risk assessment embedded with cumulative prospect theory for terrorist attacks on aquifer of karstic limestone and water supply system," PhD diss., University of Texas at San Antonio, 2009.

United States does not improve its protection policy and technology, the estimated probability of "successful" terrorist attacks is 4.33×10^{-5}, which is very high, with a very low incremental risk acceptability of 2.35×10^{-17}. If groundwater and water supply system source points are fenced and monitored by surveillance or motion detectors, the probability of successful terrorist attacks is 9.5×10^{-16}, which is low, and the incremental risk acceptability is 6.5×10^{-12}, still very low or unacceptable because there are other events that could happen, such as dumping or injection of chemical threats from the fence. If alternative number 3 in Table 9.20 is considered, the probability of successful terrorist attacks is 9.5×10^{-18}, which is very low and the incremental risk is 4×10^{-8}, a higher value that is considered *acceptable*; once the improvement of technological systems and regulatory requirements for security are employed. Therefore, several considerations for policy improvements and preventive measures against terrorist attacks should be incorporated to protect U.S. infrastructure and to achieve acceptable level of risks. Terrorism risks should not be declared as *absolutely* unacceptable to society. The methodology presented in this chapter can be utilized to obtain an acceptable level of water infrastructure terrorism risks.

9.6 Implications

Engineers, analysts, scientists, managers, and experts should integrate risk assessment based on cumulative prospect theory in policy making and technology development for U.S. water infrastructure protection. Based upon the risk assessment analysis detailed in this chapter, the Infrastructure Protection Division of the Department of Homeland Security and the U.S. EPA should consider some of the preventive measures, which are presented in Chapter 10. Additional risk acceptability analysis examples are presented in Chapter 11.

References

BBC News. 2005. *U.S.-Canada drug tunnel uncovered.* http://news.bbc.co.uk/2/hi/americas/4706339.stm (accessed May 10, 2008).

Doro-on, A. M. 2009. "Risk assessment embedded with cumulative prospect theory for terrorist attacks on aquifer of karstic limestone and water supply system." PhD diss., University of Texas at San Antonio.

Ellison, D., S. J. Duranceau, S. Ancel, and R. McCoy. 2003. *Drinking Water Distribution System: Assessing and Reducing Risk.* Washington, DC: National Academy Press.

Federation of American Scientists. 2009. *Tunnels Beneath U.S. Borders Proliferate.* http://www.fas.org/blog/secrecy/2009/02/tunnels.html (accessed May 5, 2009). FAS

Frank, R. H. 1997. The frame of reference as a public good. *Econ J* 107:1832–47.

Kahneman, D., and A. Tversky. 1979. Prospect theory: An analysis of decision under risk. *Econometrica* 47(2):263–92.

Kahneman, D., and A. Tversky. 2000. *Choices, Values and Frames.* New York: Russell Sage Foundation and Cambridge University Press.

Nogami, G., and S. Streufort. 1973. *Time Effects on Perceive Risk Taking*. Purdue University Technical Report No. 11, Lafayette, Indiana.

O'Connor, T. 2008. *Border Security*. http://drtomoconnor.com/3430/3430lect05.htm (accessed February 3, 2008).

Otway, H. 1975. *Risk Assessment and Societal Choices*, IIASA RM-75-2. International Institute for Applied Systems Analysis, p7. Laxenburg, Austria.

Rowe, W. 1977. *An Anatomy of Risk*. New York: John Wiley & Sons.

Shih, C., S., A. M. Doro-on, and G. A. Arroyo. 2007. *Risk Assessment of Terrorism Based on Prospect Theory for Groundwater Protection. Vol. 1 Environmental Science and Technology*. Houston, TX: American Science Press.

Shih, C., and A. Riojas. 1990. Risk and Its Acceptability for Groundwater Contamination by Hazardous Wastes. Risk Assessment for Groundwater Pollution Control. Editors: McTernan, W., and Kaplan, E., American Society of Civil Engineers, pp. 126–157. ASCE.

Tversky, A., and D. Kahneman. 1986. Rational Choice and the Framing of Decisions. *Journal of Business* 59:S251–S278.

Tversky, A., and D. Kahneman. 1992. Advances in Prospect Theory: Cumulative Representation of Uncertainty. *Journal of Risk and Uncertainty* 5:297–323. Kluwer Academic Publisher. http://3xfund.com/images/article009.pdf (accessed January 5, 2006).

U.S. Environmental Protection Agency. 1973. The Quality of Life Concept, a Potential New Tool for Decision Makers. Office of Research and Development, Washington D.C. http://nepis.epa.gov/Exe/ZyNET.exe/20016RYI.PDF?ZyActionP=PDF&Client=EPA&Index=Prior%20to%201976&File=D%3A\ZYFILES\INDEX%20DATA\70THRU75\TXT\00000005\20016RYI.txt&Query=000R73002%20or%20The%20or%20Quality%20or%20Life%20or%20Concept%20or%20A%20or%20potential%20or%20New%20or%20Tool%20or%20for%20or%20Decision%20or%20Makers%20or%20USEPA&SearchMethod=1&FuzzyDegree=0&User=ANONYMOUS&Password=anonymous&QField=pubnumber^%22000R73002%22&UseQField=pubnumber&IntQFieldOp=1&ExtQFieldOp=1&Docs (accessed May 10, 2009).

U.S. Environmental Protection Agency 2009. Quality of Life Indicators. http://www.epa.gov/reva/seql_qol.html (accessed May 10, 2009).

Velimirovic, H. 1975. An Anthropoligical View of Risk, IIASA RM-75-55. International Institute for Applied Systems Analysis, p. 17. Laxenburg, Austria.

Wilson, R. 1975. Examples of Risk-Benefit Analysis. *CHEMTECH*-Journal of the American Chemical Society. October 1975, p. 604–607.

Chapter 10

Emergency Preparedness, Response, and Preventive Measures

10.1 Introduction

This chapter introduces potential plans for emergency preparedness and response before or in the event of a disaster or terrorist attack against water infrastructures such as dams and reservoirs. According to Radvanovsky and McDougall (2010), based on the response levels of first responders to a given emergency situation, environment, or hazardous condition, the following groups are representative of classification of departments and agencies based on their function by various governments: (1) law enforcement; (2) fire services; (3) emergency medical services; (4) emergency management; (5) hazmat team; (6) explosives team; and (7) search and rescue. Fire fighters and police usually respond to terrorism and accidents or other incidents involving the release of hazardous materials similar to Figures 11.2b and 11.7 (when terrorists decided to dump cyanide or another type of poison before blasting the dam) discussed in Chapter 11. Responding to such events requires knowledge of the nature of chemicals so that suitable methods can be used; decisions about evacuations or traffic diversion can be made; and danger of injury, death, or property damage can be minimized. Figures 10.1 through 10.7 provide some conceptual designs of preventive measures.

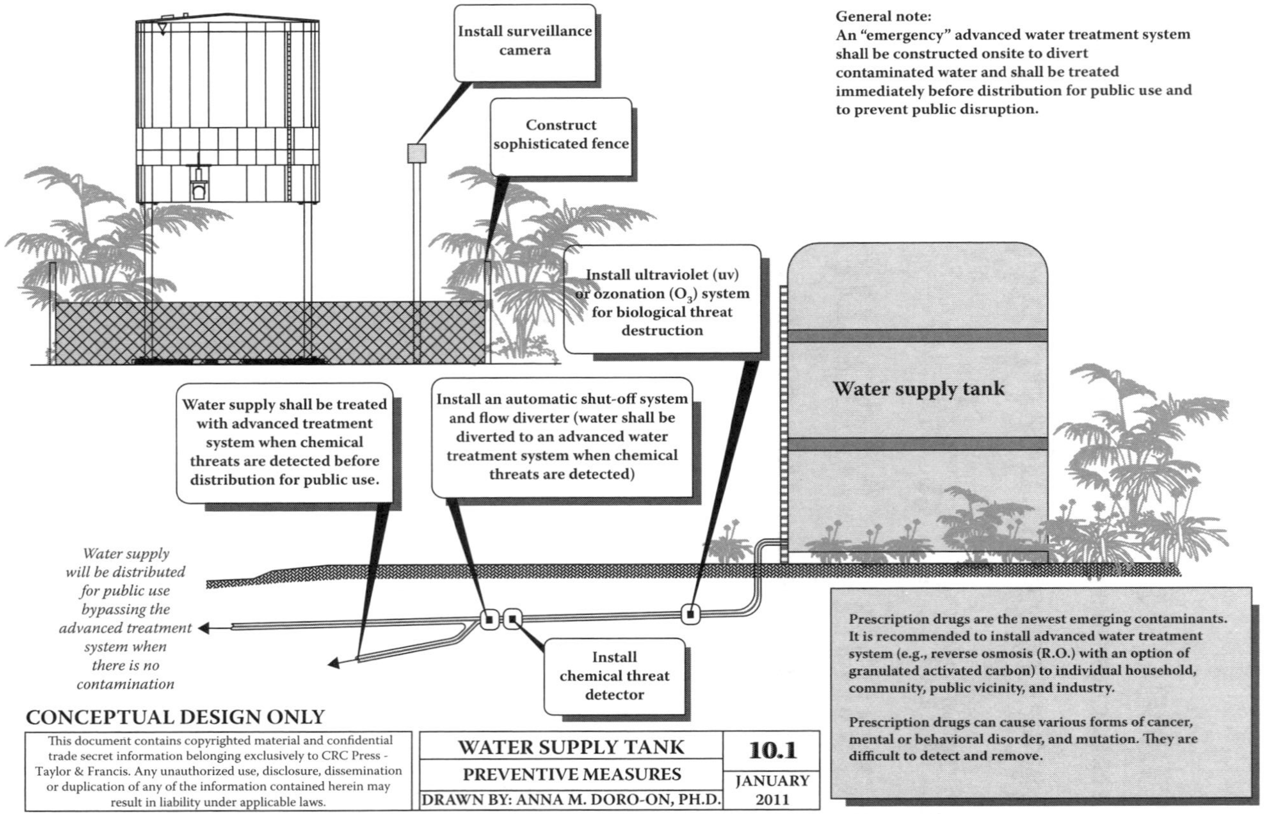

Figure 10.1 Preventive measures for water storage tanks.

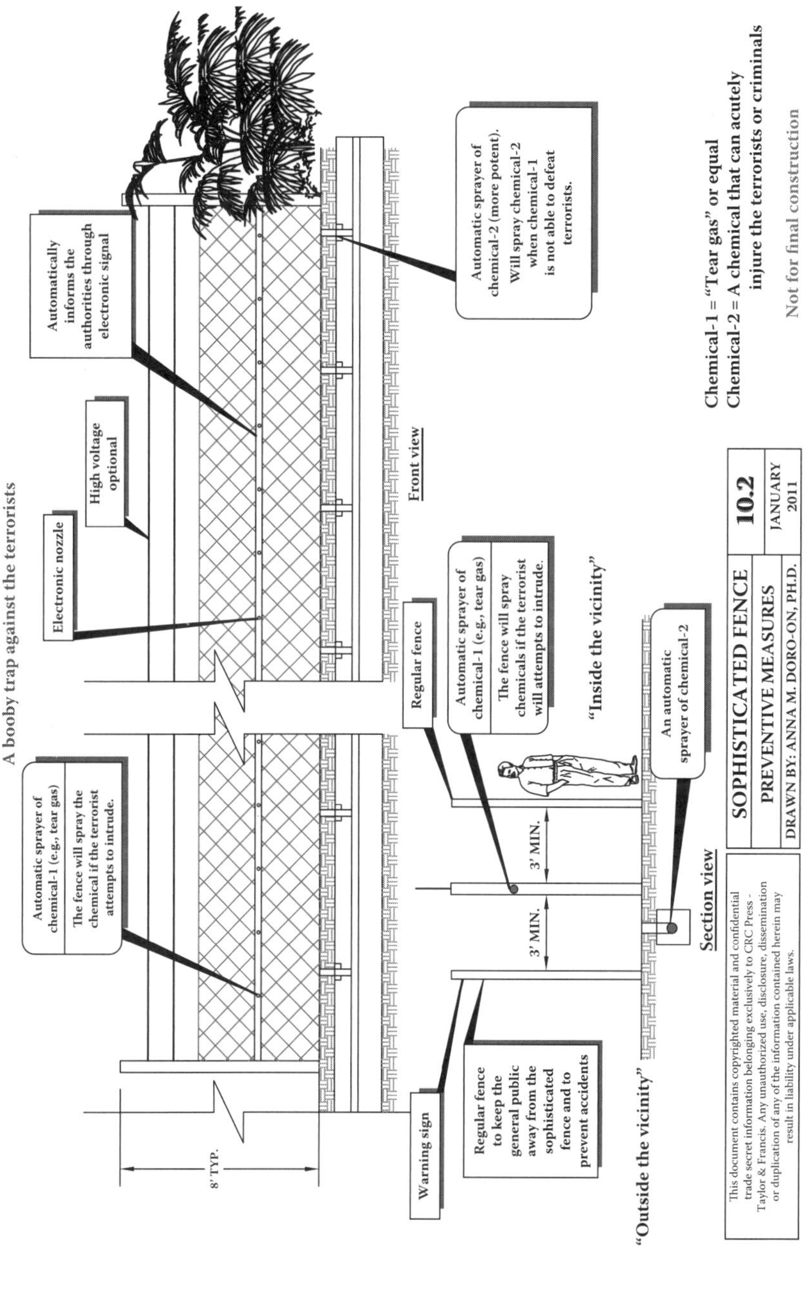

Figure 10.2 Sophisticated fence conceptual design.

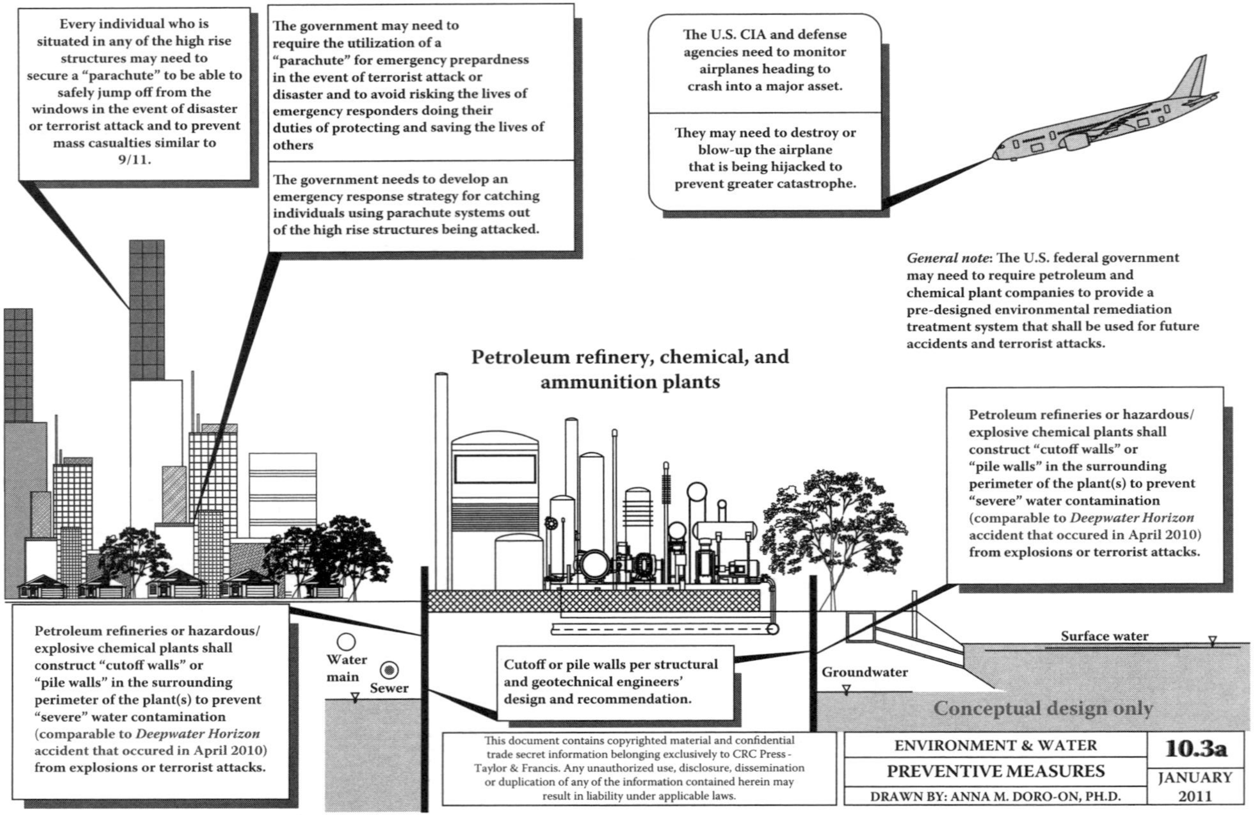

Figure 10.3a Exhibit A: Catastrophe prevention.

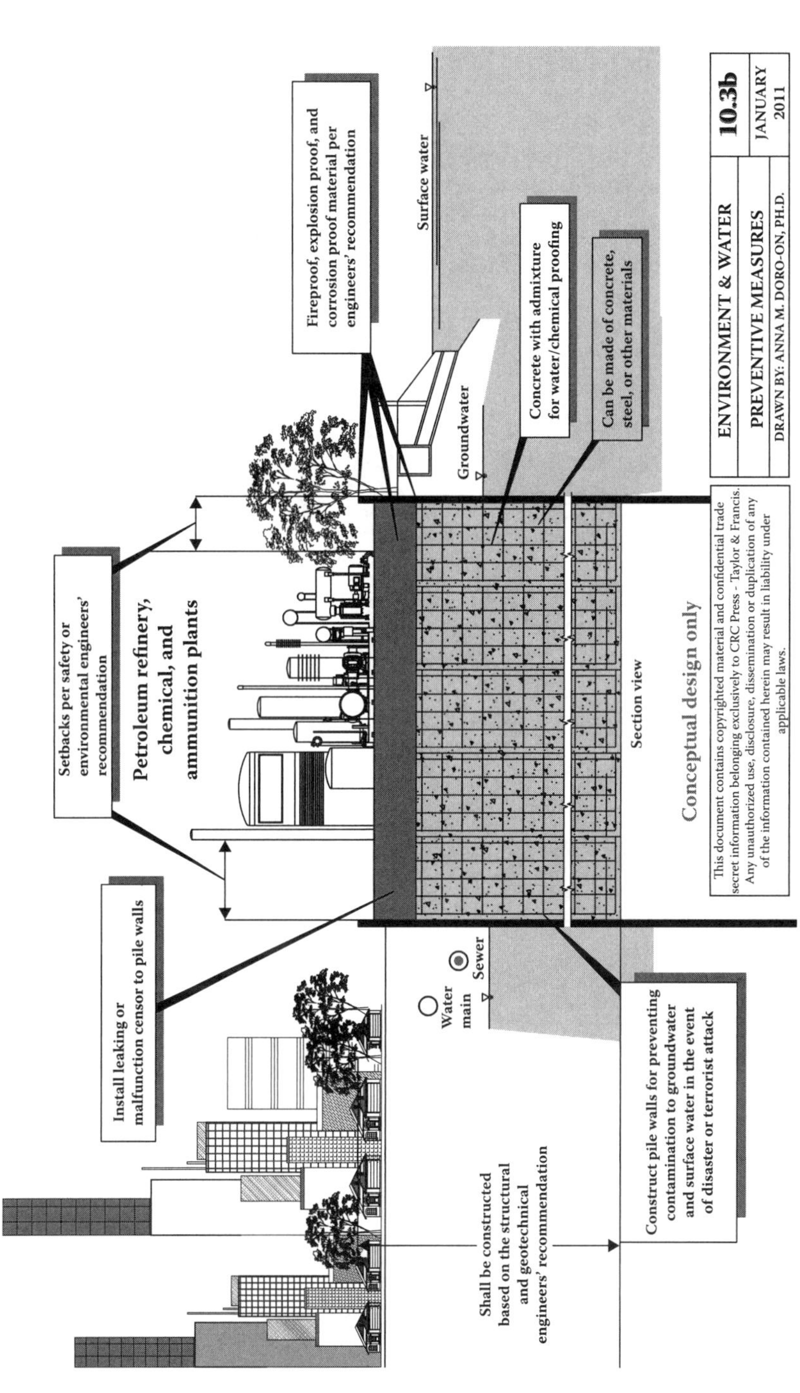

Figure 10.3b Exhibit B: Catastrophe prevention.

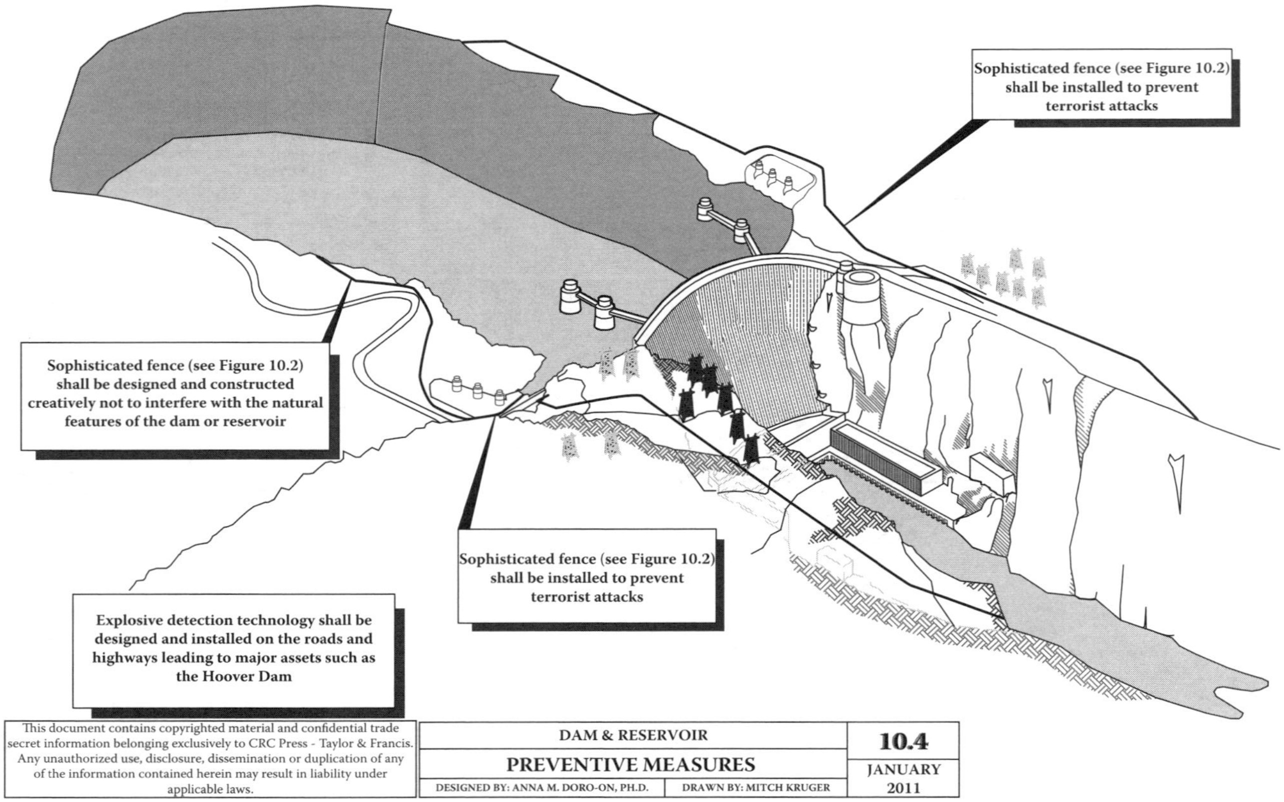

Figure 10.4 Dam and reservoir protection.

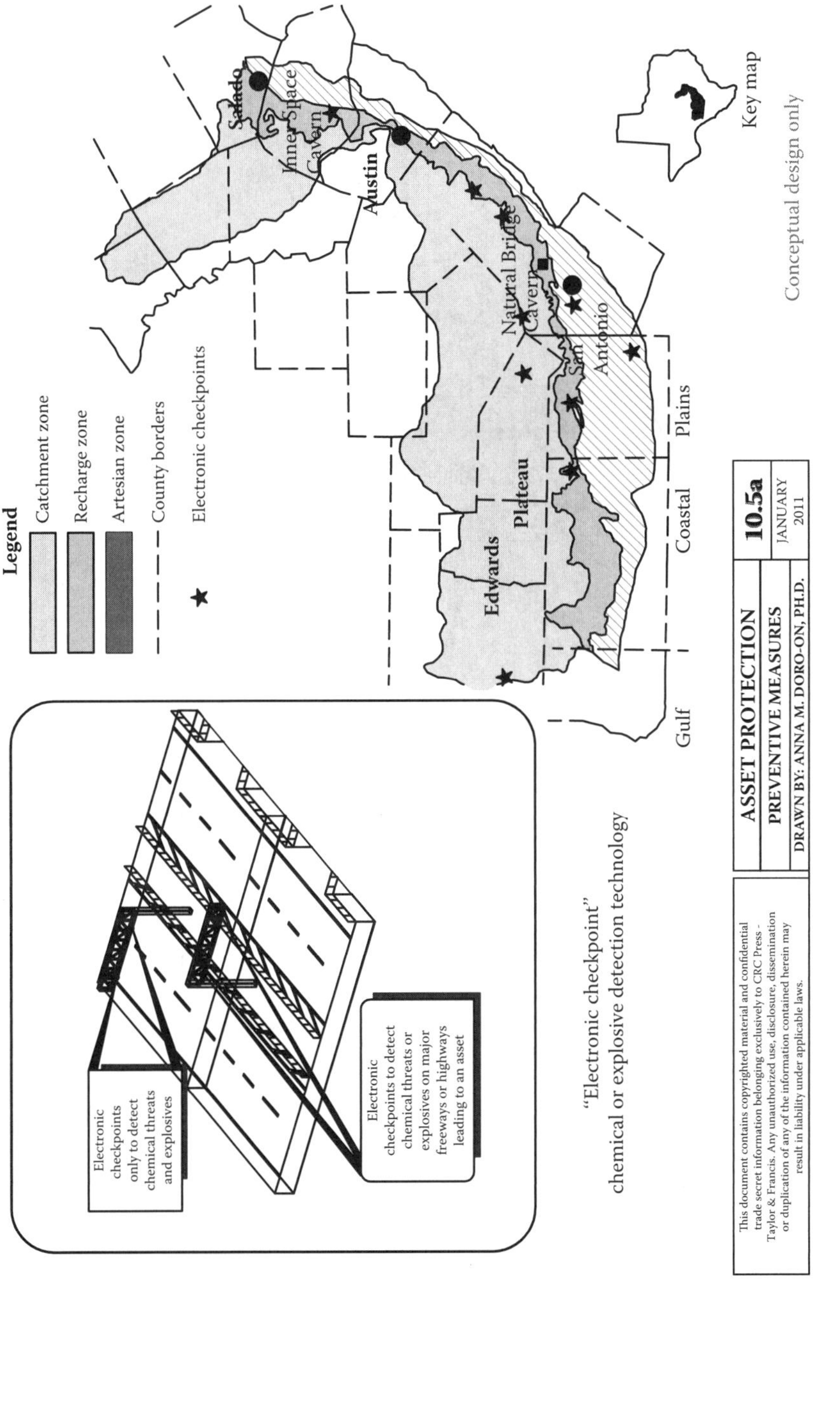

Figure 10.5a Exhibit A: Chemical threat and explosive detection.

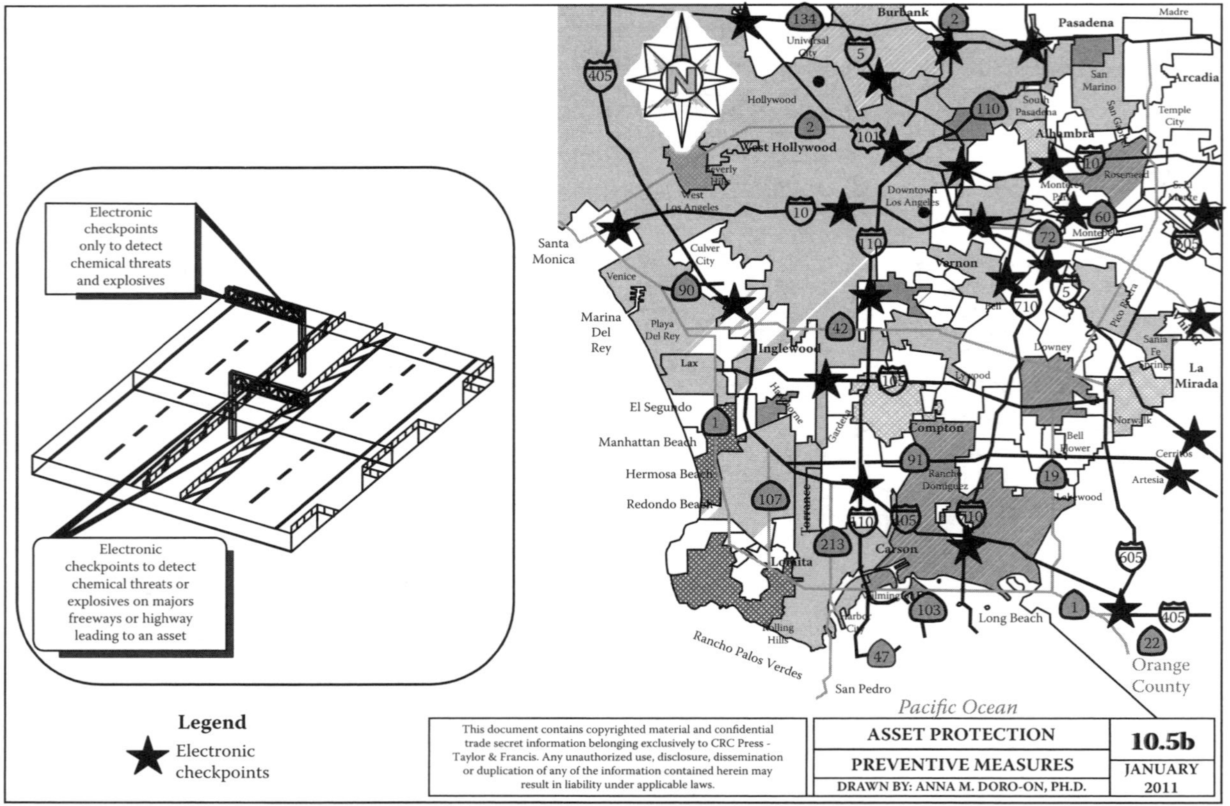

Figure 10.5b Exhibit B: Chemical threat and explosive detection.

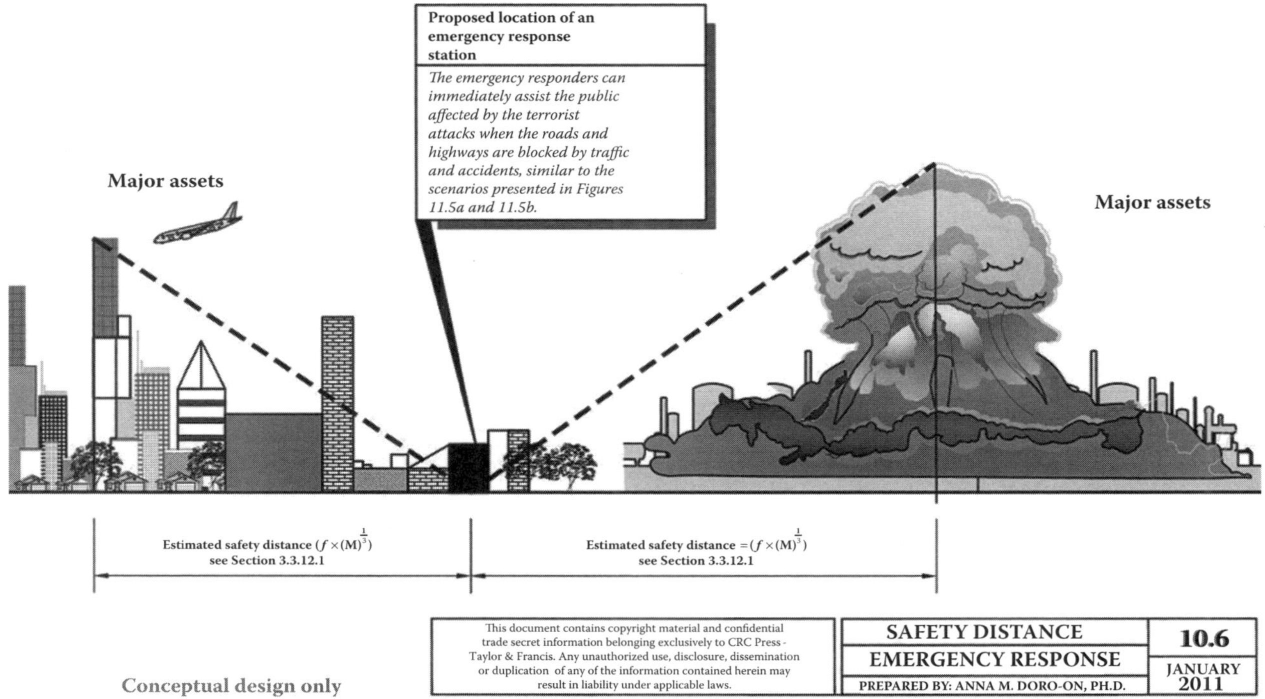

Figure 10.6 Exhibit A: Safety distance estimation of the location for the emergency response station.

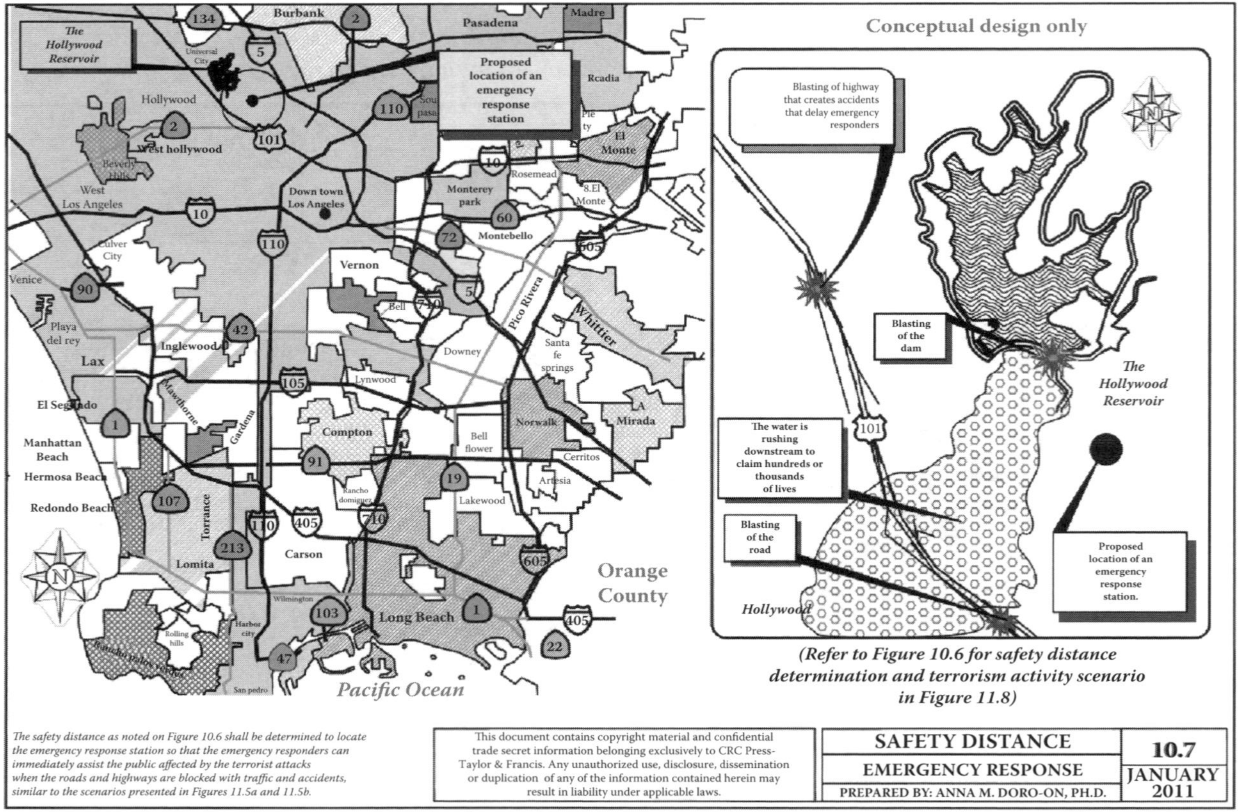

Figure 10.7 Exhibit B: Safety distance estimation of the location for the emergency response station.

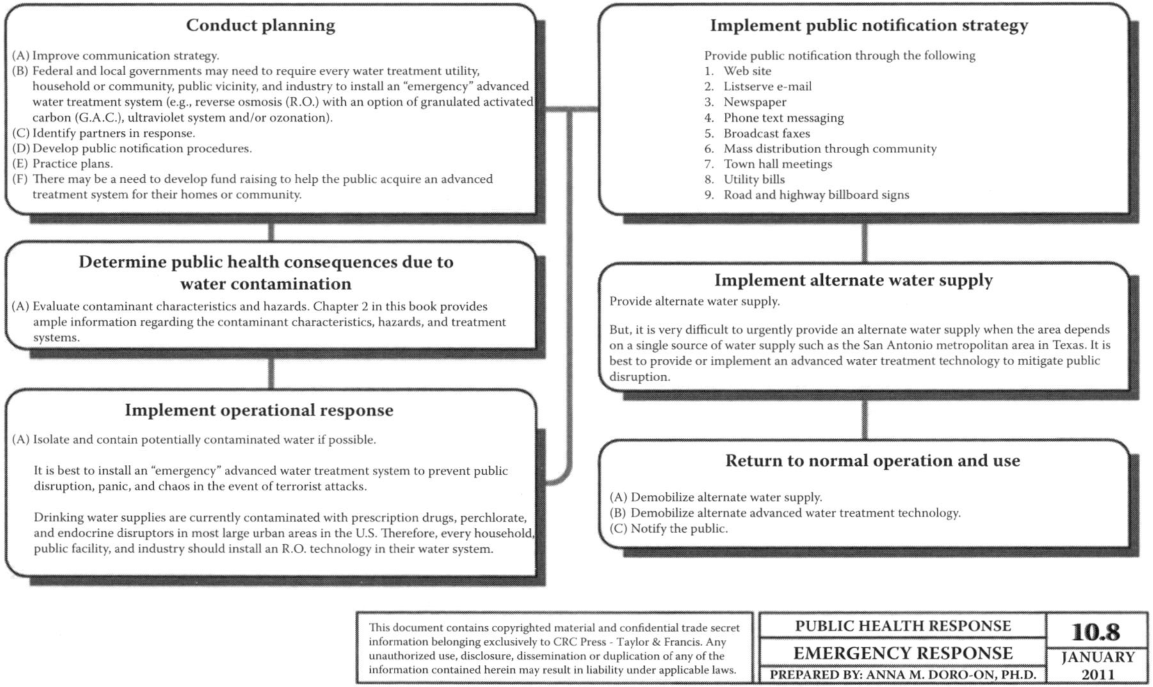

Conduct planning
(A) Improve communication strategy.
(B) Federal and local governments may need to require every water treatment utility, household or community, public vicinity, and industry to install an "emergency" advanced water treatment system (e.g., reverse osmosis (R.O.) with an option of granulated activated carbon (G.A.C.), ultraviolet system and/or ozonation).
(C) Identify partners in response.
(D) Develop public notification procedures.
(E) Practice plans.
(F) There may be a need to develop fund raising to help the public acquire an advanced treatment system for their homes or community.
Determine public health consequences due to water contamination
(A) Evaluate contaminant characteristics and hazards. Chapter 2 in this book provides ample information regarding the contaminant characteristics, hazards, and treatment systems.
Implement operational response
(A) Isolate and contain potentially contaminated water if possible.
It is best to install an "emergency" advanced water treatment system to prevent public disruption, panic, and chaos in the event of terrorist attacks.
Drinking water supplies are currently contaminated with prescription drugs, perchlorate, and endocrine disruptors in most large urban areas in the U.S. Therefore, every household, public facility, and industry should install an R.O. technology in their water system.
Implement public notification strategy
Provide public notification through the following
1. Web site
2. Listserve e-mail
3. Newspaper
4. Phone text messaging
5. Broadcast faxes
6. Mass distribution through community
7. Town hall meetings
8. Utility bills
9. Road and highway billboard signs
Implement alternate water supply
Provide alternate water supply.
But, it is very difficult to urgently provide an alternate water supply when the area depends on a single source of water supply such as the San Antonio metropolitan area in Texas. It is best to provide or implement an advanced water treatment technology to mitigate public disruption.
Return to normal operation and use
(A) Demobilize alternate water supply.
(B) Demobilize alternate advanced water treatment technology.
(C) Notify the public.
This document contains copyrighted material and confidential trade secret information belonging exclusively to CRC Press - Taylor & Francis. Any unauthorized use, disclosure, dissemination or duplication of any of the information contained herein may result in liability under applicable laws.
PUBLIC HEALTH RESPONSE
EMERGENCY RESPONSE
PREPARED BY: ANNA M. DORO-ON, PH.D.
10.8
JANUARY
2011

Figure 10.8 Public health response planning.

The key components of public health response planning

Public health response planning into emergency response plans (ERPs)

(A) State ERPs

The states normally have ERPs that are designed to provide standard procedures for the protection of critial infrastructure such as the water supply system at the state level.

(B) State or local department of public health ERPs and the centers for disease control and prevention (CDC)

The department of public health provides ERPs concerning public health protection, while the CDC aids the public health ERPs related to biological and chemical terrorism.

(C) Regional or city ERPs

The regional and city ERPs are created to meet a variety of emergencies related to natural and man-made disasters, terrorism, and violence, etc. The ERPs define the responsibilities of every agency engaged in emergency response.

(D) Utility ERPs

The utility ERPs are provided protocols on how to perform vulnerability/risk assessment and to outline ERPs for a variety of emergencies. The USEPA generally provides guidelines to assist the water sector.

Agencies involved in public health response

(A) Drinking water primacy agency

The USEPA is the drinking water primacy agency. The drinking water primacy agency is the agency that has primary enforcement responsibility for national drinking water regulations, namely, the Safe Drinking Water Act as amended (USEPA, 2004).

(B) Water utility

The water utility should have the extensive technical expertise and knowledge in the operation and protection of the water supply, treatment technologies and aspects, and preparedness and recovery strategies.

(C) State and local department of public health

The state public health ERPs provide plans to all public health related emergencies. The local department of public health may assist and coordinate in the investigation and public health protection.

(D) CDC

The CDC assist in diagnosing disease related to terrorism and disaster.

Agencies involved in public health response

(E) Health care sectors and poison control centers

In accordance with individual state laws, health care personnel report to the appropriate designated agency information on respiratory infections, skin rashes, diarrhea, and other syndromes that may provide early warning of an outbreak of disease such as anthrax, plague, and smallpox, as well as the intentional release of more common infectious agents (USEP 2004). Regional poison control centers, CDC, and Agency for Toxic Substances and Disease Registry (ATSDR) have knowledge and expertise in the diagnosis and management of poisoning.

(F) State and local emergency services

The state offices/departments of emergency services generally support the state on readiness to respond and recover from different types of emergencies (i.e., accidents, terrorism, and disasters). The local offices/departments provide training, planning, and procedures on emergency response within their particular jurisdiction.

(G) USEPA

The USEPA supports the federal response to terrorism incidents and provides assistance to water sector.

(H) ATSDR

The ATSDR provides management, surveillance, and assistance concerning protecting public health from hazardous substances in the environment.

(I) Federal Emergency Management Agency (FEMA)

FEMA is the lead federal agency for assisting state and local response to the consequences of public health emergencies involving terrorism and disasters.

(J) U.S. Army Corps of Engineers (USACE)

USACE, with FEMA, serves as one of the primary agencies responsible for emergency restoration of critical public facilities such as dams and waterways.

Communication tactics

Effective communication between water utilities and public health agencies is crucial for water terrorism preparedness, emergency response, and recovery.

Operational and public health response actions

According to the USEPA (2004), public health response decisions include containment of the suspect water and public notification regarding restrictions on water use in the event of contamination to prevent exposure to the contaminated water. It is best to consider the implementation of an "emergency" advanced water treatment system to be immediately available in the event of water contamination as illustrated in Figure 10.1.

PUBLIC HEALTH RESPONSE	10.9
EMERGENCY RESPONSE	JANUARY 2011
PREPARED BY: ANNA M. DORO-ON, PH.D.	

Figure 10.9 Key components of public health response planning.

10.2 National Response Framework

The U.S. Department of Homeland Security (DHS) developed the National Response Framework (NRF) to conduct all-hazards response in the United States (FEMA 2008). The NRF is a guide for how the federal, state, local, and tribal governments, along with nongovernmental and private sector entities, will collectively respond to and recover from all disasters, particularly catastrophic disasters such as Hurricane Katrina, regardless of their cause (GAO 2008). The NRF recognizes the need for collaboration among the myriad of entities and personnel involved in response efforts at all levels of government, nonprofit organizations, and the private sector (GAO 2008). Hence, the overall guidelines of NRF are in accordance with the National Incident Management System (NIMS), which offers a systematic standard for managing incidents.

10.2.1 Local Governments

Local governments, departments (e.g., emergency medical services, police, fire, emergency management and public works), and volunteers are usually the first to respond to incidents, threats or hazards. The responsibility of the local appointed official is to ensure the public safety and welfare, organize plans, and integrate the local government's capabilities and resources with neighboring jurisdictions.

10.2.1.1 Roles of Chief Elected or Appointed Officials

Elected officials provide direction and guidance to constituents during an incident and help modify regulatory requirements and budgets for preparedness efforts, emergency management, and response plans. However, they do not regularly focus on emergency management and response efforts.

10.2.1.2 Roles of Emergency Managers

The emergency manager mainly integrates the local emergency management program and evaluates the availability of local resources needed during an incident. Other objectives and missions of the local emergency manager include the following: (1) coordinating, planning, and working cooperatively with other local agencies and private sectors; (2) establishing common aid and support agreements; (3) facilitating damage assessments during incidents; (4) advising local officials about emergency management actions during disaster and terrorism incidents; (5) providing public awareness, standard training procedures, and education programs; (6) conducting exercises to examine plans and employ evaluation; and (7) including the private sector and NGOs in planning, training, exercises and evaluation.

10.2.1.3 Roles of Department and Agency Heads

Department and agency heads are responsible for working with the emergency manager in developing local emergency standard plans and procedures to ensure public safety and security. For instance, the EPA and local water agencies coordinate with emergency managers in preparing emergency response plans to ensure public health protection in the event of contamination or when the drinking water supply is determined to be a potential hazard.

10.2.1.4 Roles of Individuals and Households

Individuals and households should be informed and educated regarding their roles in the overall emergency management strategy. They can make a big difference by preparing supplies, emergency kits, and plans for disaster and terrorism. It is obviously very difficult to make every house and individual develop emergency plans for themselves, while their focus is to survive in the current economic situation. Local government such as the county or city should provide standard household emergency plans and provide programs to systematically remind every individual and household to be prepared in advance for an unwanted event such as disaster or terrorism. They can also be persuaded to be volunteers in response and recovery with an established voluntary agency. The local government may provide credits to individuals who take part in emergency response training courses, which may encourage them to be more devoted volunteers.

10.2.2 States, Territories, and Tribal Governments

State, territory, and tribal governments have sovereign rights; unique factors are involved in working with these entities. Stafford Act assistance is available to the states and Puerto Rico, the Virgin Islands, Guam, American Samoa, and the Commonwealth of the Northern Mariana Islands, which are included in the definition of "State" in the Stafford Act (FEMA 2008a).

10.2.2.1 Roles of the Governor

The governor is responsible for activating the state resources and implementing the strategic protocols needed for different kinds of incidents. Moreover, in accordance with state law, the governor has control over certain orders or regulations associated with response. The governor's roles based on the NFR include but are not limited to the following: (1) communicating and helping the public to cope with the unwanted consequences; (2) commanding the state military forces (not in federal service and state militias); (3) facilitating emergency aid from other states under interstate mutual aid and assistance agreements; (4) seeking federal support under Stafford Act presidential declaration of an emergency when resources are determined to be insufficient; and (5) working together with affected tribal governments within the state and initiating requests for the Stafford Act presidential declaration of an emergency on behalf of the affected tribe when needed.

10.2.2.2 Roles of the State Homeland Security Advisor

The state homeland security advisor provides counseling and guidance to the governor on homeland security issues and may serve as a liaison between the governor's office, the state homeland security structure, DHS, and other organizations both inside and outside the state (FEMA 2008a).

10.2.2.3 Roles of the Director of the State Emergency Management Agency

According to Western et al. (2008), the director of the state emergency management agency safeguards the state by providing preparedness actions to deal with large-scale disaster and terrorism emergencies, aiding local governments, and providing emergency assistance with other states and the federal government. Western et al. (2008) indicated that if local resources are determined to be insufficient, officials can request additional support from the county emergency manager or the state director of emergency management. For example, the U.S. Environmental Protection Agency (USEPA) may assess or prevent water contamination without waiting for requests from state, tribal, or local officials.

10.2.2.4 Roles of Other State Departments and Agencies

State department and agency leaders together with their staffs develop plans, and train internal policies and procedures to meet response and recovery needs safely. They should be involved in interagency training and exercises to enhance and polish the necessary capabilities.

10.2.2.5 Roles of Indian Tribes

The United States has a trust relationship with Indian tribes and recognizes their right to self-government. The state governor usually requests a presidential declaration representing the tribe to seek assistance from the state or the federal government under the Stafford Act, when local resources are inadequate.

10.2.2.6 Roles of Tribal Leaders

The tribal leader ensures the safety and welfare of the people of that tribe. As authorized by the tribal government, the tribal leader (1) is responsible for coordinating tribal resources needed for preparedness, mitigation programs, and emergency management from disaster and terrorism incidents; (2) may have powers to amend or suspend certain tribal laws; (c) can request federal assistance under the Stafford Act through the governor of the state when the tribe capabilities and resources are found to be insufficient; and (d) can deal directly with the federal government under the Stafford Act through the state governor's assistance.

10.2.3 Federal Government

The president has the authority to command the federal government to take action for federal disaster assistance in large-scale disaster and terrorism incidents under presidential declarations and the Stafford Act. According to DHS (2008), when the overall coordination of federal response activities is required, it is implemented through the Secretary of Homeland Security consistent with Homeland Security Presidential Directive 5 (HSPD-5).

10.2.3.1 Role of the Secretary of Homeland Security

The Secretary of Homeland Security provides the president with an overall systematic pattern for domestic incident management, to coordinate the federal response with the support of other federal partners. The Federal Emergency Management Agency (FEMA) administrator, as the principal advisor to the president, the secretary, and the Homeland Security Council on all issues regarding emergency management, helps the secretary in meeting the HSPD-5 responsibilities (FEMA 2008a).

10.2.3.2 Law Enforcement

According to DHS (2008), the Attorney General has the leadership and authority for criminal investigations of terrorist acts in the United States or directed at U.S. citizens or institutions in foreign countries, including the coordination of the law enforcement community and intelligence community to protect the homeland from terrorist attacks.

10.2.3.3 National Defense and Defense Support of Civil Authorities

As stated by the Office of the Under Secretary of Defense for Policy (OUSDP 2010), the primary mission of the U.S. Department of Defense (DOD) and its components is national defense. Because of this critical role, resources are committed only after approval by the Secretary of Defense or by the direction of the president (OUSDP 2010). Many DOD components and agencies are authorized to respond to emergencies and to provide support. The provision of defense support is evaluated by its legality, lethality, risk, cost, appropriateness, and impact on readiness (OUSDP 2010).

10.2.3.4 International Coordination

According to the Department of Homeland Security, Presidential Directive-5 (DHS 2008), the Secretary of State provides leadership and management of international preparedness, response, and recovery actions for the protection of U.S. citizens and U.S. interests overseas.

10.2.3.5 Intelligence

According to the Office of the Director of National Intelligence (DNI) (2007), the DNI directs the U.S. intelligence community, serves as the president's principal intelligence advisor, and oversees and directs the implementation of the National Intelligence Program.

10.2.4 Private Sector and Nongovernmental Organizations

The private sector and NGOs provide services in coordination with the governmental agencies and organizations in accordance with the NIMS principles. They are allowed to provide contingency plans and protocols.

10.2.4.1 Roles of Private Sector

Based on Western et al. (2008), private sector organizations look to the welfare and protection of their employees in the working environment. In addition, emergency managers must work with businesses that entirely involve critical infrastructure services (e.g., water, security, and power). FEMA (2008a) and Western et al. (2008) pointed out that the owners and operators of certain regulated infrastructures (e.g., petroleum refineries) may be legally accountable for preparedness and response actions to a negative incident that could happen. In the event of disaster or terrorist attack, the private sector should be working together with the local emergency managers in the decision-making process to achieve an effective response and recovery operation.

10.2.4.2 Roles of Nongovernmental Organizations

NGOs offer temporary housing, provide immediate relief, support emergency food supplies, and offer other services to assist the victims of the calamities. They usually coordinate with the government for support and planning of the allocation of substantial resources.

10.2.4.3 Roles of Volunteers and Donors

Dedicated volunteers and donors can help response endeavors in different approaches, and it is essential that governments at all levels plan ahead to effectively incorporate volunteers and donated goods into their response activities.

10.3 Emergency Preparedness

Preparedness is way of mitigating unwanted outcomes and it is one of the crucial actions in achieving safety and security in the event of calamities, disasters, and terrorism. This section presents the six essential activities for responding to an incident: (1) planning, (2) organization, (3) training, (4) equipment, (5) exercises, and (6) evaluation and improvement.

10.3.1 Planning

Effective planning includes the collection and analysis of information, policy and strategy formulations, plans, and other arrangements to operate missions and goals. It also sharpens the response operation by unequivocally defining required capabilities, increasing the speed of the response to take control of an incident, and facilitating the rapid exchange of information about the situation and event. The response plans have multiple things to address, for instance evacuations face many challenges. Therefore, systematic plans must incorporate the following: (1) the lead time required for various unexpected and anticipated events; (2) weather conditions; (3) transportation and communication; (4) interdependencies between locations of shelters and transportation (US Army-CAC 2010; FEMA 2008b); and (5) provisions of special needs populations and those with household pets (CRS 2010).

10.3.2 Organization

According to FEMA (2004), NIMS provides standard command and management structures pertaining to response. This standardized approach allows responders from different disciplines to collectively operate and respond. Government agencies and other organizations shall operate an emergency response in accordance with NIMS organizational and management policy.

10.3.3 Equipment

According to FEMA (2008a), the local, tribal, state, and federal jurisdictions need to establish a common understanding of the capabilities of different types of response equipment. A critical component of preparedness is the acquisition of equipment that will perform according to established standards, including the capability to be interoperable with equipment used by other jurisdictions and participating organizations (FEMA 2004). Efficient preparedness operation needs standards to define techniques and create strategies to acquire and direct resources and appropriate equipment in sufficient quantities to accomplish assigned missions and goals. The federal government and local governmental agencies should ensure that their personnel have the necessary resources to perform assigned response missions and tasks.

10.3.4 Training

Training methods shall be in accordance with the standards of FEMA and produce qualified skills and proficiency. FEMA and other governmental and private organizations offer response and incident management training in online and classroom formats.

10.3.5 Exercises, Evaluation, and Improvement

Well-organized exercises improve interagency coordination and communications, enhance proficiency, sharpen skills, and determine opportunities for advancement or expansion. Exercises should include but not be limited to the following: (1) include multidisciplinary, multijurisdictional incidents; (2) integrate involvement of academia, private sector, and nongovernmental organizations (including international organizations); (3) refine aspects of preparedness processes, procedures, and plans; and (4) contain a system for integrating remedial actions.

10.4 Response

Emergency response is an action and operation of activating the society's resources and capabilities to save and safeguard lives; secure assets and the environment from irreversible damages; maintain public morale and confidence; and preserve the social, economic, and political structure of the jurisdiction. The key actions usually involved in support of a response are: (1) progress and maintain awareness to every situation, condition, and event; (2) activate and deploy key resources and capabilities; (3) effectively and efficiently coordinate response actions; and (4) demobilize.

10.4.1 Baseline Priorities

Situational attentiveness requires systematic screening of potential sources of information and detailed evaluation of the information should be employed. Critical information is directed through orderly reporting systems. Priorities include (1) providing the appropriate plausible and precise information at the right time; (2) enhancing and expanding the national reporting system; and (3) involving operations centers and experts.

10.4.2 Local, Tribal, and State Actions

Local, tribal, and state governments can address the inherent challenges in establishing successful information-sharing networks by (1) creating fusion centers that integrate agencies associated with homeland security, academia, intelligence, emergency management, public health, and other agencies, as well as private sector and nongovernmental organizations locally and "internationally," to expand the information-sharing strategy; (2) implementing the National Information Sharing Guidelines to improve intelligence; (3) establishing information-sharing and reporting protocols to enable effective and timely decision making during response to incidents; and (4) developing standard procedures that can provide awareness to misleading information that can cause distortion of intelligence. The local or regional Joint Terrorism Task Force should be informed immediately when potential terrorist attacks are detected.

10.4.3 Federal Actions

The National Operations Center (NOC) is responsible for collecting, assessing, and synthesizing all-source information, across all-threats and all-hazards information comprising the range of homeland security partners. Information regarding actual or potential terrorism and disaster incidents should be reported immediately by federal departments and agencies.

10.4.4 Alerts

When notified of a threat or an incident that potentially requires a coordinated federal response, the NOC analyzes and assessed the information before it goes to the senior federal officials and federal operations centers: the National Response Coordination Center (NRCC), the FBI SIOC, the NCTC, and the National Military Command Center, to assist them with effective decision making. Once the information is verified and processed, the Secretary of Homeland Security coordinates with other appropriate departments and agencies to initiate emergency plans in accordance with the *framework*. Government and agency officials should often be aware and prepared to participate in all situations (through video and teleconference). Each federal department and agency must ensure that its response personnel are knowledgeable, well-prepared, and well-trained to utilize these tools.

10.4.5 Operations Center

Federal operations centers essentially involve awareness of circumstances, current events, and communications among governmental offices all over the nation. These operations centers can provide information, assistance, and guidance and administer resources with their state, tribal, and local partners, in the event of an incident.

10.5 Activate and Deploy Resources

According to FEMA (2008a), when an incident or potential incident occurs, responders assess the situation, identify and prioritize requirements, and activate available resources and capabilities to save lives, protect property and the environment, and meet basic human needs. Usually, this includes development of incident management objectives based on incident management priorities, development of an incident management action plan by the incident management command in the field, and development of support plans by the appropriate local, tribal, state, and/or federal government entities. The key activities include activating people, teams, resources, and capabilities based on the scope, capacity, nature, and complexity of the incident. All emergency responders should frequently exercise notification systems and protocols.

10.6 Proactive Response to Catastrophic Incidents

Prior to catastrophic incidents, state and federal governments should create models of detailed terrorism and disaster activity scenarios qualitatively and quantitatively equivalent to the combat zones presented in Chapter 11. Then, they should take proactive actions to mobilize assets in anticipation of a formal request from the state for federal assistance. They should not wait until minor and major unfavorable events take place. Such deployment of federal assets would likely occur for catastrophic events involving terror threats, disasters, or high-yield explosive weapons of mass destruction or other catastrophic incidents affecting heavily populated areas such as New York and Los Angeles. The *proactive responses* are used to ensure that resources are sufficient and reach the scene in a timely manner to assist in restoring normal function of state or local governments. Figures 10.8 and 10.9 provide a summary procedure and the key components of the public health response to water contamination. Table 10.1 consists of a list of potential entities to be notified as part of public health response, as well as the purpose of the notification for each entity. Each utility should identify the appropriate entities to be notified in its ERP. It is important to note that under 40 CFR Part 141, Subpart Q [the Federal Public Notification (PN) Rule], utilities must provide public notice to persons served by the water system in situations with significant potential to have serious adverse effects on human health as a result of short-term exposure (USEPA 2004). An emergency response station should be located based on the safety distance estimation (Figure 10.6) so that the responders can easily assist the public while avoiding traffic and accidents, which can be created by adversaries as illustrated in Figures 11.5a and b found in Chapter 11.

10.7 Recovery

Once immediate lifesaving operations are accomplished, the focus changes to assisting the critical infrastructures involved in the incidents and recovery. Within recovery, actions are taken to help the public and the nation return back to normal condition. Depending on the complexity of this level, recovery and remediation efforts involve significant contributions from all sectors of our society. In terms of water supply recovery, technological treatment systems are presented in Chapter 2.

10.8 Preventive Measures

The preventive measures listed in Table 10.2 and presented in Figures 10.1 through 10.7 and 10.10 through 10.12 could be carried out at a moderately sensible cost, and would extend to an extensive approach and technique toward improving the security of U.S. water infrastructures.

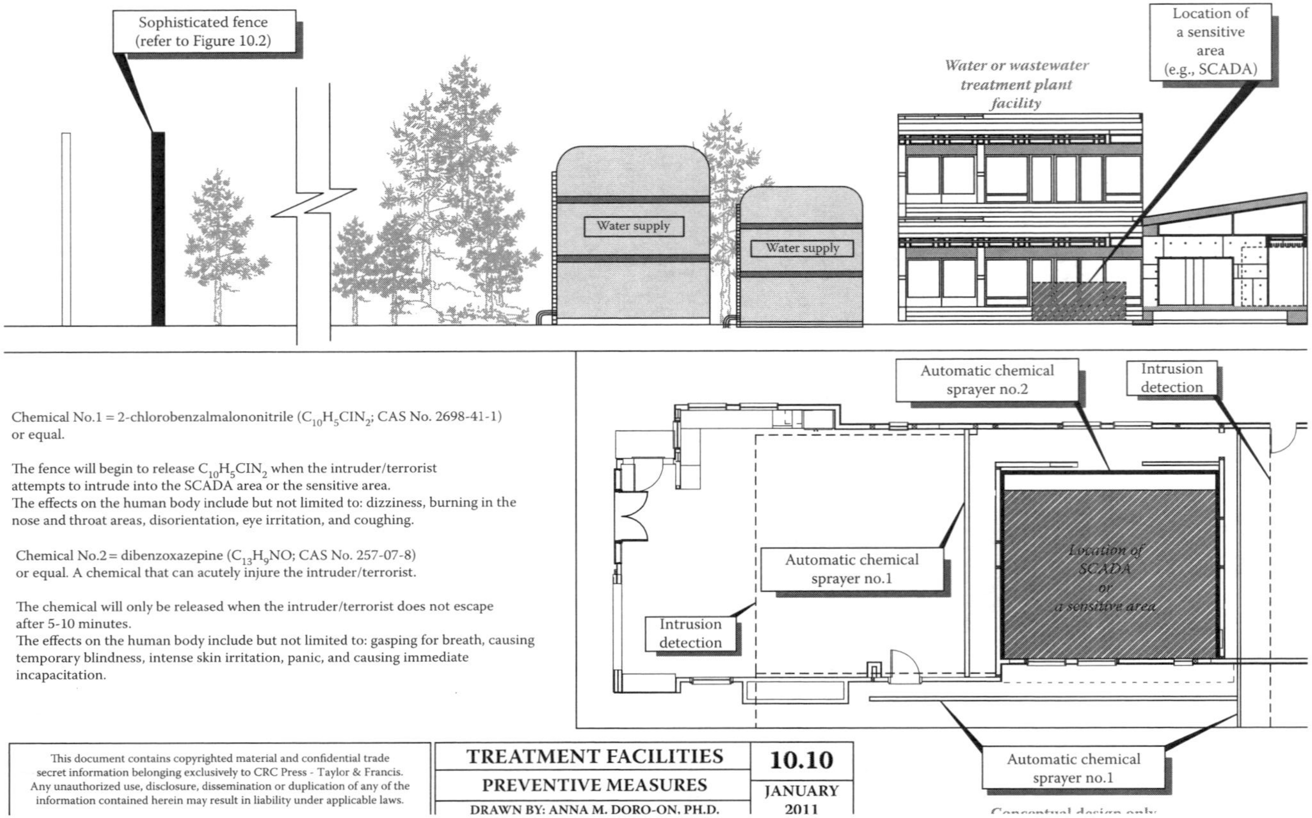

Figure 10.10 Treatment facilities protective measures.

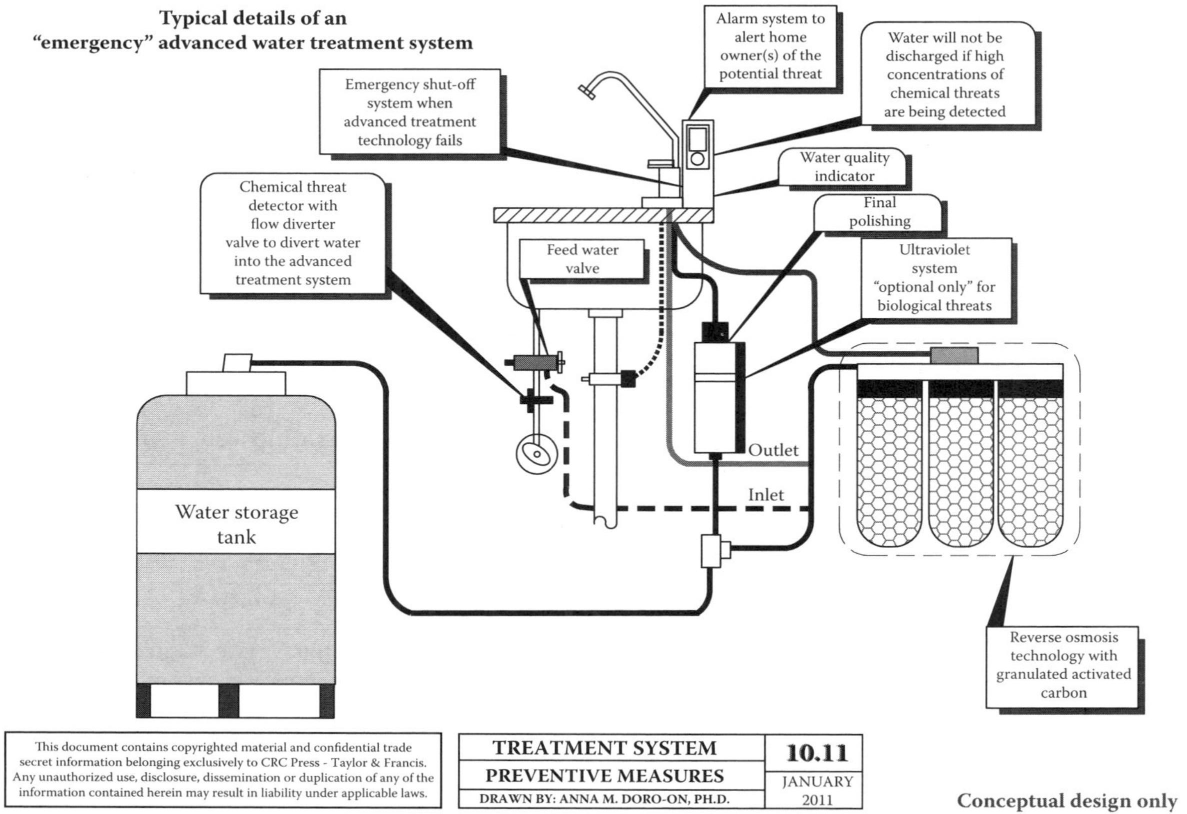

Figure 10.11 Conceptual design of an “emergency” advanced drinking water treatment system.

Table 10.1 Public Health Response—Entities That Should Be Notified

Entity		*Purpose of the Notification*
Public health agencies	State/local health and/or environmental department	To work with these officials in decision making on the distribution of "boil water," "do not drink," or "do not use" notices. These officials may be involved with public health decisions related to the proper use of the water supply, status of the water distribution system, selection of a short-term alternate water supply, and communicating the necessary public health information.
	Other associated system authorities (wastewater, water facilities)	
	Poison control centers	
Emergency responders	Emergency medical services (EMS)	To notify the organization of the need for assistance with the distribution of an alternate water supply (e.g., bottled water) and whether the contamination impacts the availability of water for firefighting. Also, these agencies should be provided with all information related to public health including information on water notices, alternate water supplies, critical care facilities, and public health notifications.
	Fire department	
	State and/or local offices of emergency services	
Law	Federal, state, and local law enforcement	Local law enforcement should be notified immediately if a malevolent act is suspected. Law enforcement agencies should also be notified of the need for assistance with getting important information out to the public and the distribution of water from the short-term alternate water supply (i.e., distribution of bottled water, etc.). Law enforcement agencies should also be contacted because the public may be contacting them through 911 regarding the incident.
Consecutive systems (i.e., public water systems that receive water from the water utility where the water contamination threat or incident occurred)		To provide information related to restrictions on the use of the drinking water supply, as well as instructions on obtaining alternate sources of drinking water, through the duration of the incident. Also, information should be provided on the status of the water supply, the potential problem, and what is being done to manage the incident.

Customers/public		To provide information related to restrictions on the use of the drinking water supply, as well as instructions on obtaining alternate sources of drinking water, through the duration of the incident. Also, information may be provided on the status of the water supply, the potential problem, and what is being done to manage the incident.
Customers with special needs	Critical care facilities (e.g., hospitals, clinics, nursing homes, dialysis centers)	These facilities should be some of the first to be notified. Information should be provided regarding the proper use of the water supply for public health purposes as well as the identity of the contaminant so these facilities can identify the symptoms of exposure as well as potential medical treatment. They may be given information on how water will be provided or how they need to obtain short-term alternate water supplies. Critical care facilities may also need to be notified of any changes in the type of chemical disinfection being used or the concentration of these chemicals in the water as this may affect some of their medical procedures.
	Schools	To provide information regarding restrictions on water use, alternate water supplies, and other public health information.
	Day care facilities	To provide information regarding restrictions on water use, alternate water supplies, and other public health information.

(*Continued*)

Table 10.1 Public Health Response—Entities That Should Be Notified (*Continued*)

Entity		*Purpose of the Notification*
	Businesses (e.g., food and beverage manufacturers, commercial ice manufacturers, restaurants, agricultural operations, power generation facilities, any other businesses identified by the utility)	To provide information regarding restrictions on water use, alternate water supplies, and other public health information. These customers may also need information regarding whether heating or superheating the water may pose a hazard.
Others	Elected officials	To provide all information related to public health, including: the status of the *threat evaluation*, information on "boil water," "do not drink," or "do not use" notices, alternate water supplies, customers with special needs, and public health notifications.

Source: Data from U.S. Environmental Protection Agency, EPA/816/R-04/002, 2004.

Table 10.2 Proposed Enhanced Preventive Measures and Strategies

Preventive Measures	*Description*
Groundwater resources and water supply systems	Groundwater resources and water supply systems *at the original source* should be fenced, be well lighted, and have a perimeter that is monitored by surveillance cameras and motion detectors with chemical threat detectors (which also includes detectors on wells). Or chemical threat detectors should be used on roads and highways leading to major recharge zone and water supply systems.
Treated water supply systems	Remaining cyber security should be enhanced, and passwords should be changed regularly. Secure and tamperproof entry points to the water distribution system. Surveillance cameras should be located on-site at key points. Sophisticated fences as illustrated in Figure 10.2 should be constructed to protect the water supply systems. Emergency advanced water treatment should be implemented at the water treatment facility to avoid public disruption and to protect public health in the event of terrorism through water contamination. Individual homes should install an advanced water treatment technology similar to Figure 10.8 to protect public health from notorious contaminants that are difficult for the traditional water treatement plant to treat and detect.
Grants/funding for technology and protection policy development	The U.S. government through USEPA should give grants for technology and protection policy development for groundwater resources and water supply systems.

(Continued)

Table 10.2 Proposed Enhanced Preventive Measures and Strategies (*Continued*)

Preventive Measures	*Description*
"Raw" water supply system	Residential and public facilities and the industry that depends on the *raw water system* (untreated water) need to install detectors and automated control systems for notifying authorities and administrators of the facilities of detection of chemical threats in the water main pipeline systems. Note: Shut-off valves and structures should be constructed or installed to avoid transport of contaminants in the water main pipeline systems, reservoirs, and water supply channels. An emergency advanced water treatment technology (e.g., reverse osmosis with granulated activated carbon) should be constructed on-site to prevent disruption and protect public health.
Protection for water supply storage and treatment facilities (including wastewater treatment facilities)	Chemical threat and explosive detection devices should be installed on roads and highways leading to the water supply storage facility (e.g., aquifer storage and recover [ASR] facility), water treatment plant, and wastewater treatment plants. An "emergency" advanced water treatment technology and sophisticated fence should be constructed as depicted in Figure 10.2. The sensitive area of the treatment facility will be protected with sophisticated intrusion detectors and security as shown in Figure 10.12. This area should only be accessed by authorized personnel. Unauthorized personnel and terrorists will be detected and will be attacked by chemical sprayers no. 1 and no. 2 if necessary as presented in Figure 10.12.
Aquifer recharge zone, artificial aquifer recharge and storage systems	Thorough background investigation of any individual or group who will purchase or rent properties adjacent or located above aquifer recharge and storage system areas. Thorough background investigation of any individual or group who are proposing to rent and to own properties within proximity of the aquifer recharge and storage system areas. Mandatory inspection by governing agencies of any ongoing construction or other major activities on properties and open land above aquifer recharge and water supply system areas.

	Monitoring wells with chemical threat detectors should be installed. Automated control systems with chemical threat detectors to notify governing agencies or authorities on underground storage systems (e.g., water and wastewater treatment tanks) located above or near major groundwater resources should be installed. Mandatory inspection by governing agencies and authorities on any underground storage tanks installed above or near groundwater resources. Shut-off systems and structures to prevent discharge, flow, or transport of contaminants and contaminated water should be installed or constructed. Surveillance cameras and motion detection should be located on-site at key points, such as the groundwater aquifer recharge zone, aquifer artificial recharge facilities, and treated water supply storage systems. A chemical threat or explosive detector should be installed on roads and highways leading to the aquifer recharge zone and water supply system as illustrated in Figures 10.5a and b.
Surface water system	Automated control systems with chemical threat detectors/sensors should be installed on dams, reservoirs, aqueducts, and to any major surface water systems that can potentially be used for urban water supply. Mandatory inspection by governing agencies on any underground storage tanks installed near a surface water system. Shut-off systems and structures should be constructed or installed to prevent discharge or transport of chemical threats and contaminated water to downstream water supply systems. A sophisticated fence as illustrated in Figure 10.2 should be constructed to protect the water supply systems.

(Continued)

Table 10.2 Proposed Enhanced Preventive Measures and Strategies (*Continued*)

Preventive Measures	*Description*
Protection for aquifers, surface waters, and water storage systems located on U.S. borders	A sophisticated fence as illustrated in Figure 10.2 should be constructed to protect the water supply systems. An emergency advanced water treatment should be implemented at the water treatment facility to avoid public disruption and to protect public health in the event of terrorism through water contamination. Chemical threat (e.g., cyanide and arsenic) detectors with automated control systems should be installed to any large above ground water supply storage tanks, surface water systems, and major groundwater recharge systems located at the U.S.–Mexico and U.S.–Canada borders.
Security against illegal inflow of chemical threats or weapons along U.S. borders (see Figure 9.3 of Chapter 9)	Motion and chemical threat detectors should be installed on U.S. borders. Thorough background investigation of any individual or group who are proposing to rent and to own properties on U.S. borders. A sophisticated fence as illustrated in Figure 10.2 should be constructed to protect water supply systems. Mandatory inspection with surveillance by governing agencies and authorities to any ongoing construction or other major activities on properties and open land located on U.S. borders. Laser beams, and wireless surveillance (accessible on the Internet) with intrusion detection should be installed on U.S. borders.

Chemical threats	Legal documentation should be required for purchasing large quantities of chemicals. A sophisticated security system or sophisticated fence as depicted in Figures 10.2 and 10.12 should be installed on chemical plants and facilities. Arsenic, endocrine disruptors, and prescription drugs should be regulated and standardized in the national maximum contaminant level (MCL) standards.
Dams, reservoirs, and aqueducts	A sophisticated fence as illustrated in Figure 10.2 should be constructed to protect the water supply systems. Chemical threat and explosive detectors should be installed on roads and highways to prevent destruction by terrorists to major critical infrastructure such as the Hoover Dam (Figure 10.4).
Petroleum refineries, hazardous chemical plants and explosive plants located near or adjacent to water resources	Pile walls and/or cutoff walls should be constructed in the surrounding perimeter of the plant to prevent severe contamination to water resources as shown on Figures 10.3a and b.
High-rise structures	A parachute will be required to every individual situated in high-rise structures as shown in Figure 10.9.
U.S. intelligence	U.S. intelligence should be improved to maintain water infrastructure security and protection. Refer to Chapter 11 for an *intelligence analysis* presentation.

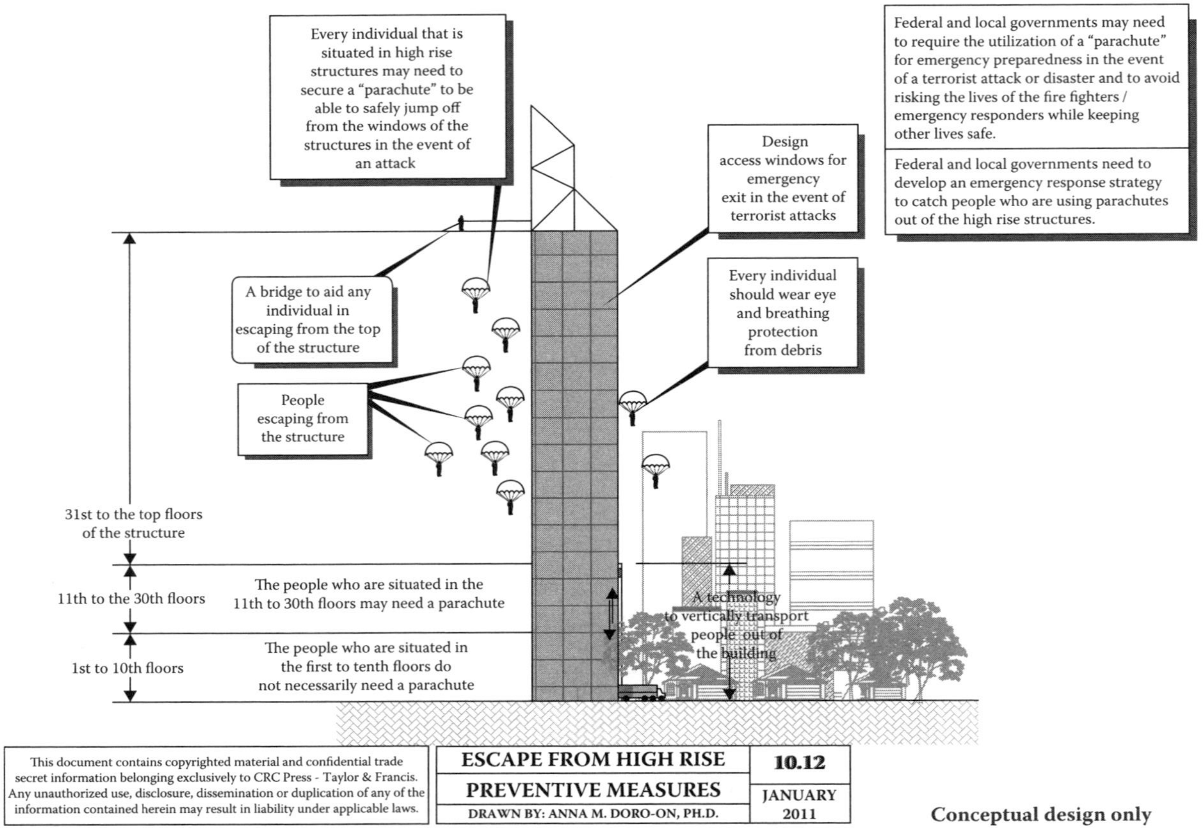

Figure 10.12 Escape strategies from high-rise structures in the event of terrorist attacks.

References

Lindsay, B. 2010. Federal Evacuation Policy: Issues for Congress. U.S Congressional Research Service. http://www.dtic.mil/cgibin/GetTRDoc?Location=U2&doc=GetTRDoc.pdf&AD=ADA620742 (accessed June 20, 2010).

Radvanovsky, R., and A. McDougall. 2010. *Critical Infrastructure-Homeland Security and Emergency Preparedness.* 2nd ed. Boca Raton, FL: CRC Press-Taylor & Francis Group.

U.S. Army Combined Arm Center. 2010. Disaster Response Staff Officer's Handbook. http://usacac.army.mil/cac2/call/docs/11-07/app_g.asp (accessed November 15, 2010).

U.S. Department of Homeland Security. 2008. Homeland Security Presidential Directive 5. http://www.dhs.gov/xabout/laws/gc_1214592333605.shtm (accessed June 20, 2010).

U.S. Environmental Protection Agency. 2004. Small and Medium Water System Emergency Response Plan Guidance to Assist Community Water Systems in Complying with the Public Health Security and Bioterrorism Preparedness and Response Act of 2002. EPA/816/R-04/002.

U.S. Federal Emergency Management Agency. 2004. National Incident Management System. http://www.fema.gov/good_guidance/download/10243 (accessed October 26, 2010).

U.S. Federal Emergency Management Agency. 2008a. Mobile National Response Framework. http://www.fema.gov/pdf/emergency/nrf/nrf-core.pdf (accessed October 1, 2010).

U.S. Federal Emergency Management Agency. 2008b. The Mass Evacuation Incident Annex. http://www.fema.gov/pdf/emergency/nrf/nrf_massevacuationincidentannex.pdf (accessed October 1, 2010).

U.S. Government Accountability Office. 2008. *National Response Framework: FEMA Needs Policies and Procedures to Better Integrate Non-Federal Stakeholders in the Revision Process.* http://www.gao.gov/new.items/d08768.pdf (accessed October 25, 2010).

U.S. Office of the Director of National Intelligence. 2007. An overview of the United States Intelligence Community. http://www.dni.gov/who_what/061222_DNIHandbook_Final.pdf (accessed August 22, 2010). DNI.

U.S. Office of the Under Secretary of Defense for Policy. 2010. Homeland Defense and America's Security Affairs: FAQ. http://policy.defense.gov/hdasa/faq.aspx (accessed August 23, 2010).

Western. J.L., Contestabile, J. and J.M. Englot. *Transportation Security Roles and Responsibilities.* Wiley Handbook of Science and Technology for Homeland Security. Wiley and Sons Inc. http://onlinelibrary.wiley.com/doi/10.1002/9780470087923.hhs143/pdf (accessed October 5, 2010).

Chapter 11

Strategic Intelligence Analysis for Water Infrastructure Terrorism Prevention

11.1 Introduction

This chapter provides a brief and concise *intelligence analysis* embedded with *cumulative prospect theory* to significantly screen valuable approaches or alternatives and to improve the effectiveness of the intelligence enterprise. In addition, this chapter will present illustrative practical examples for the approach using a series of extreme terrorism activity scenarios related to water infrastructure (including dams, aqueducts, and reservoirs), involving other critical infrastructure, *event tree analysis*, definition of a new strategic goal for intelligence, and the development of an effective information-sharing model based on cumulative prospect theory. Meanwhile, the terrorists and their leaders think in terms of a long time frame for achieving their goals, while they also carry out their own intelligence measures to identify the best target or the right timing for an attack. A higher threshold of destruction that can equal or exceed the level of the 9/11 attacks requires a degree of planning; the terrorist leaders are using highly intelligent people who do the planning for them. Therefore, there is an urgent need for systematic information-sharing strategies for terrorism threat assessment and warnings; identification of terrorism scenarios that can be used for collection of information; and rapid adaptation to changes in terrorists' tactics so that speed, risk acceptability, and accuracy of operations can be achieved with an improvement in the value of intelligence analysis focusing on water infrastructure security.

11.2 Intelligence Analysis

Intelligence is a key element of combating terrorism effectively, and it helps to identify targets deemed important to the adversary for mission accomplishment. The remarkable developments in intelligence collection methods have increased the availability of combat zone information from many different sources. Combat zone information is of only partial value until it is analyzed and exploited. Through analysis, this information becomes intelligence. Generally, the intelligence analyst coordinates the bits of information from diverse sources to manufacture a complete and accurate picture of the combat zone. Some examples of combat zone plans are presented in Figures 11.1a through c; detailed terrorism activity scenarios are given in these figures. Thus, analysis produces the intelligence that is needed to win the combat against the adversary.

Meanwhile, for many years the sharing of intelligence and law enforcement information was circumscribed by administrative policies and statutory prohibitions. The failure to deliver a tough-minded and objective assessment of Iraqi weapons of mass destruction (WMD) was the latest in a long series of Central Intelligence Agency (CIA) blunders (Goodman 2004). The 9/11 intelligence failure insinuated the need to remodel the entire intelligence structure. U.S. intelligence has been particularly weak on the issue of terrorism, and it frequently politicized intelligence (Goodman 2004). The 9/11 terrorist attack exposed the CIA's incompetence in preventing terrorist operations in the United States and to anticipate commercial airplanes being used as terrorist weapons, as presented in Figures 11.2a and b. In the meantime, what the CIA and the intelligence community members should be doing, and what they should do in the future, is of more concern today than at any time since the Cold War. The intelligence community must produce an independent source of intelligence for policy or decision makers. Furthermore, the CIA and other intelligence agencies must strengthen their intelligence-sharing networks without jeopardizing the public's privacy. Unfortunately, these agencies place greater focus on the compartmentalization of intelligence and the *need-to-know*, which can be a hindrance to effective information sharing.

11.3 Traditional Intelligence Cycle

Intelligence operations follow a five-phase process known as the *intelligence cycle*. It is a concept that describes the fundamental cycle of intelligence processing in a civilian or military intelligence agency or in law enforcement as a closed path

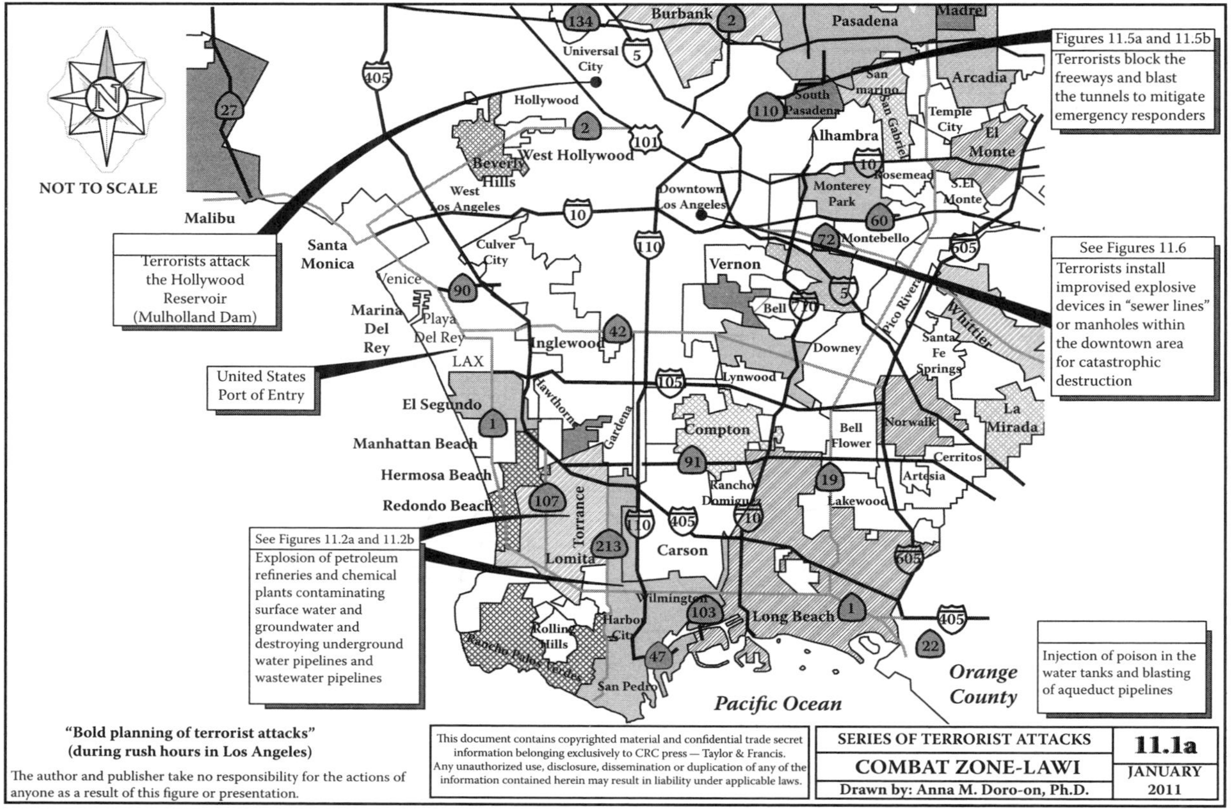

Figure 11.1a Map 1: Combat zone Los Angeles water infrastructure.

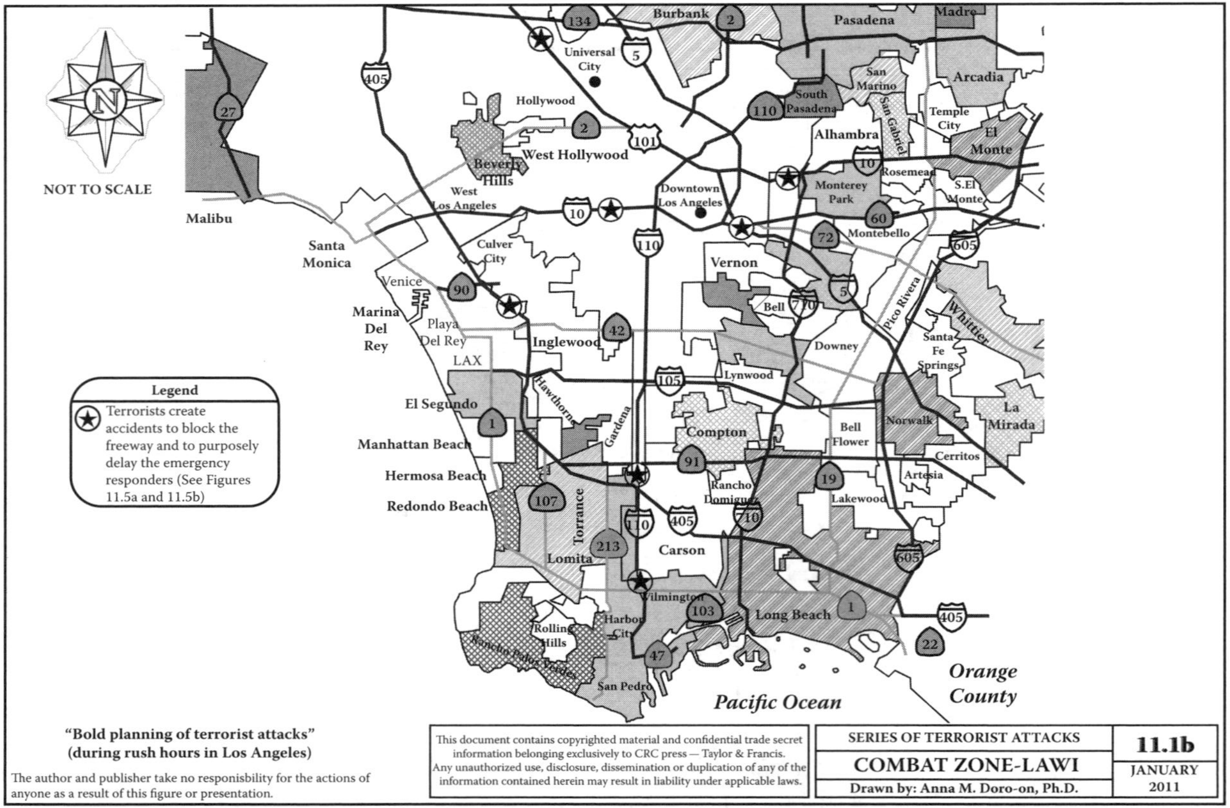

Figure 11.1b Map 2: Combat zone Los Angeles water infrastructure.

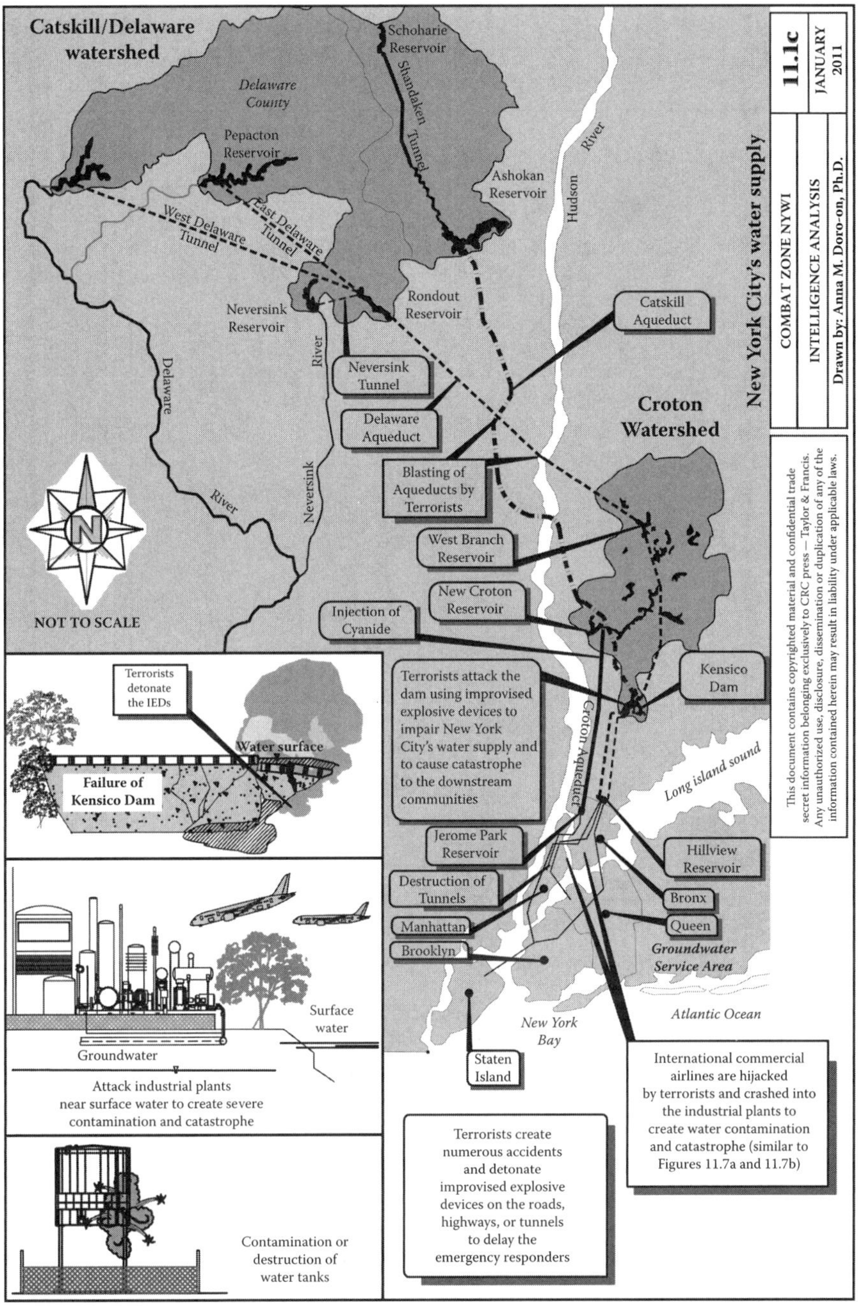

Figure 11.1c Combat zone New York water infrastructure. (Adapted from New York City Department of Environmental Conservation, http://www.dec.ny.gov/docs/water_pdf/nycsystem.pdf, 2010.)

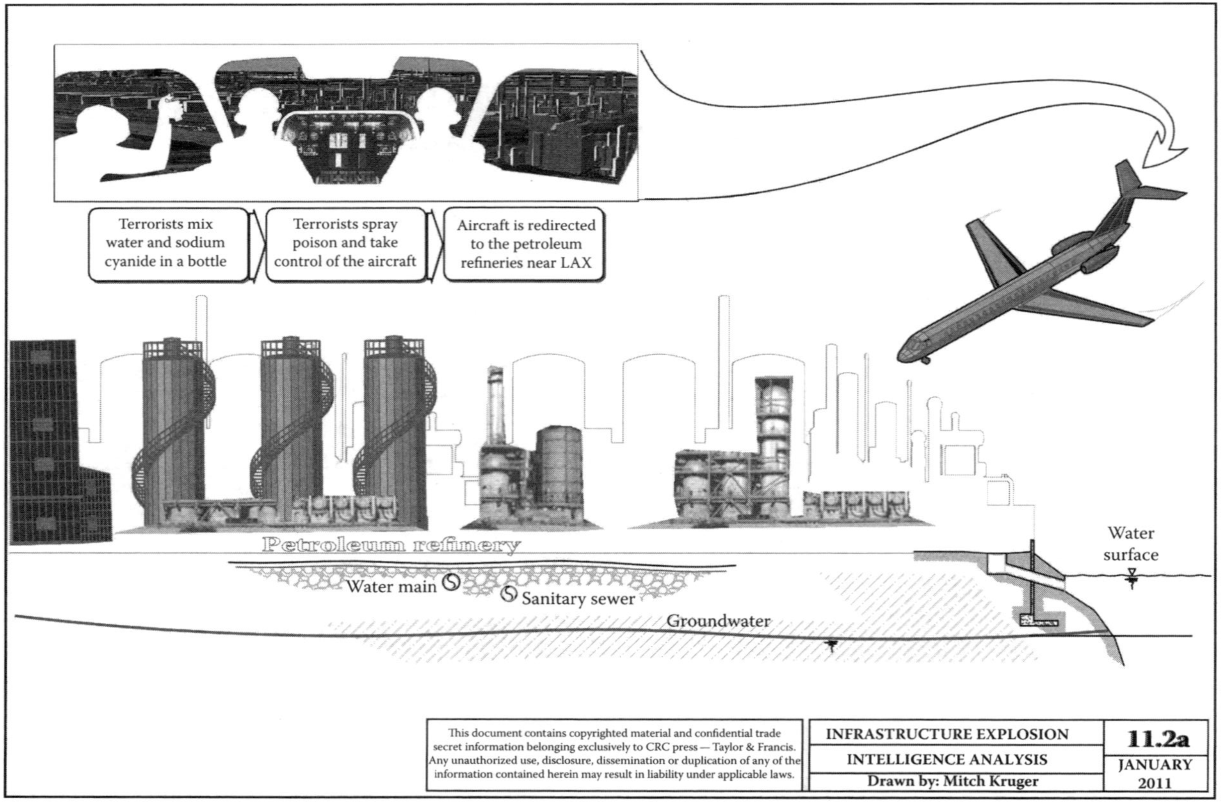

Figure 11.2a Explosions of petroleum refineries creating surface water contamination.

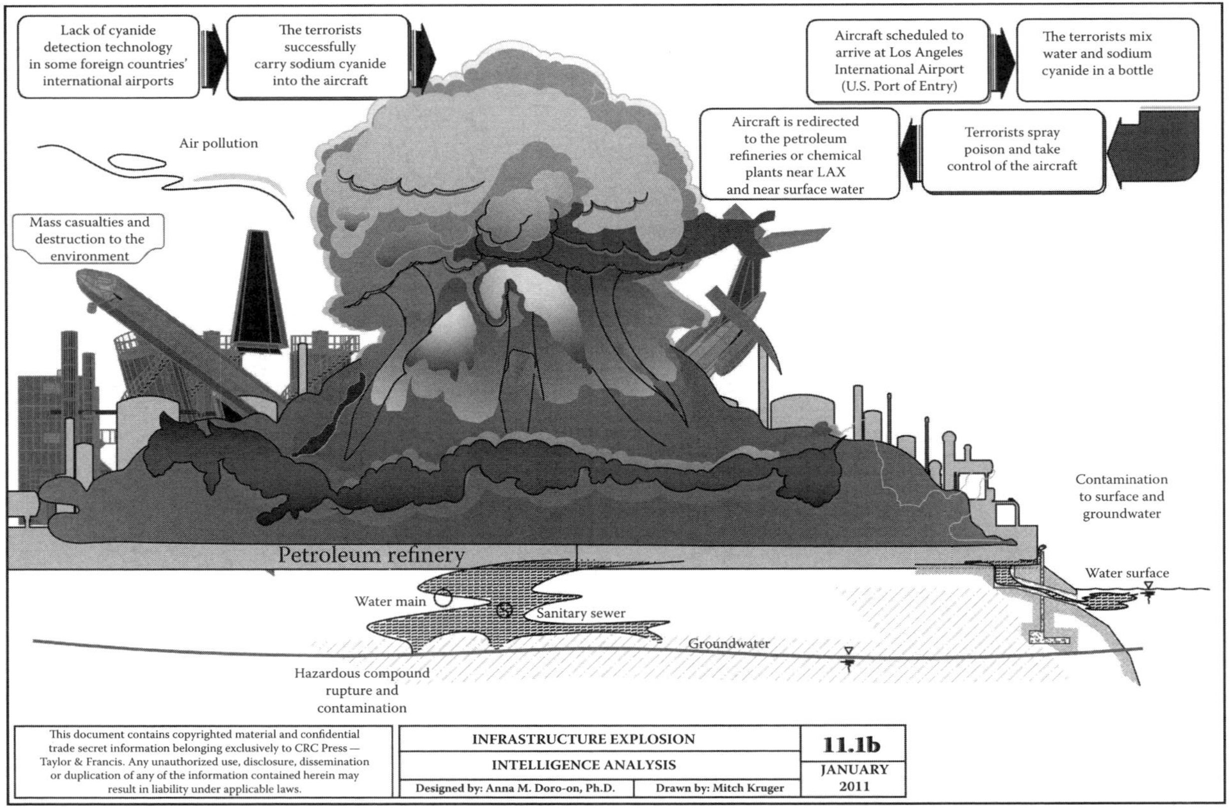

Figure 11.2b Large petroleum refinery and explosive chemical plant explosions near water systems and urban areas.

consisting of repeating nodes (CIA 2010a). The stages of the intelligence cycle include the issuance of requirements by decision makers and the collection, processing, analysis, and publication of intelligence. The circuit is completed when decision makers provide feedback and revised requirements. The traditional intelligence cycle is presented in Figure 11.3.

According to Rob Johnston (2005), the traditional intelligence cycle model should be redesigned to depict accurately the intended mission. Teaching with an inaccurate aid merely leads to misconceptions that can result in poor performance, confusion, and a need for re-teaching. If the objective is to capture the entire intelligence process from the request for a product to its delivery, including the roles and responsibilities of intelligence community members, then something more is required. This should be a model that pays particular attention to representing accurately all the elements of the process and the factors that influence them (Johnston 2005; CIA 2010b). The proposed modified intelligence analysis is presented in Figure 11.4.

11.4 Quantitative Risk Estimation Model to Aid Intelligence Analysis

The risk analysis methodology discussed in this chapter, which is also presented in Chapter 8, is a systematic approach that is integrated into the intelligence analysis for producing terrorism threat assessments and warnings. Also, a comprehensive *risk estimation model* will be developed based on event tree analysis and a *probabilistic model*. The five steps of the risk estimation process to aid intelligence analysis are also presented here utilizing Los Angeles and New York as examples for terrorism combat zone scenarios.

11.4.1 Process of Risk Estimation for Water Infrastructure Threats for Intelligence Analysis

There are multiple reasons why Los Angeles and New York water infrastructures are vulnerable to terrorist attacks. Figures 11.1a-c, 11.2a-b, 11.5a-b, 11.6, and 11.7 illustrate various examples of bold planning by terrorist leaders against the United States. The following includes some of the reasons why New York and Los Angeles are attractive to terrorists:

- Los Angeles and New York are the top cities in the United States experiencing an economic meltdown. The local governments are more focused on job creation, maintaining energy resources, tax problems, property foreclosures prevention, and economic recovery than on security. Local agencies (e.g., water resources department, water works department, and local

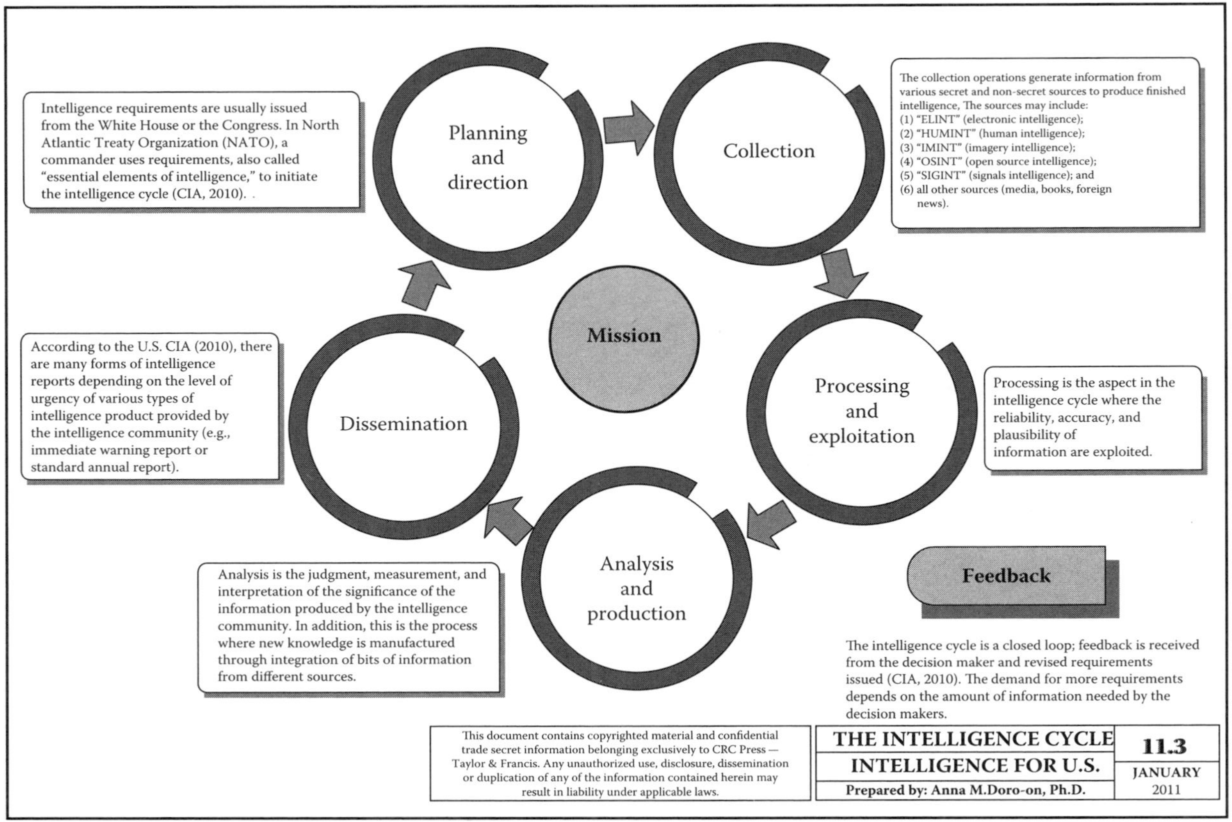

Figure 11.3 Traditional intelligence cycle.

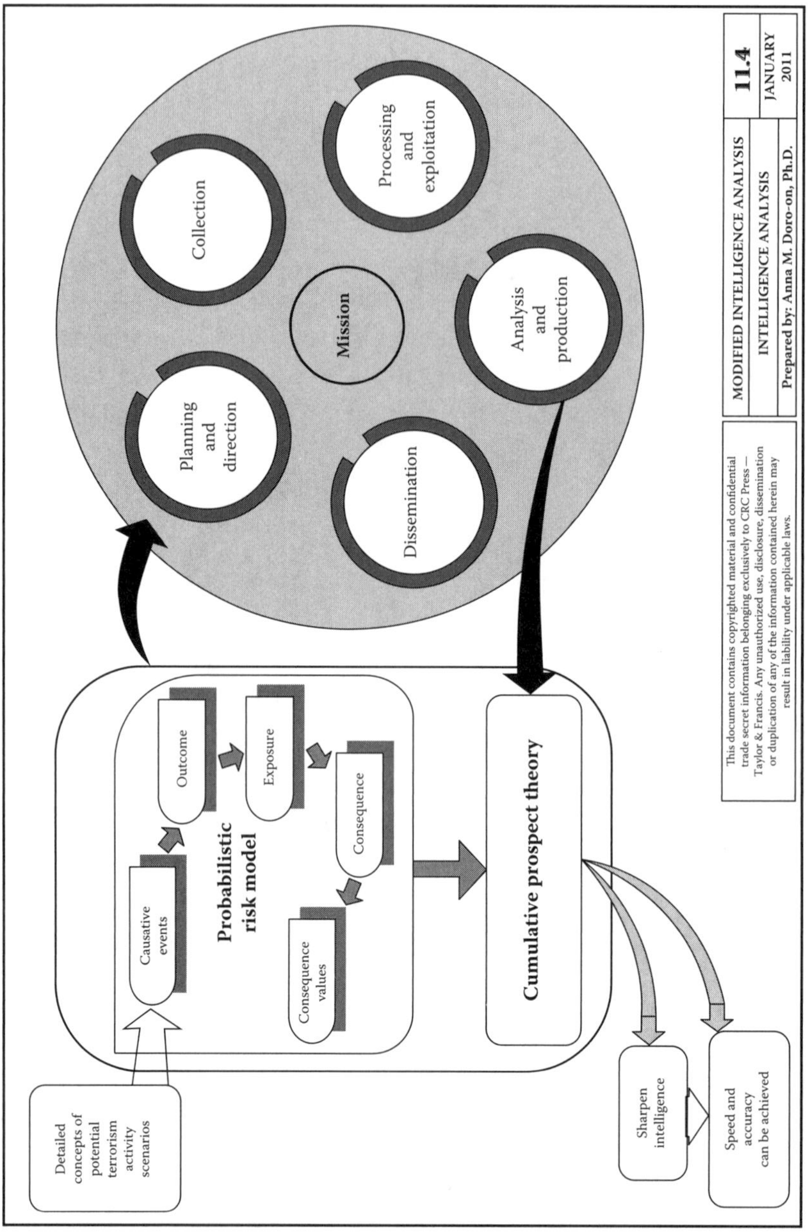

Figure 11.4 Modified intelligence analysis.

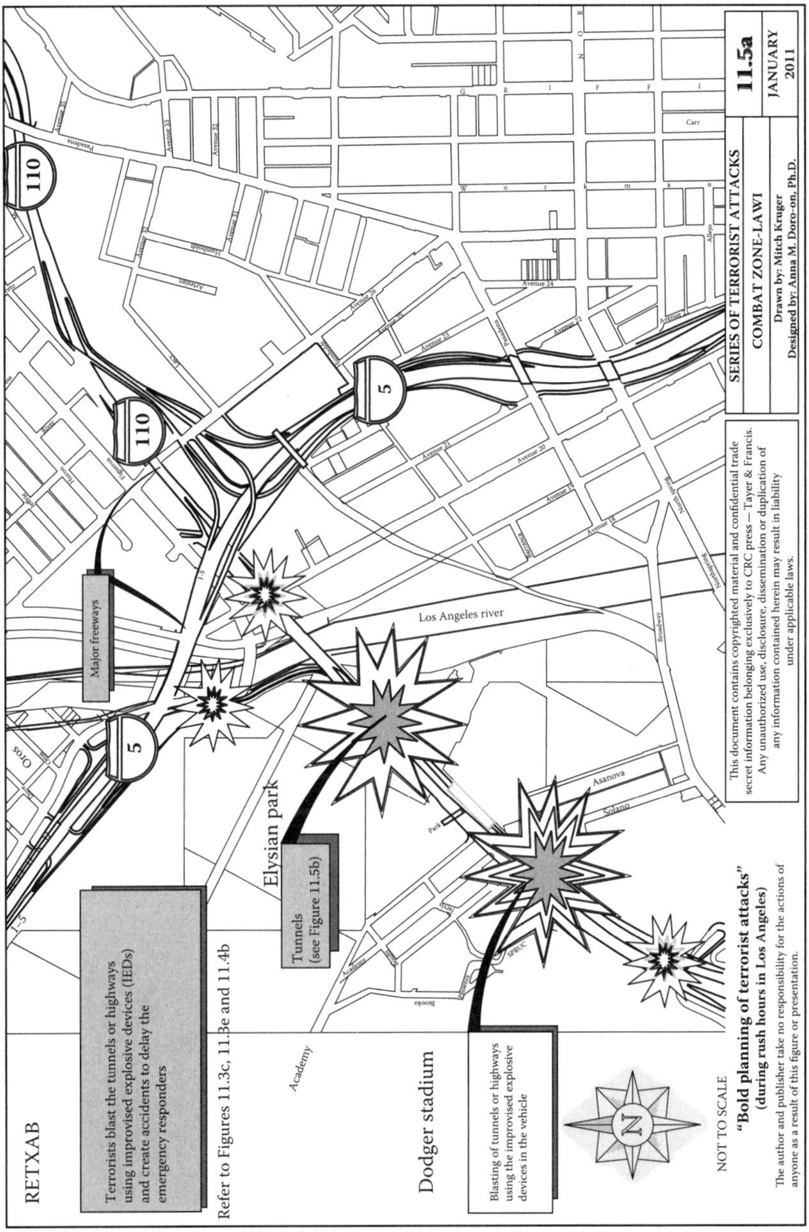

Figure 11.5a Accidents generated by terrorists on highways delaying emergency responders for water infrastructure attacks.

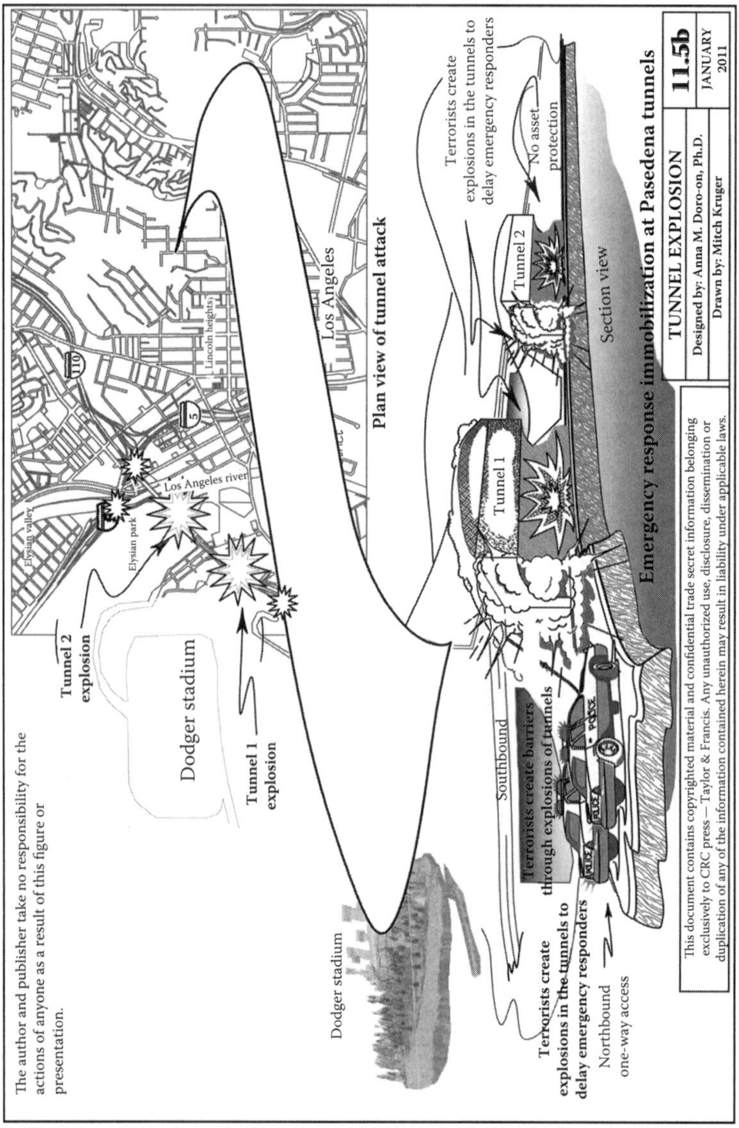

Figure 11.5b Accidents generated by terrorists in tunnels delaying emergency responders for water infrastructure attacks.

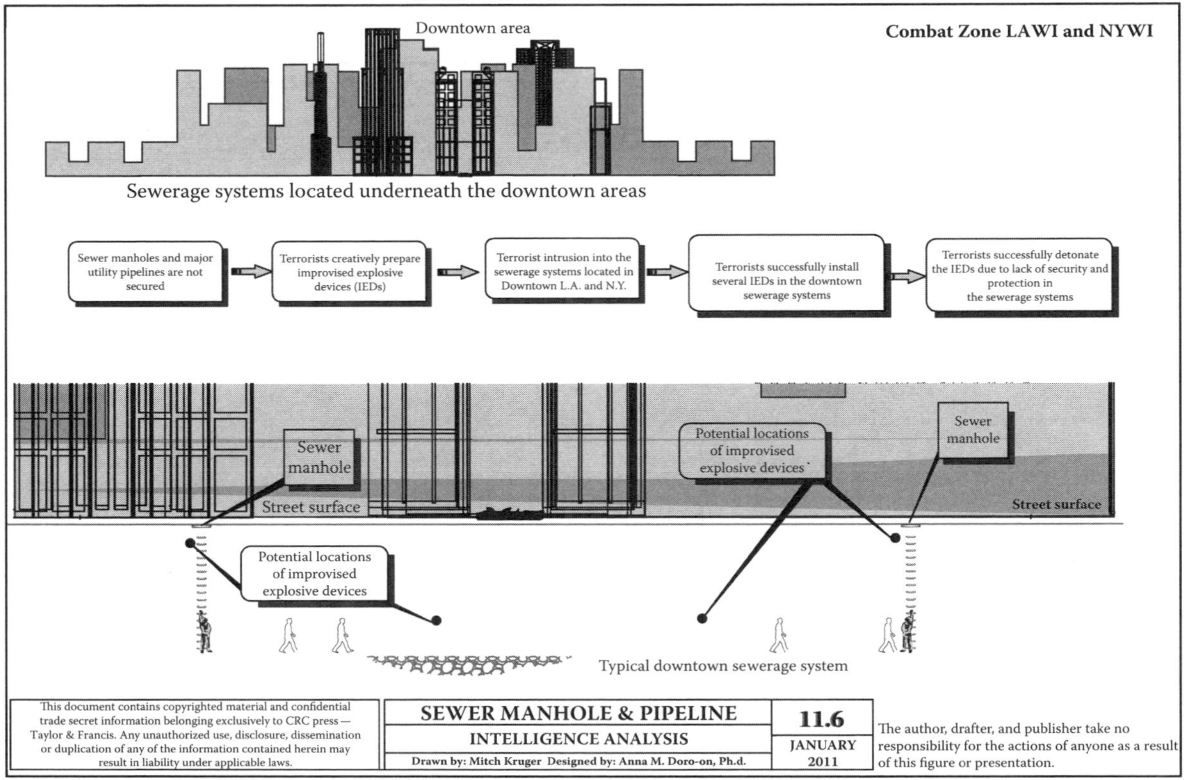

Figure 11.6 Installation of improvised explosive devices in sanitary sewer manholes and sewer pipes located in the center point of large metropolitan areas.

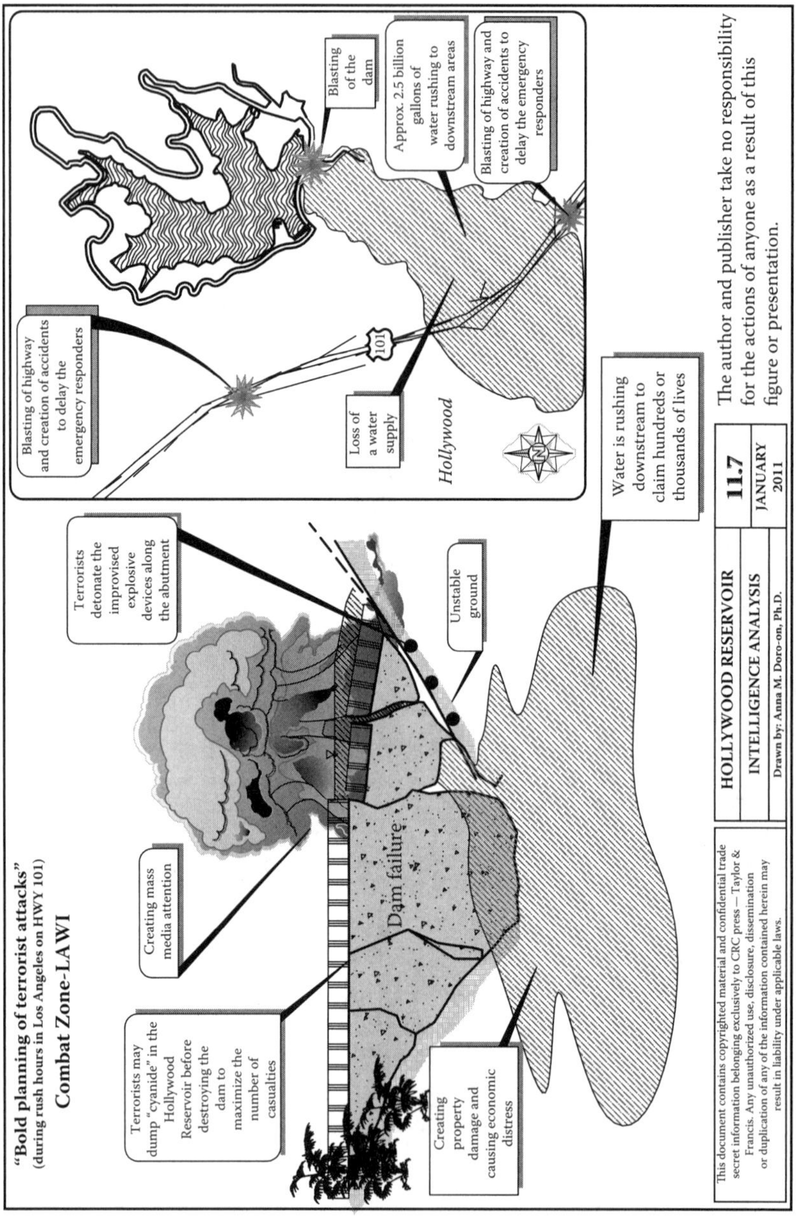

Figure 11.7 Destruction of the dam and its reservoir.

environmental protection agency) are presently diminishing their workforce due to the economic crisis. In addition, their focus is on improving green technology to mitigate the depletion of energy resources. Therefore, there is a lack of financial resources and workforce to support water infrastructure security improvements in these areas.

- There are currently no sophisticated technologies for ensuring security of water infrastructure, including dams, aqueducts, and reservoirs, against terrorism.
- Los Angeles and New York are the most traveled to cities in the United States; both can be perfect candidates for creating massive media attention when attacks happen similar to 9/11.
- Both cities are the most densely populated cities in the United States; terrorists can achieve the maximum number of casualties, catastrophe, and economic aftershocks comparable to 9/11 by attacking these places.
- Kensico Dam is the receiving point of the New York City drinking water supply from the Catskill aqueducts. If the terrorists effectively destroy the Kensico Dam, it will not only impair the New York City water supply and create water outages but also generate a catastrophe for the downstream communities. Figure 11.1c provides a combat zone of New York City's water infrastructure for intelligence analysis.
- Hollywood, in Los Angeles County, California, is one of the most popular places around the globe; destruction of its beautiful reservoir can generate mass media attention worldwide and can also create casualties, cause water supply shortages, and destroy the downstream communities and industries.
- The sanitary sewer manholes and pipelines are not secured in the downtown area of New York and Los Angeles; terrorists can utilize these utility components as accessories for launching their attacks. They can install improvised explosive devices (IEDs) randomly within the downtown areas without being detected, as shown in Figure 11.6.
- Some of the largest petroleum/industrial plants in the United States are located in Los Angeles and New York. The petroleum refineries are usually near or adjacent to surface water and potential groundwater resources. Explosion of these refineries (Figures 11.2a-b) can cause catastrophe to the surface water and the environment comparable to the Deepwater Horizon accident that occurred on April 22, 2010. Mass casualties and destruction of industries can be generated, and they can worsen the economic status of the entire nation.

The risk estimation models for Los Angeles and New York are detailed in Tables 11.1 through 11.5. And the risk estimation model for intelligence analysis is presented in Table 11.6.

Table 11.1 Risk Estimation Model: Step 1—Causative Events

Los Angeles—Causative Events	
(a)	Terrorist intrusion into Hollywood Reservoir (Mulholland Dam).
(b)	Terrorist intrusion into water supply system facilities (e.g., water supply storage tanks and water treatment facilities).
(c)	Terrorists purchase foreclosure homes or rent homes adjacent to underground pipelines of the aqueducts and near reservoirs.
(d)	Terrorist intrusion into sewer lines and manholes in the downtown Los Angeles area.
(e)	Terrorists with IEDs situate themselves on major highways and freeways leading into and out of Los Angeles.
(f)	International flights with undetected terrorists head to Los Angeles International Airport (LAX).
New York—Causative Events	
(a)	Terrorist intrusion into Kensico Dam in New York.
(b)	Terrorist intrusion into Croton system and Catskill Aqueduct.
(c)	Terrorists purchase or lease homes adjacent to underground pipelines of the aqueducts.
(d)	Terrorist intrusion into sewer lines and manholes in New York City.
(e)	Terrorists with IEDs situate themselves on major highways and freeways leading into and out of New York City and/or in the areas targeted.
(f)	International flights with undetected terrorists head to John F. Kennedy (JFK) International Airport, New York.

Table 11.2 Risk Estimation Model: Step 2—Outcome

Los Angeles—Outcome	
(a)	Terrorists blast the Hollywood Reservoir (Mulholland Dam).
(b)	Terrorists inject or dump deadly chemical threats (e.g., arsenic-/cyanide-based pesticides) into the water supply tanks and aqueducts, and blast major aqueduct pipelines.
(c)	Terrorists inject cyanide-based pesticides into the aqueducts.
(d)	Terrorists blast downtown Los Angeles area through bombs installed in the sewer lines and manholes.

Table 11.2 Risk Estimation Model: Step 2—Outcome (*Continued*)

Los Angeles—Outcome	
(e)	Terrorists with IEDs create accidents and explosions on major highways and freeways leading into and out of Los Angeles and/or in the areas targeted.
(f)	Terrorists take over international jets, land them on petroleum refineries and chemical plants located near LAX, and create high explosions. Hazardous chemicals create contamination to surface water. Explosions create destruction to underground water mains and sanitary sewers.
New York—Outcome	
(a)	Terrorists blast Kensico Dam.
(b)	Terrorists blast Croton system and Catskill Aqueduct.
(c)	Terrorists inject poison into the underground pipelines of the aqueducts.
(d)	Terrorists blast downtown New York through bombs installed in sewer lines and manholes.
(e)	Terrorists with IEDs create accidents and explosions on major tunnels, highways, and freeways leading into and out of New York and/or in the areas targeted.
(f)	Terrorists take over international jets, land them on petroleum refineries near New York airport, and create high explosions. Hazardous chemicals cause contamination to surface water. Explosions create destruction to underground water mains and sanitary sewers. Terrorists land jets directly on airport structures and create a massive explosion. This destroys other aircrafts and contaminates the surrounding surface water.

Table 11.3 Risk Estimation Model: Step 3—Exposure

Los Angeles—Exposure	
(a)	About 2.5 billion gallons of water from Hollywood Reservoir (Mulholland Dam) flood the downstream infrastructure and communities (101 Freeway, residential and commercial properties in West Hollywood).
(b)	Water mixes with poison in the water tanks and other water supply system facilities.
(c)	Water mixes with poison in the aqueduct pipelines.
(d)	Sewer lines and commercial building structures are destroyed, and massive fires are created.

Table 11.3 Risk Estimation Model: Step 3—Exposure (*Continued*)

	Los Angeles—Exposure
(e)	Traffic cannot be mobilized within the highways and freeways. Massive fires block the highways.
(f)	Release of hazardous chemicals into the groundwater, surface water, and environment. Massive fire from explosion.
	New York—Exposure
(a)	The 30.6 billion gallons of water from Kensico Dam flood many communities in Westchester, New York, and diminish the water supply of New York.
(b)	Blasting of Croton system and Catskill Aqueduct diminishes a portion of New York's water supply.
(c)	Water mixes with poison in the water tanks and other water supply system facilities.
(d)	Sewer lines and commercial building structures are destroyed, and massive fires are created.
(e)	Traffic cannot be mobilized within the highways and freeways. Massive fires block the highways.
(f)	Release of hazardous chemicals into the groundwater, surface water and environment. Massive fire from explosion.

Table 11.4 Risk Estimation Model: Step 4—Consequences

	Los Angeles—Consequences
(a)	Property damages
(b)	Mass casualties
(c)	Health effects due to exposure to hazardous chemicals
(d)	Economic aftershocks due to the following: Groundwater and surface water remediation Environmental cleanup Reconstruction of infrastructure (e.g., highways, building structures, utilities) Public health recovery Loss of some of the major commerce and industry sections (with long-term recovery) Water shortage Loss of tourism

Table 11.4 Risk Estimation Model: Step 4—Consequences (*Continued*)

(e)	Irreversible damage to water system
(f)	Injury to the environment
(g)	Destruction of some part of the Los Angeles harbor, creating further contamination
(h)	Disruption of international and national flights, affecting commercial airline companies
(i)	Damage to the beautiful Pacific Ocean near LAX, affecting business along the coast
(j)	Disruption of agricultural industry and livestock dependent on clean water supply
(k)	Abrupt crime rate increase—fighting for water supply
(l)	Surging of unemployment rate in Los Angeles due to loss of industry
(m)	Sudden price increase on goods/produce due to limited water supply
New York—Consequences	
(a)	Property damages
(b)	Mass casualties
(c)	Health effects due to exposure to hazardous chemicals
(d)	Economic aftershocks due to the following: Groundwater and surface water remediation Environmental cleanup Reconstruction of infrastructure (e.g., highways, building structures, utilities) Public health recovery Loss of some of the major commerce and industry (with long-term recovery) Water shortage
(e)	Irreversible damage to water system infrastructure
(f)	Diminishment of a large portion of the New York water supply
(g)	Injury to the environment
(h)	Damages to JFK International Airport
(i)	Disruption of international and national flights, affecting commercial airline companies

(*Continued*)

Table 11.4 Risk Estimation Model: Step 4—Consequences (*Continued*)

New York—Consequences	
(j)	Abrupt crime rate increase—fighting for water supply
(k)	Surging of unemployment rate in New York City

Table 11.5 Risk Estimation Model: Step 5—Consequence Values

Los Angeles and New York—Consequence Values	
(a)	Improve U.S. intelligence
(b)	Improve preventive measures

Table 11.6 Risk Estimation Process for Intelligence Analysis

Step 1: Causative Events	
(a)	The terrorist leaders and supporters created misleading information. The media announced that the terrorist leader has a major illness. The media announced that the terrorist group has no financial support. Foreign intelligence informed the United States that WMDs are manufactured in the designated site or country (e.g., wrongful information about Iraq's WMDs). The media announced that the principal terrorist leader is located in the designated country (the information was not verified to be plausible). The United States received intelligence information that the Iraqi government was involved in the 9/11 attacks. The media announced that the new, strong leader of the terrorist group is going to attack U.S. cyberspace. For several years, the terrorist leader (Osama bin Laden) rarely appeared in the media, and then only through videotaped talks, particularly during the peak of Iraq and Afghanistan wars. The media announced that the terrorist group had disintegrated and that the leader might be secretly dead. Those claims were proved unfounded in May of 2011 when U.S. special forces cornered Bin Laden in a compound in Pakistan and killed him there. Once again, the media announced that Al-Zawahiri may step up to replace bin Laden but would not be able to unite the al Qaeda members due to his lack of charisma and that he is not respected within the organization. Therefore, al Qaeda may not be able to strongly attack U.S.
(b)	Terrorists generated misleading events to divert U.S. intelligence attention.

Table 11.6 Risk Estimation Process for Intelligence Analysis (*Continued*)

	Step 1: Causative Events
	The terrorists created minor attacks on another infrastructure but did not cause any casualties on others such as cyberspace. The amateur terrorists created minor attacks on airports or buildings for media attention and intentionally got caught by authorities.
	Step 2: Outcome
(c)	The U.S. government exhausted its financial resources for the "wrongful war"; therefore, water security is of less priority while intelligence continues to utilize the same source of information (*business-as-usual*).
(d)	The federal government allocated a large amount of financial support for cyber and airport protection; therefore, water infrastructure has less priority. It focuses on protecting water and wastewater treatment plants only based upon the budget.
(e)	After the events of minor cyber and airport attacks, the U.S. intelligence community informed and advised the U.S government. The government allocated more support to develop sophisticated technology against cyberterrorism; however, there is less priority placed on protecting original water sources (e.g., aqueducts, aquifer, reservoirs including dams). The United States can only prioritize water treatment facilities for financial support at the moment.
(f)	The general public believes that the terrorists will not attack the original water source due to the requirement of a large quantity of chemicals; the terrorists do not have the financial resources and a large quantity of chemicals is too apparent for public to recognize. Some conservative intelligence officials validate this perception on the media.
(g)	The terrorists are currently planning for high-profile attacks, improving their own intelligence, and gaining more financial support while the United States continues to exhaust its financial resources and the CIA and other intelligence agencies maintain business-as-usual information sharing strategy.
	Step 3: Exposure
(h)	The terrorists blast dams and create destruction to reservoirs. They contaminate aquifers, aqueducts, and water supply tanks without being detected.
(i)	The terrorists hijack commercial airplanes from foreign international airports to attack petroleum refineries and explosive chemical plants near water bodies and populated areas.

(*Continued*)

Table 11.6 Risk Estimation Process for Intelligence Analysis (*Continued*)

	Step 4: Consequence
(j)	The terrorists achieve their ultimate mission, while the U.S. intelligence community failed to provide warnings and threat assessments to the U.S. leaders. The general public lost their confidence in the U.S. CIA as well as other intelligence agencies, and demanded the reform or replacement of the agency.
	Step 5: Consequence value
(k)	CIA needs to have an independent source of information; there is a need to reform or replace the U.S. intelligence community.

11.5 Event Tree Analysis Model

Event tree analysis provides a logical tracing of sequential events resulting in consequential outcomes. The event tree analysis for the development of intelligence analysis on terrorism events, their time frame, and potential terrorism warnings are presented in Figures 11.8a through g. The designed probability scales based on the author's scientific and engineering judgment for critical infrastructure analysis and intelligence analysis are provided in Tables 11.7 and 11.8, respectively. The risk rates used in the event tree analysis for intelligence analysis are shown in Table 11.9.

The dominating concern in comparative risk assessment in intelligence analysis is the question "Is the risk acceptable?" This is true especially when public and national security issues are involved. Risk can readily be quantified based on a stochastic model describing all the events leading to negative consequences. However, public risk acceptability demands a thorough understanding of the risk characteristics and a meticulous anatomy of the perception of risks as detailed in Chapter 9. Section 11.6 utilizes the cumulative prospect theory concept in social psychology to illustrate how risk acceptability can be assessed and achieved.

11.6 Perspectives of Risk Acceptability in Strategic Intelligence Analysis

Strategic intelligence development in the United States now places an increasing emphasis on consideration of risk and its consequences. Incorporating risk considerations into a strategic intelligence analysis is not as simple as it appears at first. In the public sector, there are several concepts of risk. One common approach is to view risk quantitatively in terms of the probability that some undesirable event with a negative consequence will occur. This is especially true for the risk associated with accidents or toxic substances. Preferably, risk can also be defined in terms

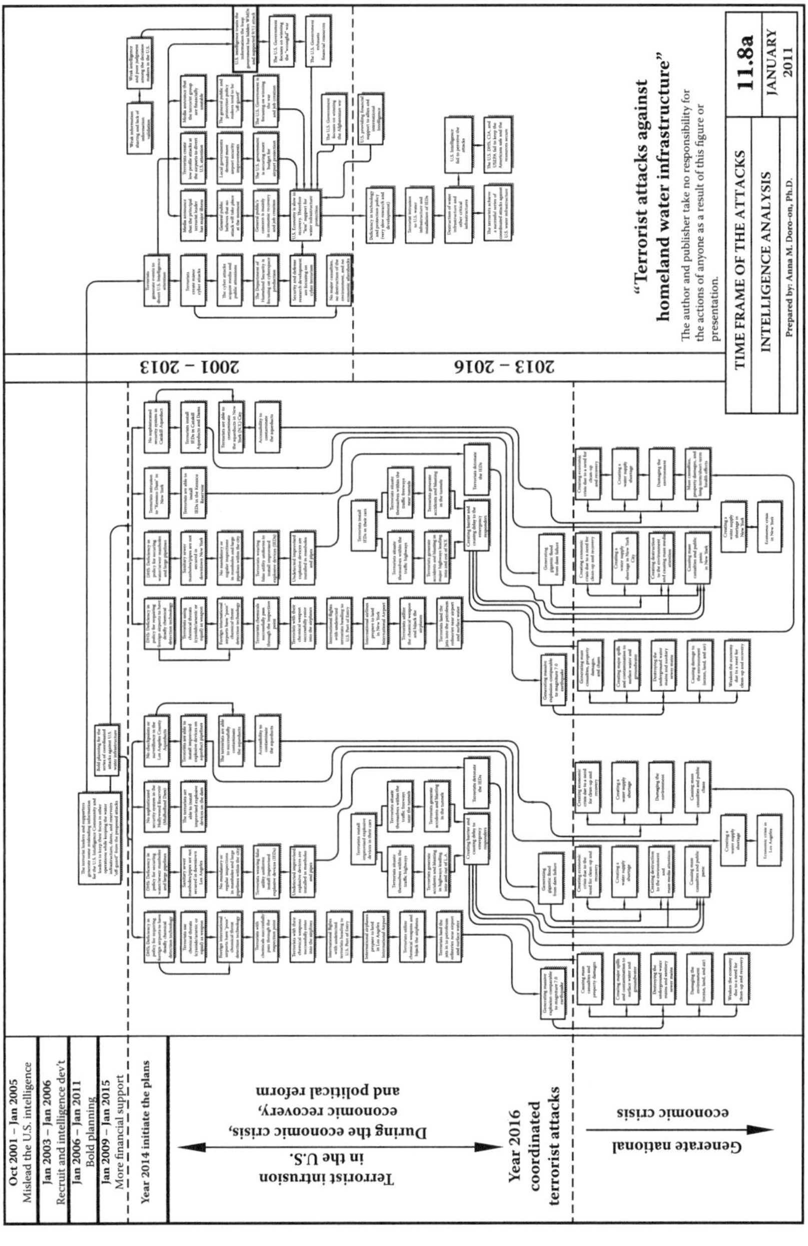

Figure 11.8a Estimated time frame for the series of coordinated terrorist attacks.

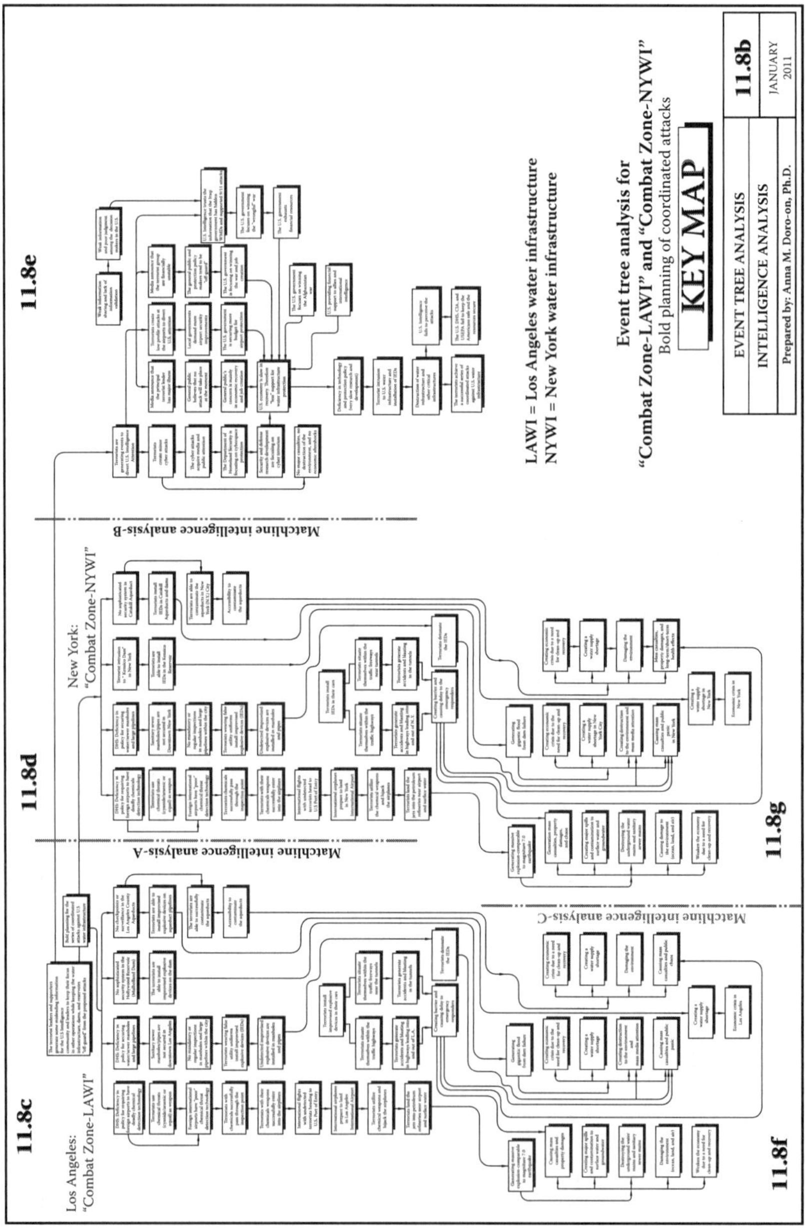

Figure 11.8b Key Map: Event tree analysis for combat zones and intelligence analysis.

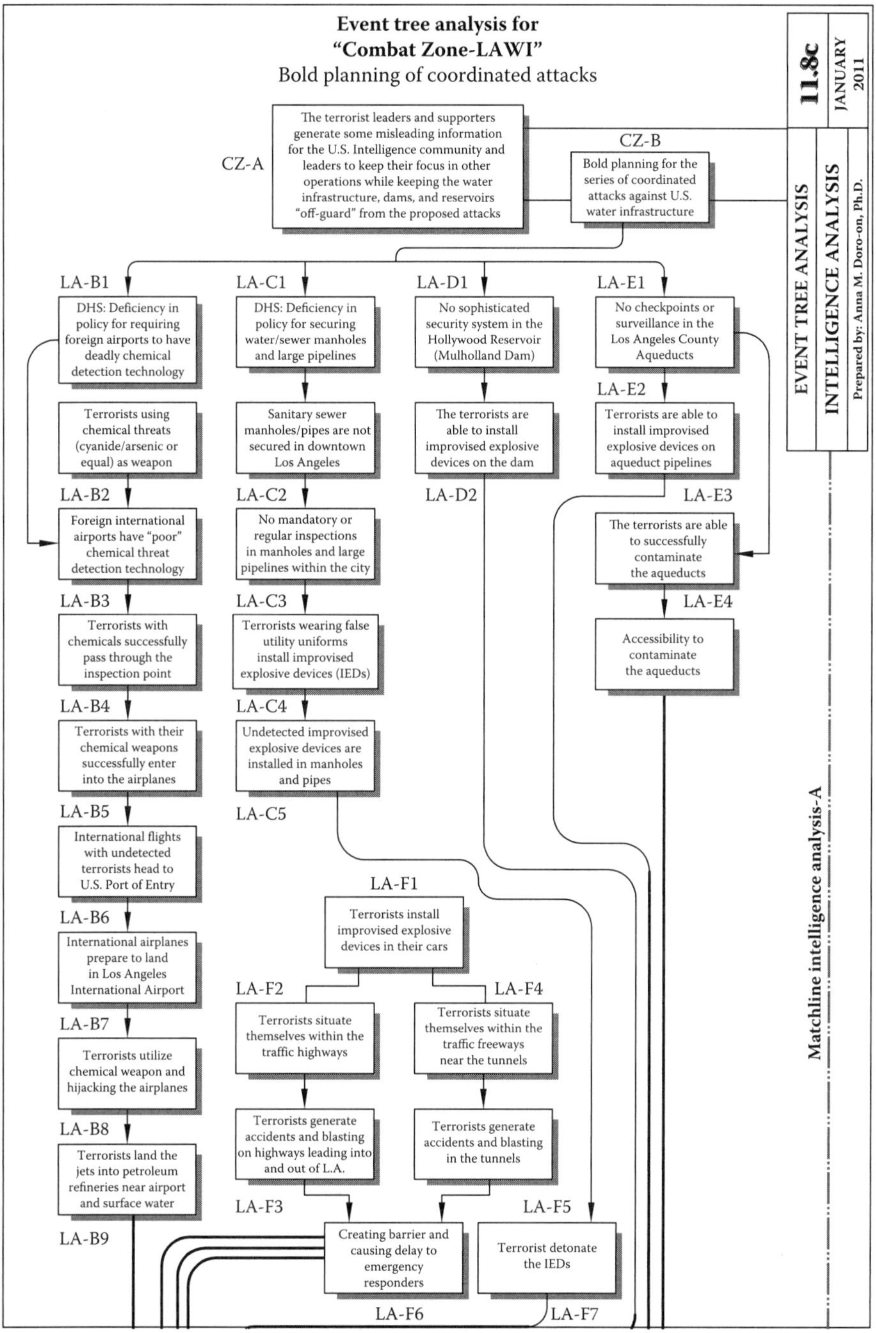

Event tree analysis for
"Combat Zone-LAWI"
Bold planning of coordinated attacks
11.8c
JANUARY 2011
EVENT TREE ANALYSIS
INTELLIGENCE ANALYSIS
Prepared by: Anna M. Doro-on, Ph.D.
CZ-A
The terrorist leaders and supporters generate some misleading information for the U.S. Intelligence community and leaders to keep their focus in other operations while keeping the water infrastructure, dams, and reservoirs "off-guard" from the proposed attacks
CZ-B
Bold planning for the series of coordinated attacks against U.S. water infrastructure
LA-B1
DHS: Deficiency in policy for requiring foreign airports to have deadly chemical detection technology
Terrorists using chemical threats (cyanide/arsenic or equal) as weapon
LA-B2
Foreign international airports have "poor" chemical threat detection technology
LA-B3
Terrorists with chemicals successfully pass through the inspection point
LA-B4
Terrorists with their chemical weapons successfully enter into the airplanes
LA-B5
International flights with undetected terrorists head to U.S. Port of Entry
LA-B6
International airplanes prepare to land in Los Angeles International Airport
LA-B7
Terrorists utilize chemical weapon and hijacking the airplanes
LA-B8
Terrorists land the jets into petroleum refineries near airport and surface water
LA-B9
LA-C1
DHS: Deficiency in policy for securing water/sewer manholes and large pipelines
Sanitary sewer manholes/pipes are not secured in downtown Los Angeles
LA-C2
No mandatory or regular inspections in manholes and large pipelines within the city
LA-C3
Terrorists wearing false utility uniforms install improvised explosive devices (IEDs)
LA-C4
Undetected improvised explosive devices are installed in manholes and pipes
LA-C5
LA-D1
No sophisticated security system in the Hollywood Reservoir (Mulholland Dam)
The terrorists are able to install improvised explosive devices on the dam
LA-D2
LA-E1
No checkpoints or surveillance in the Los Angeles County Aqueducts
LA-E2
Terrorists are able to install improvised explosive devices on aqueduct pipelines
LA-E3
The terrorists are able to successfully contaminate the aqueducts
LA-E4
Accessibility to contaminate the aqueducts
LA-F1
Terrorists install improvised explosive devices in their cars
LA-F2
Terrorists situate themselves within the traffic highways
Terrorists generate accidents and blasting on highways leading into and out of L.A.
LA-F3
LA-F4
Terrorists situate themselves within the traffic freeways near the tunnels
Terrorists generate accidents and blasting in the tunnels
LA-F5
Creating barrier and causing delay to emergency responders
Terrorist detonate the IEDs
LA-F6
LA-F7
Matchline intelligence analysis-A

Figure 11.8c Event tree analysis for the intelligence analysis.

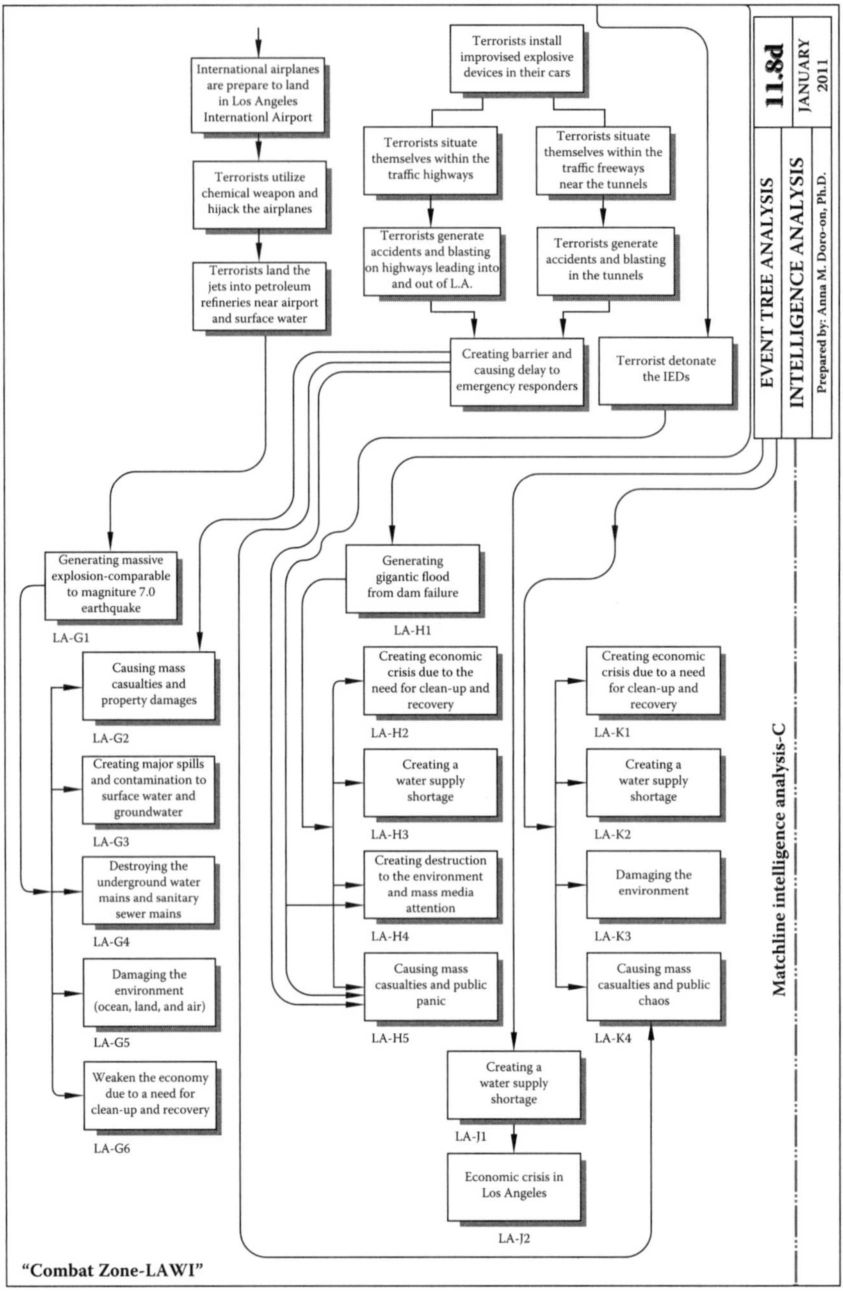

Figure 11.8d Event tree analysis for the intelligence analysis.

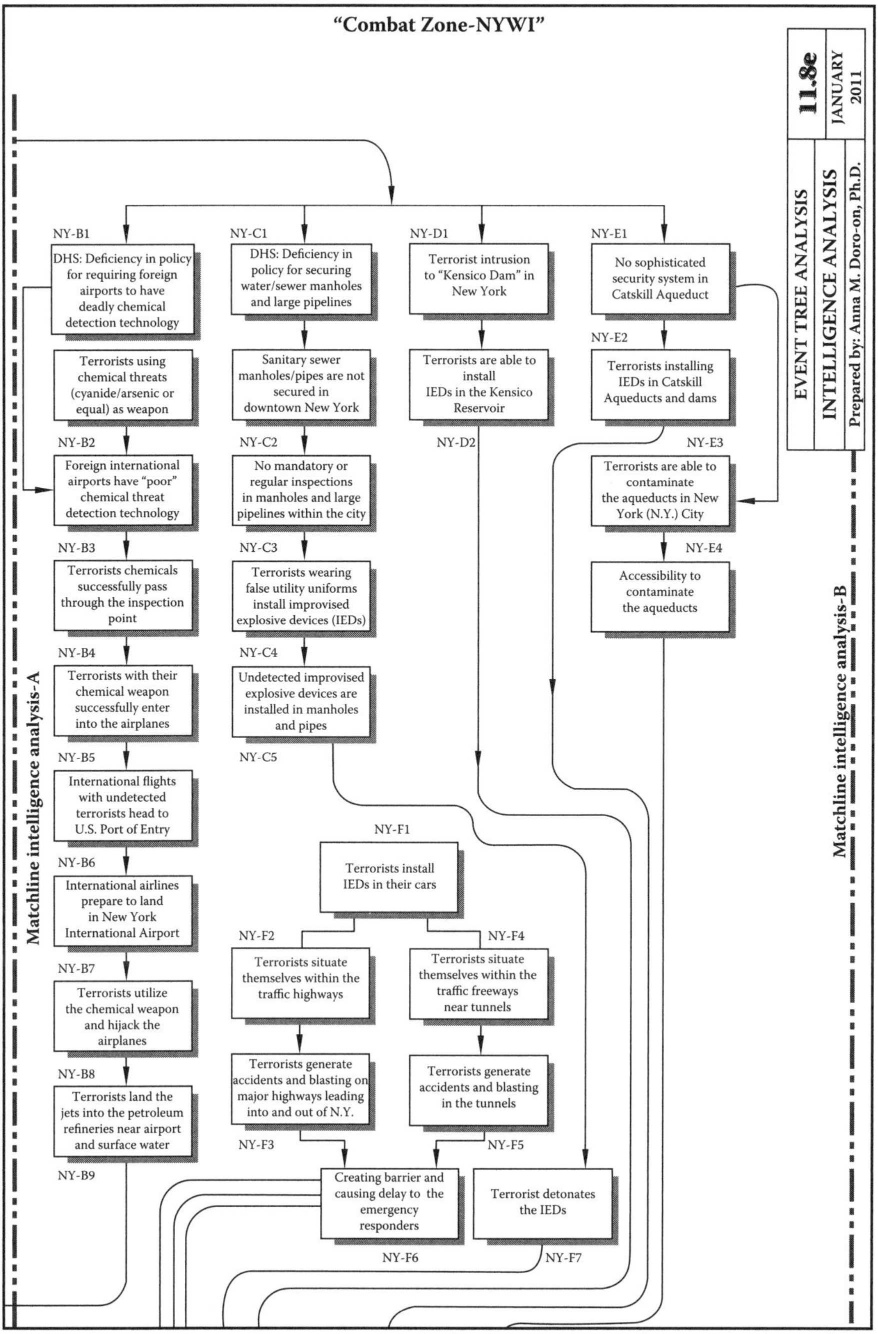

Figure 11.8e Event tree analysis for the intelligence analysis.

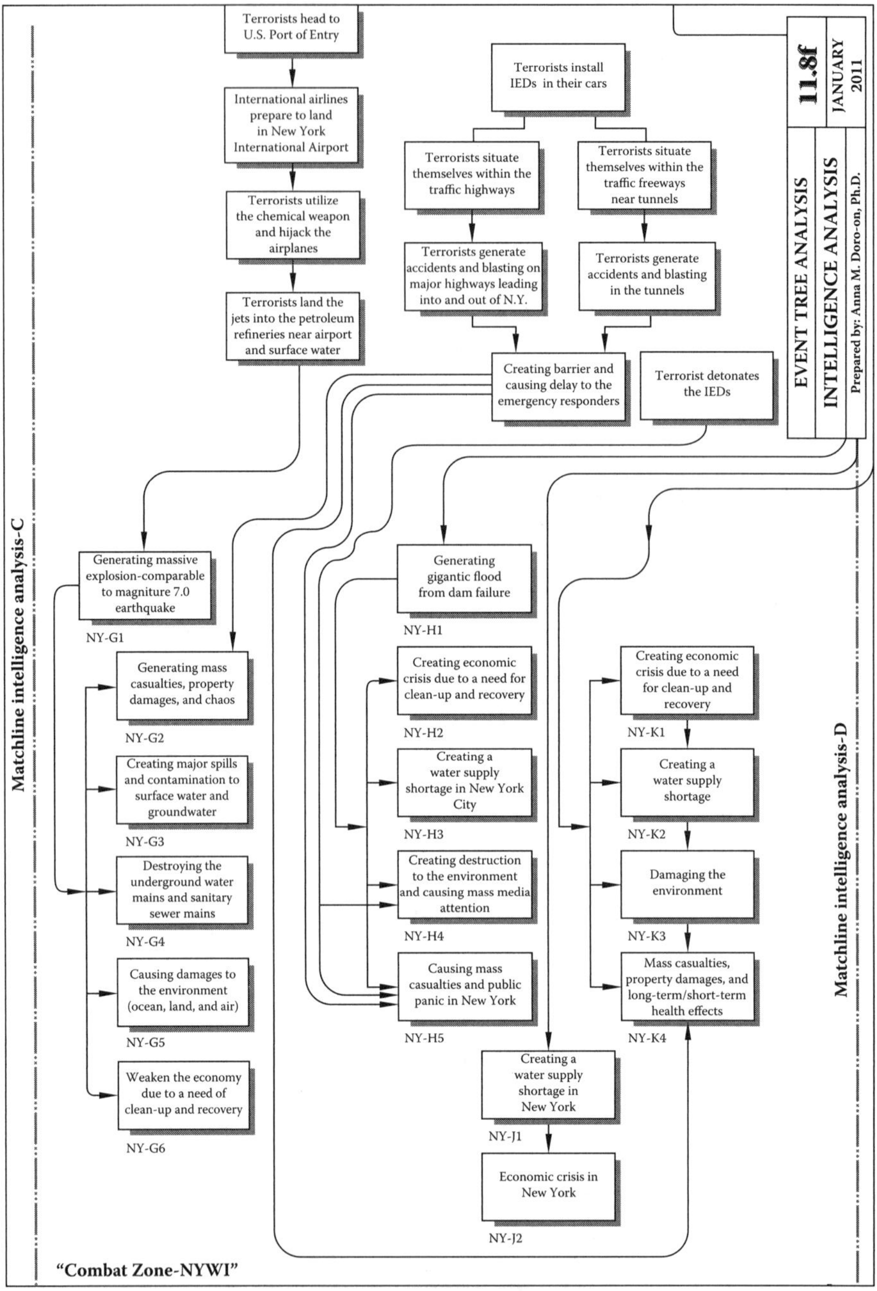

Figure 11.8f Event tree analysis for the intelligence analysis.

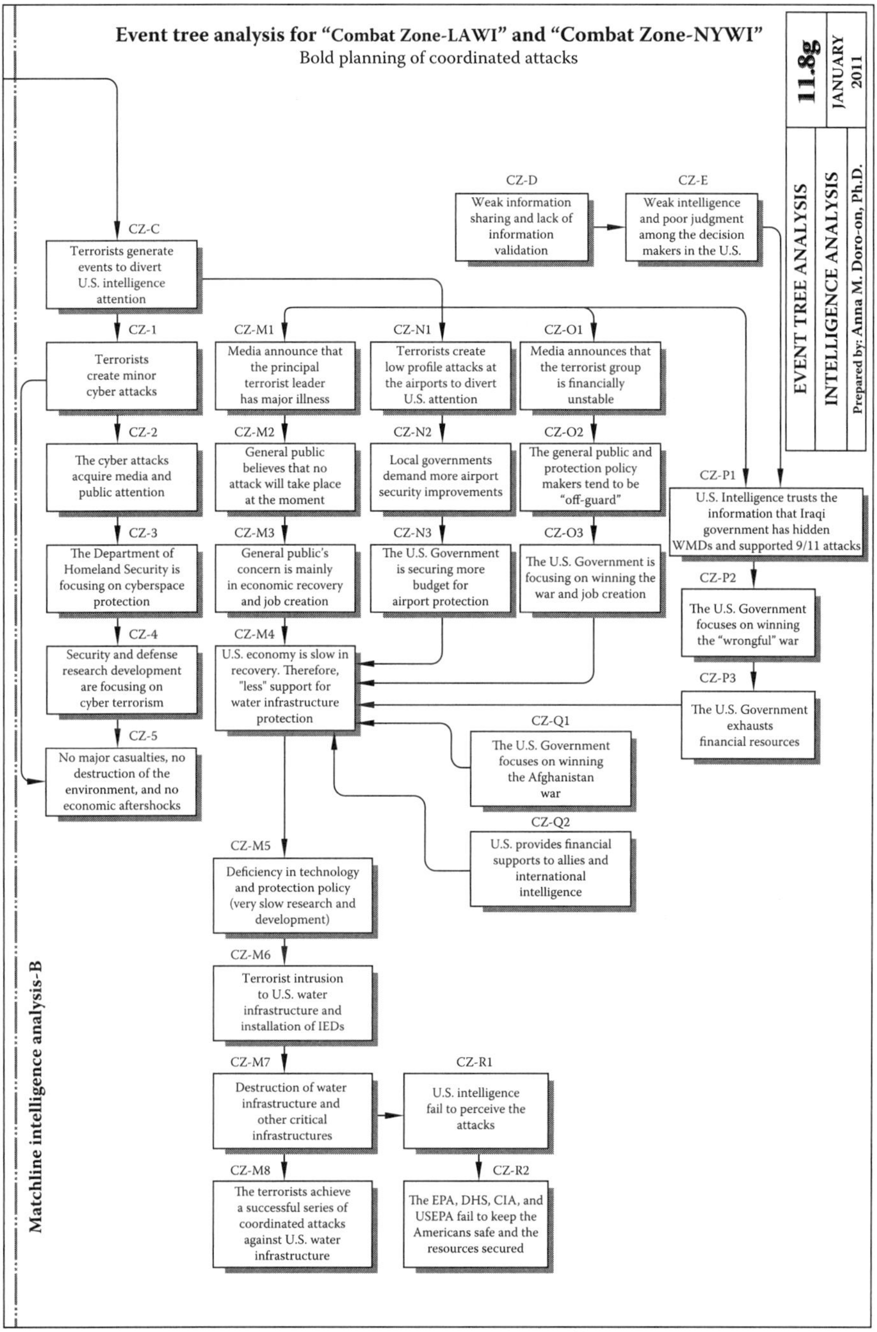

Event tree analysis for "Combat Zone-LAWI" and "Combat Zone-NYWI"
Bold planning of coordinated attacks
11.8g
JANUARY 2011
EVENT TREE ANALYSIS
INTELLIGENCE ANALYSIS
Prepared by: Anna M. Doro-on, Ph.D.
CZ-D
Weak information sharing and lack of information validation
CZ-E
Weak intelligence and poor judgment among the decision makers in the U.S.
CZ-C
Terrorists generate events to divert U.S. intelligence attention
CZ-1
Terrorists create minor cyber attacks
CZ-2
The cyber attacks acquire media and public attention
CZ-3
The Department of Homeland Security is focusing on cyberspace protection
CZ-4
Security and defense research development are focusing on cyber terrorism
CZ-5
No major casualties, no destruction of the environment, and no economic aftershocks
CZ-M1
Media announce that the principal terrorist leader has major illness
CZ-M2
General public believes that no attack will take place at the moment
CZ-M3
General public's concern is mainly in economic recovery and job creation
CZ-M4
U.S. economy is slow in recovery. Therefore, "less" support for water infrastructure protection
CZ-M5
Deficiency in technology and protection policy (very slow research and development)
CZ-M6
Terrorist intrusion to U.S. water infrastructure and installation of IEDs
CZ-M7
Destruction of water infrastructure and other critical infrastructures
CZ-M8
The terrorists achieve a successful series of coordinated attacks against U.S. water infrastructure
CZ-N1
Terrorists create low profile attacks at the airports to divert U.S. attention
CZ-N2
Local governments demand more airport security improvements
CZ-N3
The U.S. Government is securing more budget for airport protection
CZ-O1
Media announces that the terrorist group is financially unstable
CZ-O2
The general public and protection policy makers tend to be "off-guard"
CZ-O3
The U.S. Government is focusing on winning the war and job creation
CZ-P1
U.S. Intelligence trusts the information that Iraqi government has hidden WMDs and supported 9/11 attacks
CZ-P2
The U.S. Government focuses on winning the "wrongful" war
CZ-P3
The U.S. Government exhausts financial resources
CZ-Q1
The U.S. Government focuses on winning the Afghanistan war
CZ-Q2
U.S. provides financial supports to allies and international intelligence
CZ-R1
U.S. intelligence fail to perceive the attacks
CZ-R2
The EPA, DHS, CIA, and USEPA fail to keep the Americans safe and the resources secured
Matchline intelligence analysis-B

Figure 11.8g Event tree analysis for the intelligence analysis.

Table 11.7 Probability Scale Used for Critical Infrastructure Analysis Risk Estimation Model

Probability Scale for Critical Infrastructure and Key Resources Protection Analysis	
Description	*Probabilities in Decimal Description*
Very high (*indicates that there are no effective policy or protective measures currently in place to deter, detect, delay, and respond to the threat*)	0.90–1.00
High (*there are some policy and protective measures to deter, detect, delay, or respond to the asset but not a complete or effective application of these security strategies*)	0.80–0.89
	0.71–0.79
Medium high	0.61–0.70
Medium low (*indicates that although there are some effective policy and protective measures there is not a complete and effective application of these security strategies*)	0.40–0.60
Low (*indicates that there are effective protective measures in place; however, at least one weakness exists such that an adversary would be able to defeat the countermeasure*)	0.20–0.398
	0.10–0.198
Less likely—very low probability (*indicates no credible evidence of capability*)	0.05–0.099
	0.025–0.049
	<0.024

Table 11.8 Probability Scale Used for Intelligence Analysis Risk Estimation Model

Probability Scale for Intelligence Analysis	
Description	*Probabilities in Decimal Description*
Very high (*indicates that there are no effective strategies current in place to accurately validate the information, or business-as-usual*	0.90–1.00
High (*there are some strategy and procedures to acknowledge the validity of the information but not an effective application of these strategies*)	0.80–0.89
	0.71–0.79

Table 11.8 Probability Scale Used for Intelligence Analysis Risk Estimation Model (*Continued*)

Probability Scale for Intelligence Analysis	
Description	*Probabilities in Decimal Description*
Medium high (*indicates that although there are some effective strategies and procedures there is not an effective application of these strategies*)	0.61–0.70
Medium low (*indicates that there are effective strategies in place; however, at least one weakness exists that defeats the strategy to validate the accuracy of the information*)	0.40–0.60
Low (*indicates that there are effective strategies and sophisticated technologies but at least one piece of information that is not totally clear*)	0.20–0.398 0.10–0.198
Very low probability (*there are effective strategies and sophisticated technologies that validates the information being provided*)	<0.099

Table 11.9 Risk Rates for the Event Tree Analysis

Event Tree List of Events		
Symbol	*Description*	*Risk Rate*
CZ-A	The terrorist leaders and supporters generate some misleading information for the U.S. intelligence community and leaders to keep them focused on other operations while keeping the water infrastructure, dams, and reservoirs off guard from the proposed attacks.	0.90
CZ-B	Bold planning for the series of coordinated attacks against U.S. water infrastructure.	1/18 or 0.056
CZ-C	Terrorists generate events to divert the U.S. intelligence community's attention.	0.90
CZ-D	Weak information sharing and lack of information validation.	0.80
CZ-E	Weak intelligence and poor judgment among decision makers in the United States.	0.80

(*Continued*)

Table 11.9 Risk Rates for the Event Tree Analysis (*Continued*)

Event Tree List of Events		
Symbol	*Description*	*Risk Rate*
LA-B1	U.S. government/DHS: Deficiency in policy for requiring foreign international airports (or the foreign aviation department) to have chemical threat detection technology for flights heading to the United States.	0.50
LA-B2	Terrorist uses chemical threat, cyanide\arsenic, or equal as weapons.	0.95
LA-B3	Foreign international airports have poor or deficient chemical threat detection technology.	0.20
LA-B4	Terrorists with their chemical weapon successfully pass through the inspection point.	0.20
LA-B5	Terrorists with their chemical weapon successfully enter the airplanes.	0.90
LA-B6	International flights with undetected terrorists head to U.S. port of entry.	0.95
LA-B7	International commercial airplanes prepare to land at LAX.	1.0
LA-B8	Terrorists utilize the chemical weapon and hijack international commercial airplanes.	0.90
LA-B9	Terrorists land the international commercial airplanes onto the petroleum refineries near the airport and surface water (ocean or river).	0.95
LA-C1	U.S. government/DHS: Deficiency in policy for securing water and sewer manholes and pipelines.	0.80
LA-C2	Sanitary sewer manholes and pipelines are not secured in downtown Los Angeles.	0.95
LA-C3	No mandatory or regular inspections in the manholes or large pipelines within the downtown area.	0.80
LA-C4	Terrorists wear false utility company uniforms and install IEDs.	0.85
LA-C5	Undetected IEDs are successfully installed in manholes, water, or sewers, or stormwater utility pipelines.	0.95

Table 11.9 Risk Rates for the Event Tree Analysis (*Continued*)

Event Tree List of Events		
Symbol	*Description*	*Risk Rate*
LA-D1	No sophisticated security system in the Hollywood Reservoir or Mulholland Dam.	0.60
LA-D2	The terrorists were able to install IEDs on the dam's abutment.	0.65
LA-E1	No checkpoints or surveillance in Los Angeles County's aqueducts.	0.95
LA-E2	Terrorists were able to install IEDs on the aqueduct pipelines.	0.50
LA-E3	Terrorists contaminate the water supply in the aqueducts.	0.10
LA-E4	Terrorists' accessibility to contaminate the aqueducts.	0.05
LA-F1	Terrorists successfully install IEDs in their vehicle.	1.0
LA-F2	Terrorists situate themselves in the traffic highways/ freeways.	0.98
LA-F3	Terrorists generate accidents and blasts on major highways leading into and out of Los Angeles (see Figure 11.5 as an illustrative example).	0.98
LA-F4	Terrorists situate themselves in the traffic freeways near the tunnels.	0.98
LA-F5	Terrorists generate accidents and blasting of the tunnels.	0.98
LA-F6	The accidents and blasts create barriers and delay for the emergency response team.	0.90
LA-F7	Terrorists detonate the IEDs.	0.85
NY-B1	U.S. government/DHS: Deficiency in policy for requiring foreign international airports (or the foreign aviation department) to have chemical threat detection technology for flights heading to the United States.	0.50
NY-B2	Terrorist uses chemical threat, cyanide\arsenic, or equal as weapons.	0.95

(*Continued*)

Table 11.9 Risk Rates for the Event Tree Analysis (*Continued*)

Event Tree List of Events		
Symbol	*Description*	*Risk Rate*
NY-B3	Foreign international airports have poor or deficient chemical threat detection technology.	0.20
NY-B4	Terrorists with their chemical weapon successfully pass through the inspection point.	0.20
NY-B5	Terrorists with their chemical weapon successfully enter the airplanes.	0.90
NY-B6	International flights with undetected terrorists head to U.S. port of entry.	0.95
NY-B7	International commercial airplanes prepare to land in the New York International Airport.	1.0
NY-B8	Terrorists utilize the chemical weapon and hijack international commercial airplanes.	0.90
NY-B9	Terrorists land the international commercial airplanes into the petroleum refineries near the airport and near the surface water (ocean or river).	0.95
NY-C1	U.S. government/DHS: Deficiency in policy for securing water and sewer manholes and pipelines.	0.80
NY-C2	Sanitary sewer manholes and pipelines are not secured in downtown New York.	0.95
NY-C3	No mandatory or regular inspections in the manholes or large pipelines within the downtown area.	0.80
NY-C4	Terrorists wear false utility company uniforms and install IEDs.	0.85
NY-C5	Undetected IEDs are successfully installed in the manholes, water, or sewers, or stormwater utility pipelines.	0.95
NY-D1	No sophisticated security system in the Kensico Dam vicinity.	0.40
NY-D2	The terrorists were able to install IEDs on the dam's abutment.	0.65
NY-E1	No checkpoints or surveillance in the Catskill aqueducts/reservoirs/dams.	0.75

Table 11.9 Risk Rates for the Event Tree Analysis (*Continued*)

Event Tree List of Events		
Symbol	*Description*	*Risk Rate*
NY-E2	Terrorists were able to install IEDs on the aqueduct pipelines.	0.50
NY-E3	Terrorists successfully contaminate the water supply in the aqueducts.	0.05
NY-E4	Terrorists' accessibility to contaminate the aqueducts.	0.05
NY-F1	Terrorists successfully install IEDs in their vehicle.	1.0
NY-F2	Terrorists situate themselves in the traffic highways/ freeways.	0.98
NY-F3	Terrorists generate accidents and blasts on major highways leading into and out of New York City.	0.98
NY-F4	Terrorists situate themselves in the traffic freeways near the tunnels.	0.98
NY-F5	Terrorists generate accidents and blasting of tunnels.	0.98
NY-F6	The accidents and blasts create barriers and delay for the emergency response team.	0.90
NY-F7	Terrorists detonate the IEDs.	0.85
LA-G1	Terrorists generate a massive explosion, comparable to magnitude 7.0, through the attacks on petroleum refineries.	0.95
LA-G2	Cause mass casualties and property damages.	0.80
LA-G3	Create major spills and contamination to surface water and groundwater comparable to the Deepwater Horizon accident.	0.98
LA-G4	Destruction of the underground water mains and sanitary sewer pipelines.	0.85
LA-G5	Damage the environment by contaminating the ocean, river, groundwater, land, and air.	1.0
LA-H1	Generate a gigantic flood.	1.0
LA-H2	Create economic crisis due to the need for cleanup and recovery.	1.0

(*Continued*)

Table 11.9 Risk Rates for the Event Tree Analysis (*Continued*)

Event Tree List of Events		
Symbol	*Description*	*Risk Rate*
LA-H3	Create water supply shortage.	0.80
LA-H4	Cause destruction to the environment and generate mass media attention.	0.95
LA-H5	Cause mass casualties and public panic/chaos.	0.90
LA-K1	Create economic crisis due to the need for cleanup and recovery.	0.20
LA-K2	Create water supply shortage.	0.20
LA-K3	Cause mass casualties and public panic/chaos.	0.20
NY-G1	Terrorists generate massive explosion, comparable to magnitude 7.0, through the attacks on petroleum refineries.	0.95
NY-G2	Cause mass casualties and property damages.	0.80
NY-G3	Create major spills and contamination to surface water and groundwater comparable to the Deepwater Horizon accident.	0.98
NY-G4	Destroy the underground water mains and sanitary sewer pipelines.	0.85
NY-G5	Damage the environment by contaminating the ocean, river, groundwater, land, and air.	1.0
NY-H1	Generate a gigantic flood.	1.0
NY-H2	Create economic crisis due to the need for cleanup and recovery.	1.0
NY-H3	Create a water supply shortage.	0.80
NY-H4	Cause destruction to the environment and generated mass media attention.	0.95
NY-H5	Cause mass casualties and public panic/chaos.	0.90
NY-K1	Create economic crisis due to the need for cleanup and recovery.	0.20
NY-K2	Create water supply shortage.	0.20

Table 11.9 Risk Rates for the Event Tree Analysis (*Continued*)

Event Tree List of Events		
Symbol	*Description*	Risk Rate
NY-K3	Cause mass casualties and public panic/chaos.	0.20
NY-K4	Cause mass casualties and public panic/chaos.	0.20
CZ-1	Terrorists create minor cyber attacks.	0.50
CZ-2	The cyber attacks acquire media and public attention.	0.80
CZ-3	The DHS focuses on cyberspace protection.	0.75
CZ-4	Security and defense research development focus on cyberterrorism.	0.95
CZ-5	No major casualties, no destruction of the environment, and no economic aftershocks comparable.	0.20
CZ-MI	Media announces that the principal terrorist leader has a major illness.	1.0
CZ-M2	General public believes that no major attack will take place at the moment.	0.70
CZ-M3	General public's concern is mainly on economic recovery and job creation.	1.0
CZ-M4	U.S. economy is slow in recovery. Therefore, there is less support for water infrastructure protection.	0.98
CZ-M5	Deficiency in technology and protection policy (slow in research and development for water infrastructure protection and safety improvements).	0.95
CZ-M6	Terrorists intrude into U.S. water infrastructure (including dams, aqueducts, and reservoirs) and successfully install the IEDs.	0.90
CZ-M7	Water infrastructure destruction involving other critical infrastructure.	0.90
CZ-M8	The terrorists achieve a successful series of coordinated attacks against U.S. water infrastructure with the involvement of other critical infrastructure such as highways, petroleum refineries, and aviation.	0.90

(*Continued*)

Table 11.9 Risk Rates for the Event Tree Analysis (*Continued*)

Event Tree List of Events		
Symbol	*Description*	*Risk Rate*
CZ-N1	Terrorists create low-profile attacks at the airports to divert U.S. attention on aviation protection only within the U.S. mainland.	1.0
CZ-N2	The local governments demand additional airport security improvements.	0.90
CZ-N3	The U.S. government secures additional budget for airport protection.	0.90
CZ-O1	The media announces that the terrorist group is financially unstable.	0.90
CZ-O2	The general public and protection policy makers tend to be off guard.	0.50
CZ-O3	The U.S. government is focused on winning the war (with more financial support for the wrongful war) and job creation.	0.90
CZ-P1	The U.S. intelligence community trusts the information regarding Iraq's hidden WMDs and that it supported the 9/11 terrorist attacks.	0.90
CZ-P2	The U.S. government is focused on winning the wrongful war.	0.70
CZ-P3	The U.S. government exhausts its financial resources.	0.70
CZ-Q1	The U.S. government is focused on winning the Afghanistan war.	0.80
CZ-Q2	The United States provides financial support to allies and the international intelligence communities, and a lower budget for water security.	1.0
CZ-R1	The U.S. intelligence community fail to anticipate the attacks.	0.90
CZ-R2	The DHS, CIA, and U.S. Environmental Protection Agency (EPA) fail to keep the Americans safe and the resources secured.	0.90

of the total potential loss that would result if the undesirable event occurred. This second viewpoint places much less emphasis on probability and much more on the potential level of impact.

The first step in intelligence formulation involving risk is to establish the various factors that will be affected and determine the potential effects if these processes do not proceed as desired. The probabilities of occurrence can then be estimated using a stochastic model describing the events and the perceived risk. One of the most powerful techniques to present the alternative outcomes of a situation that involves risk is the event tree analysis, as shown in Figures 11.8b through g. It not only allows the analysts to isolate and examine the potential of various parts of a complicated process for creating negative outcomes but also is an effective mechanism to communicate these risks to decision makers in the intelligence process. It translates a situation with risk potential into a sequence of individual steps or subprocesses.

11.6.1 Risk Estimation and Risk Acceptability

Based on the event tree analysis for a series of coordinated terrorist attacks against U.S. water infrastructure that involves other critical infrastructure, as shown in Figures 11.8b through g, the comparison of risk estimation and risk acceptability is presented in Table 11.10. The detailed incremental risk acceptability calculations and results are presented in Table 11.11. Finally, the comparison of strategic alternatives for U.S. intelligence and infrastructure defense against terrorism is presented in Table 11.12.

11.7 Implications

U.S. intelligence is the core of all the effort of the Department of Homeland Security (DHS). The U.S. Intelligence Community (2008) and DHS (2008a,b) acknowledged and emphasized that improving the strategy and policy relating to the intelligence enterprise can be a lead toward enhancing the security of U.S. critical infrastructures. Based on the presentation and assessment presented in this chapter, there is a very urgent need to improve the intelligence analysis and information-sharing strategy within the intelligence community to defeat terrorism. Figure 11.9 provides a schematic illustration of a modified or improvised intelligence enterprise strategic plan embedded with cumulative prospect theory. The improved intelligence enterprise strategic plan may potentially make it more difficult for attacks to succeed or decrease the impact of attacks that may take place.

Table 11.10 Risk Estimation and Risk Acceptability Analysis Comparison

Sl. No.	*Description*	*Risk Estimation*	*Risk Acceptability* $V(f) = \omega(p) \cdot v(x)$
1	Terrorist leaders indirectly provide misleading information to the U.S. intelligence community using local supporters and the media	7.3×10^{-1} (Can lead to very high risk event if U.S. CIA and other intelligence communities do not have independent source of information and fail to produce strategic plans)	1.9×10^{-4} (Unacceptable risk)
2	Terrorist leaders create misleading information through foreign intelligence communities or agencies to keep the United States focused on the wrongful war	3.9×10^{-1} (Very high risk)	2.1×10^{-10} (Unacceptable risk)
3	Terrorists use cyberterrorism as a strategy to keep the United States from protecting other major infrastructure such as the water supply system and dams	4.6×10^{-2} (Very high risk)	2.9×10^{-15} (Very unacceptable risk)
4	Terrorist group allows media to announce that the principal leader of terrorists has a major illness to mislead U.S. public belief	5.7×10^{-1} (Can pose very high risk if the intelligence community does not have its own independent source of information and will not produce strategic plans)	5.9×10^{-7} (Unacceptable risk)
5	Media's credibility as a source of information regarding the terrorist group's financial and stability status	3.3×10^{-1} (Very high risk)	1.03×10^{-12} (Unacceptable risk)

6	Foreign or global (non-United States) intelligence, an effective information source to locate the terrorist leader	4.7×10^{-1} (Very high risk)	1.2×10^{-10} (Unacceptable risk)
7	Business as usual, U.S. intelligence strategy	5.1×10^{-1} (Very high risk)	3.2×10^{-15} (Very unacceptable risk)
8	Terrorists attack petroleum refineries using commercial airplanes as weapons to create massive contamination to surface water and create mass casualties	6.6×10^{-4} (Very high risk)	2.3×10^{-17} (Very unacceptable risk)
9	Generation of accidents by terrorists on major highways and freeways to delay the emergency response team during coordinated terrorist attacks in Los Angeles and New York	3.4×10^{-2} (Very high risk)	2.3×10^{-17} (Very unacceptable risk)
10	Destruction of Hollywood Reservoir by terrorists	1.7×10^{-2} (Very high risk)	2.3×10^{-17} (Very unacceptable risk)
11	Destruction of Kensico Dam in New York by terrorist	1.2×10^{-2} (Very high risk)	2.3×10^{-17} (Very unacceptable risk)
12	Blasting of aqueducts in Los Angeles by terrorists	1.9×10^{-2} (Very high risk)	5.9×10^{-8} (Acceptable risk; does not lead to catastrophe)

(Continued)

Table 11.10 Risk Estimation and Risk Acceptability Analysis Comparison (*Continued*)

Sl. No.	*Description*	*Risk Estimation*	*Risk Acceptability* $V(f) = \omega(p) \cdot v(x)$
13	Blasting of Catskill Aqueduct in New York by terrorists	1.5×10^{-2} (Very high risk)	5.9×10^{-8} (Acceptable risk; does not lead to catastrophe)
14	Contamination of aqueduct in Los Angeles	4.8×10^{-5} (High risk)	2.3×10^{-17} (Very unacceptable risk)
15	Contamination of aqueduct in New York	2.3×10^{-5} (High risk)	2.3×10^{-17} (Very unacceptable risk)
16	Terrorists utilizing manholes and pipelines within downtown Los Angeles and New York for location and installation of IEDs for destruction of infrastructure	2.1×10^{-2} (Very high risk)	2.3×10^{-17} (Very unacceptable risk)

Table 11.11 Risk Acceptability Analysis

Sl. No.	*Description*	*Risk Reference*	*Proportionality by Degree of Voluntarism*	*Derating*	*Controllability*	*Risk Acceptability* $V(f) = \omega(p) \cdot v(x)$
1	Terrorist leaders indirectly provide misleading information to U.S. intelligence community using local supporters, the media, and minor terrorism events that cause the United States to exhaust its financial resources	0.38 Man-originated Catastrophic Voluntary	1	0.09	5.6×10^{-3}	1.9×10^{-4}
2	Terrorist leaders create misleading information through foreign intelligence communities or agencies to keep the United States focused on the wrongful war	1.8×10^{-6} Man-originated Catastrophic Voluntary	1	0.09	1.3×10^{-3}	2.1×10^{-10}
3	Terrorists use cyberterrorism as a way to keep the United States from protecting other major infrastructure such as the water supply system and dams	9.8×10^{-8} Man-originated Catastrophic Involuntary	0.09	0.0081	4.1×10^{-5}	2.9×10^{-15}
4	Terrorists group allows media to announce that its principal leader has a major illness to mislead U.S. public belief	1.8 Man-originated Catastrophic Involuntary	0.09	0.09	4.1×10^{-5}	5.9×10^{-7}

(Continued)

Table 11.11 Risk Acceptability Analysis (*Continued*)

Sl. No.	*Description*	*Risk Reference*	*Proportionality by Degree of Voluntarism*	*Derating*	*Controllability*	*Risk Acceptability* $V(f) = \omega(p) \cdot v(x)$
5	Media's credibility as a source of information regarding the terrorist group's financial and stability status	9.8×10^{-8} Man-originated Catastrophic Involuntary	0.09	0.09	1.3×10^{-3}	1.03×10^{-12}
6	Foreign or global (non-United States) intelligence an effective information source to locate the terrorist leader	1.8×10^{-6} Man-originated Catastrophic Voluntary	0.09	0.0081	0.091	1.2×10^{-10}
7	Information-sharing techniques and intelligence analysis are business as usual	9.8×10^{-8} Man-originated Catastrophic Involuntary	0.09	0.000065	5.6×10^{-3}	3.2×10^{-15}
8	Terrorists explosions to petroleum refineries using commercial airplanes as weapons, to create massive contamination to surface water and mass casualties as illustrated	9.8×10^{-8} Man-originated Catastrophic Involuntary	0.09	0.000065	4.1×10^{-5}	2.3×10^{-17}

9	Generation of accidents by terrorists on major highways and freeways to delay emergency response team during the coordinated terrorist attacks in Los Angeles and New York	9.8×10^{-8} Man-originated Catastrophic Involuntary	0.09	0.000065	4.1×10^{-5}	2.3×10^{-17}
10	Destruction of Hollywood Reservoir by terrorists	9.8×10^{-8} Man-originated Catastrophic Involuntary	0.09	0.000065	4.1×10^{-5}	2.3×10^{-17}
11	Destruction of Kensico Dam in New York by terrorists	9.8×10^{-8} Man-originated Catastrophic Involuntary	0.09	0.000065	4.1×10^{-5}	2.3×10^{-17}
12	Blasting of aqueducts in Los Angeles by terrorists	1.8 Man-originated Catastrophic Involuntary	0.09	0.000065	5.6×10^{-3}	5.9×10^{-8}
13	Blasting of Catskill Aqueduct in New York by terrorists	1.8 Man-originated Catastrophic Involuntary	0.09	0.000065	5.6×10^{-3}	5.9×10^{-8}

(*Continued*)

Table 11.11 Risk Acceptability Analysis (*Continued*)

Sl. No.	*Description*	*Risk Reference*	*Proportionality by Degree of Voluntarism*	*Derating*	*Controllability*	*Risk Acceptability* $V(f) = \omega(p) \cdot v(x)$
14	Contamination of aqueduct in Los Angeles	9.8×10^{-8} Man-originated Catastrophic Involuntary	0.09	0.000065	4.1×10^{-5}	2.3×10^{-17}
15	Contamination of aqueduct in New York	9.8×10^{-8} Man-originated Catastrophic Involuntary	0.09	0.000065	4.1×10^{-5}	2.3×10^{-17}
16	Terrorists utilizing manholes and pipelines within downtown Los Angeles and New York for location and installation of IEDs for destruction of infrastructures	9.8×10^{-8} Man-originated Catastrophic Involuntary	0.09	0.000065	4.1×10^{-5}	2.3×10^{-17}

Note: Refer to Figures 11.8b through 11.8g for event tree analysis. Refer to Figure 11.8a for the designed time frame of terrorism in Los Angeles and New York. Refer to Table 11.9 for the risk rates assigned in the event tree analysis relating to infrastructure protection, and risk rates in Table 11.9 are used for the intelligence-related analysis.

Table 11.12 Comparison of Strategic Alternatives for U.S. Intelligence Improvements and Homeland Preventive Measures

Alternatives	*Risk Estimation*	*Risk Acceptability* $V(f) = \omega(p) \cdot v(x)$
Business as usual	5.1×10^{-1} (Very high risk)	3.61×10^{-16} (Unacceptable)
A proposed independent local/international media to solely aid and serve the U.S. intelligence community while proving valid information to the public and protecting sensitive information. Media with local workforce that can blend into the designated area of interest. Media that can distort the terrorist leaders' and their supporters' goal to provide misleading information to U.S. intelligence and the general public.	9.0×10^{-9} (Low risk)	6.0×10^{-3} (Acceptable)
A method shall be developed to scrutinize information and the source of information from foreign intelligence communities. May need to replace and reform U.S. intelligence in local and foreign countries. The intelligence reformation shall not defeat critical infrastructure protection particularly in the current economic recovery situation in the United States.	4.0×10^{-4} (High risk) There is a high risk of terrorism even if there is a reformation in the U.S. intelligence community if the U.S. critical infrastructure such as the water supply system is not provided with sophisticated technology and policy for protection (in this case, it is assumed that most water supply systems are still not protected after the reformation of the intelligence system in the United States).	6.54×10^{-12} (Unacceptable)

(Continued)

Table 11.12 Comparison of Strategic Alternatives for U.S. Intelligence Improvements and Homeland Preventive Measures (*Continued*)

Alternatives	*Risk Estimation*	*Risk Acceptability* $V(f) = \omega(p) \cdot v(x)$
Recommendation: Intelligence community and risk analysts shall work together and develop all possible terrorism activity scenarios with a designed time frame of the potential attacks in detail to accurately define the areas that need more information and protection in a timely manner and defeat terrorism.	–	–
The federal government, U.S. intelligence community, and DHS decision makers shall be aware of these minor attacks (attacks that do not generate fatalities and major property damages). These could create distortion of priorities for protection. A focus on protecting the assets (e.g., water infrastructure and dams) that can cause catastrophe when being attacked by terrorists should be the main priority for improvements, and the top priority for financial support and security.	5.83×10^{-11} (Very low risk)	7.94×10^{-10} (Somewhat acceptable)

The U.S. intelligence community shall monitor all foreign international airports (see Figure 4.10 of Chapter 4) with flights to the United States as to whether they have sophisticated detection technologies for biological threats, chemical threats, and explosives. If those foreign countries do not have technologies, individuals travelling to the United States shall be inspected and profiled accurately. May require commercial airplanes to install these sophisticated technologies to monitor and detect possible threats immediately.	1.3×10^{-6} (High risk) Without sophisticated detection technologies at the checkpoint level, there is a potential risk of successful terrorist attacks.	7.5×10^{-14} (Unacceptable)
	1.6×10^{-10} (Very low risk) If the United States is not able to regulate foreign countries to install sophisticated detection technologies at the checkpoint level, the U.S. intelligence community shall monitor the commercial airplanes for hijackers. Then the United States may need to destroy or blow up the airplanes heading to attack a critical infrastructure that could cause catastrophe for U.S. defense (see Figure 4.10 of Chapter 4 as an illustrative example of the attack).	4×10^{-8} (Acceptable) This option has high incremental risk acceptability to the general public. 1.54×10^{-13} (Unacceptable) This option has low incremental risk acceptability to the owners of commercial airplanes and the families of those suffering casualties.
	1.6×10^{-11} (Very low risk) If the U.S. intelligence community monitors all foreign international airports, sophisticated detection technologies are employed, and the travelers are accurately inspected, then potential terrorism and catastrophes can be prevented.	9×10^{-9} (Acceptable)

(*Continued*)

Table 11.12 Comparison of Strategic Alternatives for U.S. Intelligence Improvements and Homeland Preventive Measures (*Continued*)

Alternatives	*Risk Estimation*	*Risk Acceptability* $V(f) = \omega(p) \cdot v(x)$
Installation of sophisticated detection technologies for explosive devices or explosive agents and surveillance on highways/roads leading to a critical asset such as dams.	6.2×10^{-6} (High risk) A risk of potential attack could still succeed even after being detected if there is no method in place to deter the attackers.	6.54×10^{-12} (Unacceptable) Low incremental risk acceptability
	1.6×10^{-8} (Low risk) Detection technologies are being utilized; authorities and emergency response teams will deter the attackers.	5.89×10^{-13} (Unacceptable) Low incremental risk acceptability if there is no strategic planning and no system to keep the surrounding assets or people from being affected during the confrontation of the attackers.
	7.7×10^{-10} (Very low risk) The U.S. intelligence community is able to detect the potential attack, inform the owner of the asset to be prepared, and secure the area. Detection technologies are being utilized; authorities and emergency response teams will assist to deter the attackers. Also, strategic planning is implemented with the use of special technologies for defeating the attack without affecting the surrounding assets and people.	1.6×10^{-2} (Acceptable)

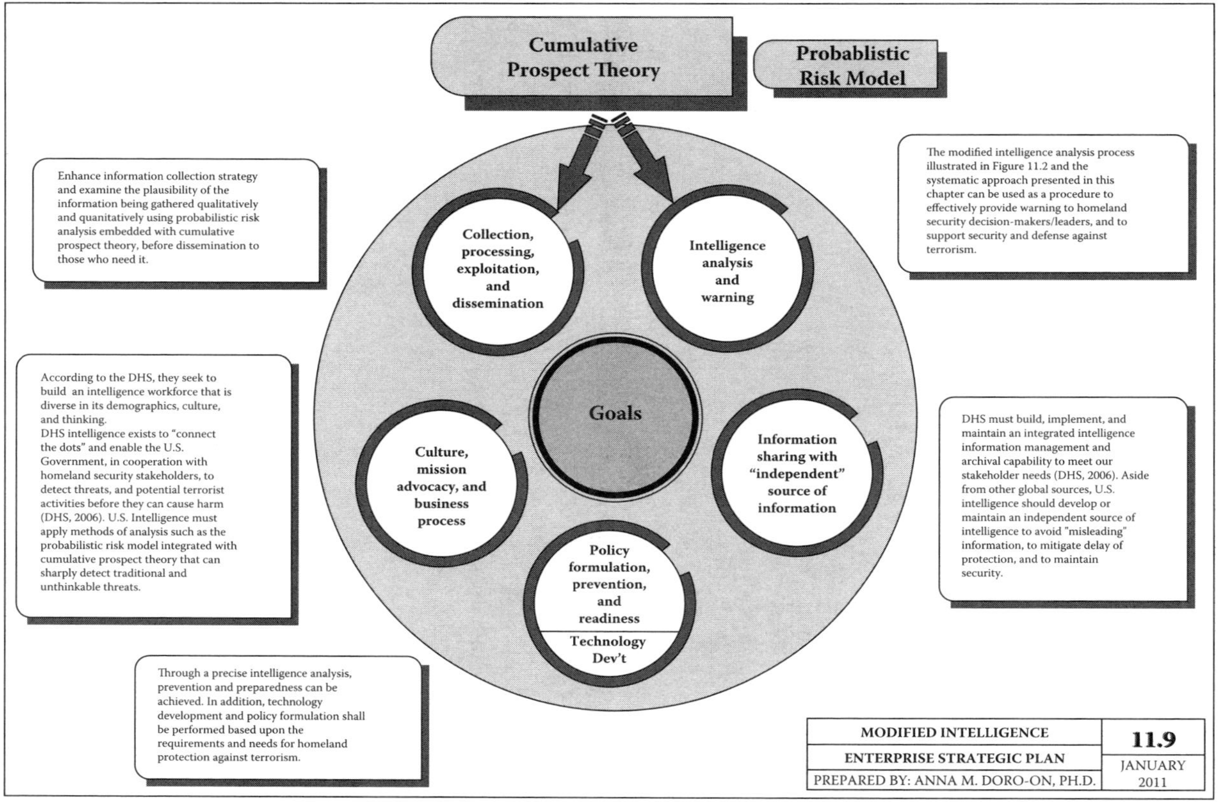

Figure 11.9 Modified intelligence enterprise strategic plan.

References

Goodman, M. A. 2004. *International Policy Report: Uses and Misuses of Strategic Intelligence*. Center for International Policy. http://www.ciponline.org/nationalsecurity/reports/jan04goodman.pdf (accessed October 10, 2010).

Heuer, R. 1999. *Psychology of Intelligence Analysis*. Washington, DC: CIA Center for the Study of Intelligence.

Johnston, R. 2005. Analytic Culture in the U.S. Intelligence Community. Center for the Study of Intelligence, Central Intelligence Agency. https://www.cia.gov/library/center-for-the-study-of-intelligence/csi-publications/books-and-monographs/analytic-culture-in-the-u-s-intelligence-community/full_title_page.htm (accessed October 16, 2010).

New York City Department of Environmental Conservation. 2010. New York City's Water Supply System. http://www.dec.ny.gov/docs/water_pdf/nycsystem.pdf (accessed October 21, 2010).

U.S. Central Intelligence Agency. 2010a. *The Intelligence Cycle*. https://www.cia.gov/kids-page/6-12th-grade/who-we-are-what-we-do/the-intelligence-cycle.html (accessed October 2, 2010).

U.S. Central Intelligence Agency. 2010b. *System Model of the Intelligence Cycle*. https://www.cia.gov/library/center-for-the-study-of-intelligence/csi-publications/books-and-monographs/analytic-culture-in-the-u-s-intelligence-community/page_52.pdf (accessed October 2, 2010).

U.S. Department of Homeland Security. 2008a. *Information Sharing Strategy*. http://www.dhs.gov/xlibrary/assets/dhs_information_sharing_strategy.pdf (accessed October 23, 2010)

U.S. Department of Homeland Security. 2008b. *Safeguarding Information Designated as Chemical-Terrorism Vulnerability Information (CVI)*, Revised Procedural Manual.

U.S. Intelligence Community. 2008. *Information Sharing Strategy*. http://www.dni.gov/reports/IC_Information_Sharing_Strategy.pdf (accessed October 22, 2010).

U.S. Department of Homeland Security. 2006. DHS Intelligence Enterprise Strategic Plan. http://www.fas.org/irp/agency/dhs/stratplan.pdf (accessed October 16, 2010).

Index

Page numbers followed by "*f*" indicates a figure and "*t*" indicates a table.

D

Q

R

S